Natural Selection

in

Human Populations

The Measurement of Ongoing Genetic
Evolution in Contemporary Societies

i

Natural Selection

in

Human Populations

THE MEASUREMENT OF ONGOING GENETIC EVOLUTION IN CONTEMPORARY SOCIETIES

Edited by

Carl Jay Bajema, Ph.D.
Department of Biology
Grand Valley State College
Allendale, Michigan

and

Research Associate
Harvard Center For Population Studies
Cambridge, Massachusetts

John Wiley & Sons, Inc., New York ● London ● Sydney ● Toronto

To Charles Robert Darwin and Alfred Russel Wallace who, in 1858, proposed the theory of natural selection that provided man with a key with which to break many of the shackles of ignorance and superstition concerning the origin and destiny of the human species and enabled man to gain a more realistic view of his origins and future.

Library of Congress Catalogue Card Number: 72-154322

ISBN 0-471-04380-x

Printed in the United States of America.

10 9 8 7 6 5 4 3 2 1

Preface

Natural selection, the mechanism by which genetic evolution takes place, was first proposed by Charles Darwin and Alfred Wallace in 1858. The theory of natural selection triggered a revolution in men's minds. The static view of nature (special creation) was rapidly supplanted by the dynamic view of nature (evolution).

The theory of natural selection provides the key to man's understanding of the history of life on this planet but, more important, it is also the key to man's understanding of his current predicament (for instance, competition and racism) and his future. Numerous books have been published that deal with man's evolutionary past. Very few books that have dealt with man's ongoing evolution have been published. This book of readings, which is concerned with man's continuing genetic evolution and his evolutionary future, should help fill a crucial gap in the literature on human evolution. The articles reprinted in this volume include papers that represent the best original research with respect to the measurement of ongoing human evolution and excellent review papers that summarize research and intelligent speculation with respect to the ongoing evolution on man and eugenics—the social control of future human evolution. In addition to helping students, teachers, and laymen gain a better understanding of ongoing human evolution (including the difficulties of measuring ongoing genetic change) and of the alternatives man must choose from as he begins to more effectively control his future evolution, I hope that the publication of these papers in one volume will stimulate further research with respect to the measurement of the ongoing genetic evolution of man. The book is part of my contribution to the 100th anniversary celebration of the publication of *The Descent of Man and Selection in Relation to Sex* (February 24, 1871).

Allendale, Michigan, 1971   *Carl Jay Bajema*

Contents

I.   **Introduction** *1*

   1   Man and Natural Selection
       T. Dobzhansky *4*

   2   Patterns of Survival and Reproduction in the United States:
       Implications for Selection
       D. Kirk *19*

   3   The Study of Mutation and Selection in Human Populations
       H. Newcombe *29*

       Selected Bibliography *55*

II.   **Natural Selection in Relation to Physical Characteristics** *57*

   4   Selection in Natural Populations. New York Babies
       L. Van Valen and G. Mellin *59*

   5   Fertility and Physique—Height, Weight, and Ponderal Index
       A. Damon and R. Thomas *84*

   6   The Operation of Natural Selection on Human Head Form in
       an East European Population
       T. Bielicki and Z. Welon *92*

   7   Factors in the Alteration of Reproductive Potential of
       Chondrodystrophics
       W. Cotter *103*

   8   Congenital Malformations in Man and Natural Selection
       L. Penrose *111*

   9   Some Factors Influencing the Relative Proportions of Human
       Racial Stocks
       F. Hulse *119*

       Selected Bibliography *143*

III. **Natural Selection in Relation to Disease** *145*

   10 ABO Blood Groups and Smallpox in a Rural Population of
       West Bengal and Bihar (India)
       F. Vogel and M. Chakravartti *147*

   11 Polymorphism and Natural Selection in Human Populations
       A. Allison *166*

12   On the Selective Advantage of Cystic Fibrosis Heterozygotes
A. Knudson, L. Wayne, and W. Hallett *191*

13   Natural Selection and Degenerative Cardiovascular Disease
E. Nye *196*

     Selected Bibliography *202*

IV.   Natural Selection in Relation to Human Behavior *205*

14   Population Dynamics of Tay-Sachs Disease. I. Reproductive
Fitness and Selection
N. Myrianthopoulos and S. Aronson *208*

15   Huntington's Chorea in Michigan. 2. Selection and Mutation
T. Reed, and J. Neel *226*

16   Selection and Schizophrenia
L. Erlenmeyer-Kimling and W. Paradowski *259*

17   Estimation of the Direction and Intensity of Natural Selection
in Relation to Human Intelligence by Means of the Intrinsic
Rate of Natural Increase
C. Bajema *276*

     Selected Bibliography *292*

**V.   Natural Selection and the Future Genetic Composition of Human
Populations** *295*

18   Do Advances in Medicine Lead to Genetic Deterioration?
P. Medawar *300*

19   The Quality of People: Human Evolutionary Changes
J. Crow *309*

20   The Population Explosion, Conservative Eugenics and Human
Evolution
L. Ornstein *321*

⚹ 21   Possible Genetic Consequences of Family Planning
E. Matsunaga *328*

22   The Tragedy of the Commons
G. Hardin *345*

23   The Genetic Implications of American Life Styles in Reproduction and Population Control
C. Bajema *359*

⁎ 24   A Return to the Principles of Natural Selection
F. Osborn *369*

25   What Genetic Course Will Man Steer?
H. Muller *378*

     Selected Bibliography *405*

The following is a list of addresses of the authors (or senior authors in the case of multiple authorship) of the papers included in this book of readings:

Dr. Anthony Allison, National Institute for Medical Research, Mill Hill, London, Great Britain

Dr. Tadeusz Bielicki, Director, Institute of Anthropology, Polish Academy of Sciences, Warsaw, Poland

Dr. William Cotter, Department of Anatomy, University of Kentucky, Lexington, Kentucky

Dr. James Crow, Department of Genetics, University of Wisconsin, Madison, Wisconsin

Dr. Albert Damon, Department of Anthropology, Harvard University, Cambridge, Massachusetts

Dr. Theodosius Dobzhansky, Department of Genetics, Rockefeller University, New York City

Dr. L. Erlenmeyer-Kimling, New York State Psychiatric Institute, 722 West 168th St., New York City

Dr. Garrett Hardin, Department of Biology, University of California at Santa Barbara, Santa Barbara, California

Dr. Frederick Hulse, Department of Anthropology, University of Arizona, Tucson, Arizona

Dr. Dudley Kirk, Food Research Institute, Stanford University, Palo Alto, California

Dr. Alfred Knudson, Jr., Dean, Graduate School of Biomedical Sciences, University of Texas at Houston, Texas Medical Center, Houston, Texas

Dr. E. Matsunaga, Director, National Institute of Genetics, Mishima, Japan

Dr. Ntinos Myrianthopoulos, National Institute of Neurological Diseases and Blindness, National Institutes of Health, Bethesda, Maryland

Dr. Peter Medawar, Director, National Institute of Medical Research, London, Great Britain

The late Dr. Hermann J. Muller was Professor of Genetics at the University of Indiana, Bloomington, Indiana

Dr. Howard Newcombe, Head, Biology Branch, Atomic Energy of Canada Ltd., Chalk River, Ontario, Canada

Dr. E.R. Nye, Department of Medicine, University of Otago Medical School, Dunedin, New Zealand

Dr. Leonard Ornstein, Division of Cell Biology, Mount Sinai Graduate School of Biological Sciences, Mount Sinai Hospital, New York City

Mr. Frederick Osborn, Treasurer, American Eugenics Society, 230 Park Avenue, New York City

Dr. Lionel Penrose, Kennedy-Galton Centre, Harperbury Hospital, St. Albans, Hertfordshire, Great Britain

Dr. T. Edward Reed, Department of Zoology, University of Toronto, Toronto, Canada

Dr. Leigh Van Valen, Department of Anatomy, University of Chicago, Chicago, Illinois

Dr. Friedrich Vogel, Director, Institute for Anthropology, and Human Genetics, University of Heidelberg, Heidelberg, Germany

PART I

Introduction

Over 100 years have elapsed since Charles Darwin and Alfred Russel Wallace first proposed the theory of natural selection to explain biological evolution. The evolution of man was mentioned very briefly by Darwin in his classic book, *The Origin of Species by Means of Natural Selection*, which was first published in 1859. Darwin considered the problem of human evolution in greater depth in his monumental work, *The Descent of Man and Selection in Relation to Sex*, which was published 12 years later. The year 1971 marks the 100th anniversary of the publication of *The Descent of Man*. It is fitting on the eve of this centennial to take stock of the advances that have been made during the past century with respect to the study of human evolution.

One of the most important advances that has taken place with respect to evolutionary theory during the past century has been the clarification of what natural selection is and what it is not. It is doubtful that any concept has been more misunderstood in the history of scientific thought than the concept of natural selection. Charles Darwin contributed to the development of this misunderstanding when he used the cliche, "survival of the fittest," to describe natural selection. This cliche conveyed the idea to most people that the fittest individuals are those individuals who survive. However, the crux of natural selection lies not merely in the success of an individual with respect to survival but in the individual's reproductive performance—the production of living offspring by the individual. In other words, the fittest individuals are those individuals who pass the most genes on to the next generation or, as Theodosius Dobzhansky, a leading evolutionary biologist states it, "the fittest are nothing more than the parents or grandparents of the greatest number of surviving descendants." Natural selection then, is nothing more than differential reproduction—the consistent production of more offspring on the average by individuals with certain genetic characteristics than those individuals without those characteristics. The modern understanding of natural selection places the emphasis where it belongs—on reproduction as opposed to the survival of the individual.

1

Natural selection, when viewed from the perspective of cybernetics (see diagram), is the feedback mechanism which favors the production of DNA codes—"programs"—which enables the species to adapt to the environment. It is via natural selection that information about the environment is transmitted to the gene pool of a population. This is accomplished by differential reproduction of different genotypes changing the frequencies of genes in populations. When viewed from this perspective natural selection is clearly a creative force capable of bringing about directional change and not just a stabilizing force maintaining the status quo by weeding out maladaptive genetic mutants.

The past century has been marked by many discoveries of the fossil remains of the progenitors of modern man. This documentation of human evolution via the fossil record has provided the data for a number of important contributions to the understanding of natural selection as it operates in the human species. We have been able to quantitatively estimate the rate of evolutionary change with respect to several skeletal characteristics and thus have become acutely aware of the mosaic nature of human evolution. For instance, it has been shown that the rates of evolutionary change for different human characteristics within so limited a complex as human cheek teeth (premolars vs. molars) differ not only with respect to rate, but also with respect to time of maximum evolutionary change (Bilsborough, 1969). Evidence concerning the cranium indicates that the maximum rate at which the hominid cranium increased in size is higher than the rates of evolution calculated for various characteristics of other pleistocene mammals and that the size of the human cranium has been quite stable over the past 40,000 years (Campbell, 1963). The fossil record, then, makes it obvious that the direction and intensity of natural selection can vary over time with respect to a given characteristic and can vary simultaneously with respect to several characteristics.

Much progress has been made with respect to the theoretical basis of evolution (Fisher, 1930; Haldane, 1932, 1964; Wright, 1931, 1932, 1968, 1969) and with respect to better understanding the origin of man (Hockett and Ascher, 1964; Coon, 1962, 1965; Campbell, 1966). What progress has been made during the past 100 years with respect to measuring ongoing evolution in human populations? It is in relation to this specific question that the papers that appear in this volume were selected.

Meaningful scientific studies of the operation of natural selection in human populations were not undertaken until more than half a century had elapsed after Darwin published his monumental work on the evolution of man. They had to await the theoretical studies of R. A. Fisher, J. B. S. Haldane, and S. Wright, which demonstrated that natural selection "can be investigated and measured almost as a physical force by means of experiments or observations properly devised to test theoretical models worked out in mathematical terms" (Montalenti, 1965).

The following articles represent some of the best attempts to measure ongoing natural selection and illustrate how much and yet how little we know about the operation of natural selection in human populations. Hopefully, they will stimulate further investigations concerning ongoing human evolution as well as help the reader attain a better understanding of human evolution. A bibliography is included at the each of each section to assist the reader in locating additional papers relevant to the topics included in that section. The bibliographies contain papers that are concerned with man's evolutionary past and with ongoing human evolution.

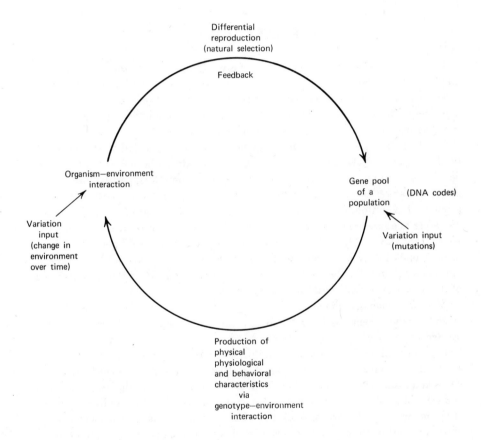

1. Man and Natural Selection*

THEODOSIUS DOBZHANSKY

Man's Evolutionary Uniqueness

By changing what man knows about the world, he changes the world he knows; and by changing the world in which he lives, he changes himself. Herein lies a danger and a hope; a danger because random changes of the biological nature are likely to produce deterioration rather than improvement; a hope because changes resulting from knowledge can also be directed by knowledge.

The human species, *Homo sapiens*, mankind, is the unique and most successful product of biological evolution, so far. This has sometimes been questioned, I suspect without too much conviction on the part of the doubters, perhaps only to mock man's pretensions or to challenge his values. But man *is* the most successful product of evolution, by any reasonable definition of biological success. Man began his career as a rare animal, living somewhere in the tropics or subtropics of the Old World, probably in Africa. From this obscure beginning, mankind multiplied to become one of the most numerous mammals, for there will soon be about three billion men living. Numbers may not be an unadulterated blessing, but they are one of the measures of biological success of a species.

Moreover, man has spread and occupied all the continents and most islands, except for the frozen wastes of Antarctica and of the interior of Greenland; he has learned to traverse seas and oceans and deserts; he is well on the way towards control or elimination of the predators and parasites which used to prey on him; he has subdued and domesticated many animal and plant species, made them serve his needs and his fancies, broadened enormously the range of utilizable food supplies, and learned to make use of a variety of energy sources. Modern man lives no longer at the mercy of wild beasts and vagaries of the climate; he has reached a status where his continuation as a species is in no danger, except perhaps as a result of man's own folly or of a cosmic accident.

The evolutionary uniqueness of man lies in that in mankind the biological evolution has transcended itself. With man commences a new, superorganic, mode of

SOURCE: T. Dobzhansky, "Man and Natural Selection," *American Scientist*, **49**: 285-299 (Princeton, New Jersey, 1961).

*A Sigma Xi-RESA National Lecture, 1960-1961.

evolution, which is the evolution of culture. Culture is a tremendously potent instrument for adaption to the environment. A very large part of the evolutionary progress, both biologically and culturally, has come from adversity. Life faces environments which are more often niggardly than bountiful, more frequently inimical than benign. For life to endure, it must develop defences and adaptions. Biological adaptation occurs through natural selection; new genes arise through mutation, sexual recombination creates new combinations of genes, and natural selection acts to multiply the successful genetic endowments and to reduce the frequencies of the unsuccessful ones. In man and in man alone, adaptation may occur also through alteration of culture. Many species of mammals have become adapted to cold climates by growing warm fur; man alone has achieved the same end by donning fur coats. Birds have mastered the air by becoming flying machines; man has conquered the air by building flying machines.

Biological and cultural evolutions of man are not independent; they are interdependent. The superorganic has an organic basis. Formation and maintenance of culture presuppose a human genotype. Even the most clever ape cannot learn human culture. Some writers have jumped to the conclusion that the genetic development of the human species was completed before culture appeared, and that the evolution of culture has replaced biological evolution. This is not true. The two evolutions go together, interacting and usually mutually reinforcing each other. There is feedback between genetics and culture. Culture is an adaptive mechanism supplemental to, but not incompatible with, biological adaptation. To be sure, adaptation by culture proved to be more efficacious, and, before all else, more rapid than adaptation by genes. This is why the emergence of the genetic basis of culture was the master stroke of the biological evolution of the human species. The genetic basis of culture should be improved or at least maintained. It should not be allowed to deteriorate.

Natural Selection, Struggle, and Fitness

Man has not only evolved; for better or for worse, he is also evolving. Is he to become a superman or a demigod? Or will the fate in store for him be like that of so many successful species of the past, which eventually declined and became extinct? Long-term prophecies do not expose the prophets to the risk of being proved wrong too soon. I nevertheless do not wish to indulge here in prophecies or in designing utopias. I wish rather to investigate some of the evolutionary forces currently at work in the human species.

According to the theory set forth by Darwin and Wallace more than a century ago, adaptation occurs in biological evolution by way of the process of natural selection. Does natural selection operate in modern mankind? This has to be answered obviously in the negative, if by "natural" you mean a world uninhabited and uninfluenced by man. But it is here that we must proceed with the greatest caution. What, indeed, is selection, and when is a selection "natural" and when is it not? Darwin said that natural selection is the outcome of the survival of the fittest in the struggle for life. "Struggle" suggests strife, contention, competition. Darwin himself wrote that " . . . from the war of nature, from famine and death, the most exalted object we are capable of conceiving, the higher animal, directly follows."

However, natural selection does not ineluctably depend on any of these things. Birch (1957) has defined competition thus: "Competition occurs when a number of animals (of the same or different species) utilize common resources the supply of which is short; or if the resources are not in short supply, competition occurs when the animals seeking these resources nevertheless harm each other in the process." Natural selection may, however, take place when resources are not limiting, if the carriers of some genes possess greater reproductive potentials than others. Some cases have been observed in experiments with animals and plants when a genotype superior under competition is inferior in the absence of competition, or vice versa.

And who is the "fittest," whose survival results in natural selection? Does natural selection make us fit for life in the society of other men, or for wisdom, or for good will, or for unselfishness? It does not necessarily ordain any of these qualities. Darwinian fitness is a measure of the reproductive proficiency. Its guiding principle is "be fruitful, and multiply, and replenish the earth." It is quite indispensable to distinguish Darwinian fitness from excellence in human estimation. The two may go hand in hand, but sometimes they may be in opposition.

It is man and man alone who can probe, scrutinize, and question the wisdom of the evolutionary process which brought him into being. He may improve what nature hath wrought. When man chooses the individuals who are to become parents of the succeeding generation he practices artificial selection. According to Lerner (1958), artificial selection is a process which has a goal that can be visualized. Any selection which is not artificial is natural selection.

Mutation and the Normalizing Natural Selection

There are several forms of natural selection operating in the human species, as well as in other organisms. Although the distinctions between these forms of selection are neither absolutely rigid nor always clear-cut, they are helpful for straight thinking about evolutionary problems. Stabilizing (Schmalhausen 1949) or normalizing (Waddington 1957) natural selection is the simplest and most obvious. It is a conservative force; it counteracts the spread in populations of detrimental mutants, hereditary diseases, and weaknesses of various kinds. Failure, or at least relaxation, of the normalizing selection in human populations leads to fears that an insidious process of genetic decay is at work in the human species. Is there a basis for these fears?

Mutations continue to arise in man, even as they have been arising since the dawn of life. Some mutants cause grave and even fatal hereditary diseases, such as retinoblastoma, hemophilia, epiloia, etc. Others cause malformations, such as achondroplasia, arachnodactyly, or brachydactyly. Still others, and probably a majority of mutants, cause small and unspectacular changes in the appearance or physiology or behavior of their carriers. Small mutations are difficult to study quantitatively. Not enough is known about them, even in Drosophila and in other organisms more favorable for experimentation than man. Large mutations are certainly more often deleterious than useful, at least in the environments in which the species normally lives, and at least in homozygous condition. A useful major mutant is like a needle in a haystack; it may be there but it is notoriously hard to find among the many harmful ones. It is tempting to suppose that all or most small mutations will also be harmful, in various small and devious ways. This is what geneticists who are adherents of the "classical" hypothesis of the population structure believe to be the case. The evidence on which this belief rests is not fully convincing (see below).

A brief consideration of some examples of the operation of the normalizing selection in man will be useful at this point. Achondroplasia or chondrodystrophy are formidable names for a rather common type of dwarfism in man, caused by the presence of a single dominant mutant gene. Achondroplasts have bodies and heads of about normal size, but their limbs are very short. Some achondroplastic dwarfs are born in families in which at least one of the parents is a dwarf. Such dwarfs evidently inherit the genes for the abnormality from the affected parents. But some dwarfs are children of parents of normal stature; these dwarfs carry the dominant gene for dwarfism newly arisen by mutation. Mørch has recorded the birth of eight dwarf mutants among about 94 thousand children in Denmark; this indicates that about one sex-cell per 24,000 sex cells produced by normal parents transports a newly arisen gene for achondroplastic dwarfism, a mutation rate of the order of 4×10^{-5} per generation.

The genes for achondroplastic dwarfism newly arisen by mutation are introduced into the gene pool of human populations; this happens relentlessly in every generation. Does it follow that the dwarfism will grow more and more widespread in the course of time? Not necessarily. The spread of the genes for dwarfism is opposed by natural selection. Mørch found that the achondroplastic dwarfs in Denmark produce only about twenty per cent as many children as do their non-dwarf brothers and sisters. This means that the adaptive value, or the Darwinian fitness, of the achondroplastic dwarfs is only 0.2 normal; it can also be said that the gene for the dwarfism is opposed by a normalizing natural selection of a magnitude of 0.8.

How does natural selection operate on this gene? Why do the dwarfs produce so few children? It turns out that this is chiefly because many of them remain unmarried. And their failure to find mates is not due to a weakened sexual drive but to their external appearance being not in accord with what is regarded in our society as handsomeness or attractiveness. Here, then, is an example of a natural selective process which operates not by causing early death of the carriers of certain genes but acting instead via certain culturally conditioned forms of human behavior. A person who remains childless may be described as genetically "dead;" but childlessness is a form of "genetic death" which does not immediately produce a cadaver.

Some mutant genes are, to be sure, eliminated by a normalizing selection acting through "genetic death" which is also real death. Retinoblastoma is a form of cancer of the eye, which afficts infants and children; unless the eyes are removed promptly by a surgical operation early death is inevitable. Neel and Falls found the mutation from the normal to the dominant mutant gene causing retinoblastoma to be about as frequent as that causing the achondroplastic dwarfism (see above). Before the invention of surgery which saves the lives of retinoblastomatic children, the normalizing selection was eliminating in every generation all the mutant genes for retinoblastoma which arose in that generation (i.e., in the sex-cells which gave rise to that generation). The surgery may be said to frustrate this normalizing selection; the retinoblastomatics, though blind, grow up and become capable of being parents of children one-half of whom will now inherit the gene for retinoblastoma. The retinoblastomatics, like the achondroplasts, are now of two kinds—those due to new mutations and those having inherited the gene from their parents.

Genes which reduce the Darwinian fitness of their carriers less dramatically than do the genes discussed above are nevertheless also opposed by normalizing selection. Predisposition to diabetes mellitus, or at least to some of its forms, seems to be inherited through a recessive gene. The same seems to be true of some forms of myopia (short-sightedness). We do not know either the mutation rates which produce these genes, nor the selection rates which oppose their spread. What is of interest for our purposes is that these selection rates doubtless vary greatly depending on the environment. Myopia is likely to be more incapacitating to people engaged in some occupations (i.e., hunters or automobile drivers) than in others (i.e., handicraftsmen or clerks), and besides myopia can be "cured" by wearing glasses. The onset of diabetes mellitus may be early or late in life; it may thus strike after the close of the reproductive age, and thus fail to reduce the number of the children produced and hence to reduce the Darwinian fitness. And some forms of diabetes may be "cured" by insulin therapy.

Now, to "cure" a hereditary disease obviously does not mean to change the genes which have caused it. Health and disease refer however to the states of well-being or infirmity of a person; a person consulting his doctor does not ask the latter to alter his genes, but only to advise how to change the environment in a way such that the genes with which the person has been born will react by producing a reasonable state of well-being. A myopic wearing glasses, or a diabetic having received his insulin injections, is no longer incapacitated, or at least less so than both were before the "cure." The relief of the incapacitation may however increase their Darwinian fitness, and mitigate the severity of the normalizing natural selection.

The "Normal" Man

The removal of a mutant from a population is called a genetic death or, less dramatically, genetic elimination. The death of a child with retinoblastoma eliminates from the population one mutant retinoblastomatic gene. When an achondroplastic dwarf fails to marry, or raises a family with fewer children than he would have if he were not a dwarf, a mutant gene for achondroplasia is eliminated. The classical hypothesis of population structure assumes that there exists for every biological species, or may appear or be produced in the future, one normal, or best, optimal genetic endowment. This genetic endowment would give the ideal, the normal, the archetypical man, or the ideal Drosophila fly, or the ideal corn plant—the man, the Drosophila, the corn as they ought to be. The way to produce this normal man is to let the normalizing natural selection remove, eliminate, purge from the population all the mutant genes. What then would remain would be the ideal constellation of genes.

This hypothesis has the advantages of simplicity, though not necessarily of accuracy. It is a product of typological thinking, which has, to be sure, venerable antecedents. It goes back to Plato's eternal ideal types. Many scientists find the appeal of the typological mode of thought irresistible. Indeed, if the classical hypothesis were justified, the problem of the guidance of the biological evolution of the human species would be theoretically simple, however difficult it might still remain in practice. We would have to arrange for the normalizing natural selection to protect the ideal normal human genes from contamination by mutations. Or, to put it in positive rather than negative terms, we might strive to obtain a mankind in which everybody would be a carrier of the normal genetic endowment. The three billion humans, or whatever numbers this "normal" mankind might contain, would then be as similar genetically as identical twins. But they would presumably be strong and healthy and happy identical multitudes!

Diversifying Natural Selection

Fortunately or unfortunately, depending upon one's point of view, the classical hypothesis of population structure is perhaps only a half-truth. That it is true to some extent is clear enough. Harmful mutants do arise, and some of them seem to be unconditionally harmful. It is hard to imagine an enviroment in which retinoblastoma could possibly be harmless, not to speak of usefulness. Accumulation of such mutants in human populations can only augment human misery. To counteract this accumulation, one could, conceivably, reduce the mutation frequencies, or else eliminate the mutants as painlessly as possible. Unfortunately, for the time being, man has learned how to increase the mutability by exposure to high-energy radiations and by other means, but not how to decrease the mutation rates (Muller 1950, 1960).

The concept of a single genetic endowment, ideal or optimal for each living species, is however a typological fiction. No living individual, and certainly no population and no species, exist in absolutely uniform environments. Environments vary in time and in space. Now does life solve the problem of adaptation to a multiplicity of environments? There are two possible solutions, and, with its habitual opportunism, evolution has used them both. The first is the genetically-conditioned physiological and developmental plasticity. The organism reacts to changes in the environment by adaptive modifications. For example, the human body has physiological mechanisms which enable it to conserve an almost constant internal temperature despite external temperature variations. In man, the most important kind of plasticity is his educability. Most humans can be trained for, and can acquire skills to perform passably in, any one of the many employments which most human societies have to offer.

Secondly, a population facing a diversity of environments may become genetically diversified. Natural selection favors different genetic endowments in different environments. This is the diversifying (also called disruptive, Mather 1955) form of natural selection. Instead of one perfect genotype, diversifying selection favors many genotypes; it favors genetic polymorphism. A population which abounds in genetic variety has a better grip on a complex environment than a genetically uniform population; polymorphism is one way of exploiting the environment more fully.

Diversity and Equality

Adaptations by way of plasticity and by way of genetic diversity are not mutually exclusive. They are complementary. Culture is by far the most potent adaptive mechanism which has emerged in the evolution of life. Its potency is due to its being transmitted by teaching and learning, instead of by genes. The ability to be trained, and to become competent in whatever occupation a person meets his opportunity, increases the social and usually also the biological fitness. Genetic specialization for one vocation may confer very high fitness for that particular vocation. But educability permits a choice of vocations, and thus confers fitness in complex and changeable environments. In an evolving culture new occupations constantly arise. Who needed aircraft pilots a century ago, but now many blacksmiths are now needed in technologically advanced countries?

Culture is, however, an adaptive contrivance to make people diversified, not to make them alike. If uniformity were advantageous, genetic fixity would most likely have emerged in evolution. Genetic diversity or polymorphism joins force with developmental plasticity. I could have probably learned many kinds of jobs other than that I actually fixed upon. But no amount of effort and training could make me a first-rate wrestler or sprinter or painter or concertmaster. I just do not possess the genetic wherewithal for these occupations. Other people do, and this genetic diversity enriches the store of man's capabilities. It is the leaven of cultural progress.

The facts of biology are compatible with the lofty vision of human equality. All men have been born equal; they certainly are not all alike. It is nonsense to think that only identical twins can be equals. Human equality is not predicated on identity, or even on identity of ability. It presupposes, however, something approaching an equality of opportunity to develop whatever socially useful gifts and aptitudes a person's genes have provided him with, and which he may choose to develop.

Culture fosters a multitude of employments and functions to be filled and served; equality of opportunity stimulates division of labor rather than sets it aside; it enables every person to choose among occupations for which he is qualified by his abilities. It is wrong to think that equality of opportunity makes genetic variation unimportant. It does precisely the opposite. It makes the differences between people reflect meaningfully their genetic differences. Inequality of opportunity acts, on the contrary, to hide, to distort, and to stultify the genetic diversity.

Balancing Natural Selection

Far from being an unfortunate deviation from the ideal state, genetic diversity is an adaptive response of life to its environment. A gene useful in one environment may cause a genetic disease in another; a mutant harmful in combination with some genes may be useful in other combinations. Some adaptively ambivalent genes increase the fitness of their carriers when in single dose, in heterozygous state, but are harmful, or even lethal in homozygotes.

The sickle-cell anemia and the Mediterranean anemia, or thalassemia, are the best known human examples. These almost invariably fatal diseases are due to homozygosis for genes which in heterozygous condition seem to confer a relative immunity to certain kinds of malarial fevers. Heterozygous carriers consequently enjoy a superior fitness, hybrid vigor, or heterosis, in countries where these particular forms of malaria are prevalent. The genes for sickle-cells and for thalassemia are common in tropical and subtropical parts of the Old World, and they have been introduced in the New World as well.

Natural selection holds the "normal" and the sickle-cell and thalassemia genes in balanced polymorphism in populations of malarial countries. This is the balancing form of natural selection. How important is balancing selection in the human species is an open issue. It may well be that many of the genes responsible for the individual differences which we observe among so-called "normal" persons, differences in facial features, body build, health, intelligence, longevity, etc., are kept in balanced polymorphism. Morton and Chung (1959) have presented evidence that the genes for the M and N blood types are maintained by balancing selection. The heterozygotes MN have a Darwinian fitness superior to MM and NN homozygotes. Just what is the advantage of the heterozygous state is unknown, nor is the nature of the drawback in homozygotes. The homozygotes are certainly viable; I am one of them. A very slight superiority in fitness in heterozygotes is, however, sufficient to maintain a balanced polymorphism.

There are grounds to suspect that the O-A-B blood types in man are also held in balanced polymorphism. The evidence for this is at present inconclusive, but recent works of Levene, Rosenfield, Morton, Crow, and others show that the genes involved are subject to a still different kind of selective pressure. This is a relative incompatability between the mother and the fetus when they differ in genetic constitution in certain ways. Such incompatibility has been known to arise because of the genes for a still different system of blood types, the Rhesus, or Rh system. An Rh positive fetus in an Rh negative mother may suffer injury or even be aborted.

Directional Natural Selection

The form of natural selection most important in the long run is directional selection. It was pivotal in Darwin's theory of evolution. Living species may respond to challenges of their environments by genetic changes. Genes once useful become inferior to others and are gradually eliminated, replaced by superior gene variants. The whole genetic system of the species is eventually rebuilt.

There is a fairly general agreement among biologists that the emergence of the human species was due in the main to directional selection. It has, however, been questioned whether directional selection, or indeed any form of natural selection, has continued to operate after the genetic basis of culture has taken shape. Now, let us admit that the success of mankind as a biological species was due precisely to man's culture being able to change ever so much faster than his genes can. Man adapts his environments to his genes more often than he adapts his genes to his environments. But, as pointed out, the two methods of adaptation are complementary and not alternative. Far from making human environments stable and uniform, culture increases the tempo of change. Given an environmental flux, the necessary and sufficient condition for genetic change is availability in the populations of genetic variants, some of which are better and others less well adapted to shifting environments. Natural selection will multiply the favorable and depress or eliminate the unfavorable variants.

We do not know exactly how much genetic change has taken place in mankind at different stages of its history. Modern man might or might not be able to survive, even if properly trained, in the environments of his ancestors of 100,000, or even 10,000 years ago. Or, if he survived, he might not be as efficient or as happy in those environments as his ancestors were. We do not know for sure. Neanderthal man may or may not have been capable of becoming a reasonably well adjusted citizen if raised in New York or in New Orleans. Perhaps some Neanderthals might have been fit to become Ph. D.'s and to be elected members of the Society of Sigma Xi. But, on the other hand, they may have been unfit for any now existing education. Modern women are alleged to experience greater difficulties in childbirth than did their great-grandmothers. All this is conjectural and not rigorously proved. It is, however, a fallacy to assert that what is unproved did not occur. In point of fact, some of the above changes probably did occur.

Are Culture and Natural Selection Compatible?

We have seen that several forms of natural selection operate in modern mankind. But they certainly do not operate as they did during the Stone Age, or even as they did a century ago. Neither does natural selection operate always in the same way in wild and "natural" species, quite "unspoiled" by culture. This is inevitable. Natural selection depends on environments, and environments change. Human environments have changed a great deal in a century, not to speak of millennia.

The real problem is not whether natural selection in man is going on, but whether it is going on towards what we, humans, regard as betterment or deterioration. Natural selection tends to enhance the reproductive proficiency of the population in which it operates. Such proficiency is however not the only estimable quality with which we wish to see people endowed. And besides, a high reproductive fitness in one environment does not even insure the survival of the population or the species when the environment changes.

Normalizing selection is, as we have seen, not the only form of natural selection. The relaxation of some of its functions is, however, a cause of apprehension. Medicine, hygiene, civilized living save many lives which would otherwise be extinguished. This situation is here to stay; we would not want it to be otherwise, even if we could. Some of the lives thus saved will, however, engender lives that will stand in need of being saved in the generations to come. Can it be that we help the ailing, the lame, and the deformed only to make our descendants more ailing, more lame, and more deformed?

Suppose that we have learned how to same the lives of persons afflicted with a hereditary disease, such as retinoblastoma, which previously was incurably fatal. In genetic terms, this means that the Darwinian fitness of the victims of the disease has increased, and that the normalizing selection against this disease is relaxed. What will be the consequence? The incidence of the disease in the population will increase from generation to generation. The increase is likely to be slow, generally no more than by one mutation rate per generation. It may take centuries or millennia to notice the difference for any one disease or malformation. However, the average health and welfare of the population are liable to show adverse effects of relaxed selection much sooner.

The process of mutation injects in every generation a certain number of harmful genes in the gene pool of the population; the process of normalizing selection eliminates a certain number of these genes. With environment reasonably stable, the situation tends to reach a state of equilibrium. At equilibrium, the mutation and the elimination are equal. If mutation becomes more frequent (as it does in man because of exposure to high-energy radiations and perhaps to some chemicals), or if the elimination is lagging because of relaxation of normalizing selection, the incidence of harmful mutant genes in the population is bound to increase. And take note of this: If the classical theory of population structure were correct, all harmful mutations would be in a sense equivalent. For at equilibrium there is one elimination for every mutation, regardless of whether the mutation causes a lethal hereditary disease like retinoblastoma, or a malformation like achondroplasia, or a relatively mild defect such as myopia (Muller 1950, 1960).

The problem is, however, more complex than this theory would suggest. It calls for research in what Wright (1960) describes neatly as "unfortunately the unpopular and scientifically somewhat unrewarding borderline fields of genetics and the social sciences." Although at equilibrium there may be one genetic elimination for every mutation, it is unrealistic to equate the human and social consequences of different mutations. The elimination of a lethal mutant which causes death of an embryo before implantation in the uterus is scarcely noticed by the mother or by anyone else. Suffering accompanies the elimination of a mutant, such as retinoblastoma, which kills an infant apparently normal at birth. Many mutants, such as hemophilia or Huntington's chorea, kill children, adolescents, or adults, cause misery to their victims, and disruption of the lives of their families. There is no way to measure precisely the amount of human anguish; yet one may surmise that the painful and slow death of the victims of so many hereditary diseases is torment greater than that involved in the elimination of a gene for achondroplasia owing to the failure of an achondroplastic dwarf to beget children.

Looked at from the angle of the costs to the society, the nonequivalence of different mutants is no less evident. Myopia may be inherited as a recessive trait. Increases of the frequencies in populations of the gene for myopia are undesirable. Yet, it may become more and more common in future generations. However, only a fanatic might advocate sterilization of the myopics or other radical measures to prevent the spread of this gene. One may hope that civilized societies can tolerate some more myopics; many of them are very useful citizens, and their defect can rather easily be corrected by a relatively inexpensive environmental change—wearing glasses. The effort needed to eradicate or to reduce the frequency of myopia genetically would exceed that requisite to rectify their defect environmentally, by manufacturing more pairs of glasses.

Diabetes mellitus is, given the present level of medicine, more difficult and expensive to correct than is myopia. Some diabetics may nevertheless be treated successfully by insulin therapy, helped to live to old age, and enabled to raise families as large as nondiabetics. The incidence of diabetes may therefore creep up slowly in the generations to come. Now, most people would probably agree that it is better to be free of diabetes than to have it under control, no matter how successfully, by insulin therapy or other means. The prospect is not a pleasant one to contemplate. Insulin injections may perhaps be almost as common in some remote future as taking aspirin tablets is at present.

Towards Guidance of Human Evolution

Are we, then, faced with a dilemma—if we enable the weak and the deformed to live and to propagate their kinds, we face the prospect of a genetic twilight; but if we let them die or suffer when we can save them we face the certainty of a moral twilight. How to escape this dilemma?

I can well understand the impatience which some of my readers may feel if I refuse to provide an unambiguous answer to so pressing a problem. Let me however plead with you that infatuation with over-simple answers to very complex and difficult problems is one of the earmarks of intellectual mediocrity. I am afraid that the problem of guidance of human evolution has no simple solution. At least I have not found one, nor has anybody else in my opinion. Each genetic condition will have to be considered on its own merits, and the solution which may be adopted for different conditions will probably be different. Suppose that everybody agrees that the genes causing myopia, achondroplasia, diabetes and retinoblastoma are undesirable. We shall nevertheless be forced to treat them differently. Some genetic defects will have to be put up with and managed environmentally; others will have to be treated genetically, by artificial selection, and the eugenic measures that may be needed can be effected without accepting any kind of biological Brave New World.

Let us face this fact: Our lives depend on civilization and technology, and the lives of our descendants will be even more dependent on civilized environments. I can imagine a wise old ape-man who deplored the softness of his contemporaries using stone knives to carve their meat instead of doing this with their teeth; or a solid conservative Peking man viewing with alarm the new fangled habit of using fire to make oneself warm. I have yet to hear anyone seriously proposing that we give up the use of knives and of fire now. Nor does anyone in his right mind urge that we let people die of smallpox or tuberculosis, in order that genetic resistance to these diseases be maintained. The remedy for our genetic dependence on technology and medicine is more, not less, technology and medicine. You may, if you wish, feel nostalgic for the good old days of our cave-dwelling ancestors; the point of no return was passed in the evolution of our species many millenia before anyone could know what was happening.

Of course, not all genetic defects can be corrected by tools or remedies or medicines. Even though new and better tools and medicines will, one may hope, be invented in the future, this will not make all genetic equipments equally desirable. It is a relatively simple matter to correct for lack of genetic resistance to smallpox by vaccination, or for myopia by suitable glasses. It is not so simple with many other genetic defects. Surgical removal of the eyes is called for in cases of retinoblastoma; this saves the lives of the victims, but leaves them blind. No remedies are known for countless other genetic defects. Human life is sacred; yet the social costs of some genetic variants are so great, and their social contributions are so small, that avoidance of their birth is ethically the most acceptable as well as the wisest solution. This does not necessarily call for enactment of Draconian eugenic laws; it is perhaps not over-optimistic to hope that spreading biological education and understanding may be a real help. Make persons whose progeny is likely to inherit a serious genetic defect aware of this fact; they may draw the conclusions themselves.

The strides accomplished by biochemical genetics in recent years have led some biologists to hope that methods will soon be discovered to induce specific changes in human genes of our choice. This would, indeed, be a radical solution of the problem of management of the evolution of our species, and of other species as well. We would simply change the genes which we do not like, in ways conforming to our desires. Now, if the history of science has any lesson to teach us, it is the unwisdom of declaring certain goals to be unattainable. The cavalier way in which the progress of science often treats such predictions should instill due humility even in the most doctrinaire prophets. The best that can be said about the possibility of changing specific genes in man in accordance with our desires is that, although such an invention would be a great boon, it is not within reach yet. And it cannot be assumed to be achievable.

Let us also not exaggerate the urgency of the problem of the genetic management of the evolution of our species. Another problem, that of the runaway overpopulation of our planet, is far more immediate and critical. If mankind will prove unable to save itself from being choked by crowding it hardly needs to worry about its genetic quality. Although the problems of numbers and of quality are not one and the same, yet they may be closely connected in practice. As steps towards regulation of the population size will begin to be taken, and this surely cannot be postponed for much longer, the genetic problem will inexorably obtrude itself before people's attention. The questions "how many people" and "what kind of people" will be solved together, if they will be solved at all.

Some people believe that all would be well with mankind, if only natural selection were permitted to operate without obstruction by medicine and technology.

Let us not forget, however, that countless biological species of the past have become extinct, despite their evolution having been directed by natural selection unadulterated by culture. What we want is not simply natural selection, but selection, natural and artificial, directed towards humanly desirable goals. What are these goals? This is the central problem of human ethics and of human evolution. Darwinian fitness is no guide here. If, in some human society, genetically duller people produce more progeny than the brighter ones, this simply means that, in the environment of that particular society, being a bit thick-headed increases the Darwinian fitness, and being too intelligent decreases it. Natural selection will act accordingly, and will not be any less "natural" on that account.

Human cultural evolution has resulted in the formation of a system of values, of *human* values. These are the values to which we wish human evolution to conform. These values are products of cultural evolution, conditioned of course by the biological evolution, yet not deducible from the latter. Where do we find a criterion by which these values are to be judged? I know of no better one than that proposed by the ancient Chinese sage: "Every system of moral laws must be based upon man's own consciousness, verified by the common experience of mankind, tested by due sanction of historical experience and found without error, applied to the operations and processes of nature in the physical universe and found to be without contradiction, laid before the gods without question or fear, and able to wait a hundred generations and have it confirmed without a doubt by a Sage of posterity."

References

Most of the ideas and arguments presented in this article will be discussed in more detail in a book entitled "The Human Species," to be published by the Yale University Press. The references specifically mentioned in the article are as follows:

Birch, L. C. 1957. The meaning of competition. *Amer. Natur.*, 91, 5-18.

Crow, J. F. and N. E. Morton. 1960. The genetic load due to mother-child incompatibility. *Amer. Natur.*, 94, 413-419.

Lerner, I. M. 1958. The genetic basis of selection. John Wiley, New York.

Levene, H. and Rosenfeld, R. E. 1961. ABO incompatibility. Progress in medical genetics. Grune and Stratton, New York.

Mather, K. 1955. Polymorphism as an outcome of disruptive selection. *Evolution*, 9, 52-61.

Morch, E. T. 1941. Chondrodystrophic dwarfs in Denmark. *Opera Domo Bio. Hered. Hum. Univ. Hafniensis*, 3, 1-200.

Morton, N. E., and C. S. Chung. 1959. Are the MN blood groups maintained by selection? *Amer. Jour. Human Genetics*, 11, 287-251.

Muller, H. J. 1950. Our load of mutations. *Amer. Jour. Human Genetics*, 2, 111-176.

——————— . 1960. The guidance of human evolution. In S. Tax's "Evolution After Darwin," 2, 423-462.

Schmalhausen, I. I. 1949. Factors of evolution. Blakeston, Philadelphia.

Waddington, C. H. 1957. The strategy of the genes. Allen and Unwin, London.

Wright, Sewall. 1960. On the appraisal of genetic effects of radiation in man. The biological effects of atomic radiation. Summary Reports, 18-24. *Nat. Acad. Sciences,* Washington.

2. Patterns of Survival and Reproduction in the United States: Implications for Selection

DUDLEY KIRK

The outstanding human biological fact of our time is the rapid multiplication of our species. In this country, and most of the Western world, man has freed himself from many of the selection pressures that have kept his numbers down since his beginnings. Now this achievement is spreading to the rest of the world. The result is what is dramatized as the "population explosion."

At the same time growth has not been unchecked. Western populations have undergone a revolution in their vital processes of birth and death, as fundamental as the industrial revolution and modernization in its impact on the individual. The vital revolution, or the demographic transition, as it is more academically described, is the transition from wastefully high birth and death rates to the much more efficient lower birth and death rates that now prevail in the more advanced countries. The following discussion relates to the United States, but similar developments are occurring in all advanced countries.

In this country, as in all developed countries, *both* deaths and births are now largely *controlled*, on the one hand by mastery of the physical environment and on the other by the voluntary choices of couples on the number of their offspring.

In mortality, of course, controlled means *postponed*, not eliminated altogether. But for present purposes, postponement beyond the ages of reproduction is equivalent to immortality as a factor in natural selection.

Similarly, the reduction in average number of offspring is chiefly the result of voluntary choice by millions of couples to restrict family size. Surveys show that over 90 per cent of white couples of proved fertility practice birth control at some time during their married life. This practice is not always efficient, but collectively in terms of statistical averages it is very effective in restricting family size. In 1800, the average American woman passing through her reproductive life had *seven* children; today she has *three* or less.

It seems reasonable to suppose that such major changes should have very important genetic effects. In what follows, demographic changes will be discussed in relation to

SOURCE: Dudley Kirk, "Patterns of Survival and Reproduction in the United States: Implications for Selection," *Proceedings of the National Academy of Sciences*, 59, 662-670 (Washington, D. C.: National Academy of Sciences, 1968).

natural selection and selection intensity.

In any society natural selection occurs because different genotypes produce different numbers of offspring and because different proportions of offspring survive to the age of reproduction. It also follows that differentials in reproduction and survival are important only to the extent that they involve important differences in genotype. Consequently I am here discussing not the actual genotypic selection, which is, of course, extraordinarily difficult to measure, but *possible* selection intensity under different demographic conditions, that is, the *opportunity* for natural selection.

I shall discuss first the relation of demographic changes to survival, and second their relation to reproduction.

Survival. There has been a dramatic reduction in mortality and hence presumably in selection intensity. While the fact is generally understood, the magnitude of the gains is probably not often realized. Historical experience and projection for the near future are shown in Table 1.

Table 1. Per Cent Surviving to Age 15, 30, and 45, United States, White Males and Females, by Year of Birth, 1840–1960

Year of Birth	Age 15		Age 30		Age 45	
	Males	Females	Males	Females	Males	Females
1840	62.8	66.4	56.2	58.1	48.2	49.4
1880	71.5	73.1	65.7	67.4	58.3	61.1
1920	87.6	89.8	83.4	88.0	79.8	85.8
1960	96.6	97.5	95.1	96.9	92.9	95.9

Data from Jacobson, Paul H., *Milbank Memorial Fund Quarterly*, 42, 36 (1964).
These are cohort data by year of birth and therefore involve projections, especially for those born in 1960. Already by 1965, current life table values had exceeded or were approximating the projections given in this table.

The reduction of mortality is most spectacular for females. In 1840 only two thirds of white females born in 1840 reached age 15 and only about one half reached age 45. According to conservative projections, 97.5 per cent of females born in 1960 will live to age 15 and 96 per cent to age 45. Fewer and fewer now die before the end of the reproductive years—only about 4 per cent instead of 50 per cent as in 1840.

Somewhat less progress has been made in reducing deaths among males. The estimates for survivorship of male cohorts born in 1960 are less than 97 per cent to age 15 and about 93 per cent to age 45. Furthermore, male reproductivity generally starts later and continues beyond age 45 with less abrupt termination than for women. The effect of mortality is greater at these higher ages. Consequently, selection intensity is somewhat greater for males, but still minimal in comparison with the universal situation in the past. Barring major catastrophes, further reduction of selection intensity for both sexes is almost certain, but of course within a narrow range—it is already so close to zero.

Most of the force of natural selection is directed at maintaining stability, i.e., cancelling the deterioration of the organism's adaptations to its environment that otherwise would occur. Has relaxation of this "protective" selection occurred? Surely it must have. The force of this is somewhat ameliorated by several factors.

(1) The major saving of life has been in the reduction of infectious and epidemic diseases through environmental means. It does not seem that genetically determined resistance to most infectious diseases (to the extent that it exists) is an adaptation of great consequence in modern society. Several of the great epidemic and endemic diseases—smallpox, cholera, plague—were mastered and largely eliminated in the 19th century. Since then, great progress has been made against those remaining: pneumonia and influenza, typhoid fever, tuberculosis, syphilis, diarrhea and enteritis, and the communicable diseases of childhood. Organic diseases are now the great killers—cardiovascular-renal diseases and cancer account for about three fourths of all deaths. These organic diseases may be more dependent on genetic inheritance, but they are not so important in earlier life and therefore do not much affect survival to and through the reproductive period.[1]

Largely because other causes of death have been so reduced, deaths by accident and violence are major causes of death prior to age 45. It is certainly nothing new to have negative selection of persons prone to accident and violence. This selective factor remains, though unhappily it seems unlikely that fatal automobile accidents will early lead to a genetically superior race of better drivers!

(2) Deaths related to specific inherited defects have not been reduced nearly so much as the total. Thus two thirds of female deaths and over 60 per cent of male deaths up to age 30 are now due to congenital malformations and diseases of early infancy, which often involve immaturity.[2] These include an important component of genetic defects. The most serious of such defects are still being removed from the population through deaths.

(3) Were it possible to do so, the survival factor should be computed from *conception* rather than from live births. Fetal wastage and stillbirth, especially of the malformed, is still high. The wastage in early pregnancy is believed to be still very high, presumably much more important than subsequent fetal wastage and mortality in eliminating gross malformations and genetic anomalies.

It would be unwise to take much comfort from these ameliorative factors. Undoubtedly many persons of genetically weak constitution are now surviving who earlier would have succumbed to disease or to a harsher environment. Furthermore, determined efforts are being made to reduce infant deaths, including those involving genetic defects. It is reasonable to expect that an increasing proportion of individuals having a weak physical constitution and/or carrying deleterious mutations will survive and reproduce. The direction is not at issue—only the specific nature and rapidity of the changes. Measurement of the latter is extremely difficult and beyond the scope of this paper.

On the other hand, environmental gains are more than compensating for any genetic deterioration, as evidenced by the gains in physical stature, by the reductions in morbidity, and by the falling death rates themselves.

Even more important, what is deteriorating is our fitness for the physical environment of the *past*, not that of the present or the future. The genetic qualities required for survival and effective performance in a peasant or pastoral society are presumably very different from those required for best performance in our primarily urban and sedentary life. We are losing our adaptability for the former but certainly much less for the life of the present, and perhaps even less for that of the future.

Reproduction. It is a commonplace that in modern societies natural selection by deaths has been replaced by the social selection of births. This frequent observation is faulty in two ways. First, selection by number of progeny is just as "natural" as selection by deaths; and second, selection by number of progeny has usually been the more important element in the past, as it is in the present. Thus Spuhler's comparisons of selective intensity from mortality and from fertility components show the latter factor to be more important in a majority of the numerous modern and premodern populations covered.[3]

Demographic changes in four areas affect the opportunity for natural selection through differential fertility. These are (1) mating and marriage, (2) childlessness, (3) number of offspring, and (4) age at childbearing and mean length of generation.

(1) *Mating and marriage:* The married state has become more prevalent in the United States over the last two generations, but particularly since 1940. The per cent of adults in the principal ages of reproduction who are married has risen substantially. Thus, of all men at ages 14-44, 50.5 per cent were currently married in 1940 and 61.2 per cent in 1960. The corresponding data for women are 58.7 per cent in 1940 and 68.9 per cent in 1960, in each case a gain of over 10 per cent.[4] This increase is due to three factors: a larger percentage of men and women marry; they marry earlier; and they spend more of their potentially reproductive years in marriage.

To illustrate, in 1966, 95.0 per cent of all women at ages 35-44 (i.e., at the later years of childbearing) had ever married, as compared with 88.6 per cent in 1910 and 91.4 per cent in 1940. The median age at first marriage of men fell from 26.1 in 1890 to 22.8 in 1960; that of women from 22.0 in 1890 to 20.3 in 1960. This earlier first marriage plus less widowhood and higher rates of remarriage have led to more years in the reproductive ages spent in marriage despite a rise in divorce.

Childbearing is not entirely restricted to married couples. Illegitimacy in the United States has risen rapidly. Since 1940, the estimate of per cent of illegitimate births has doubled for whites and increased by over 60 per cent for nonwhites. National estimates, based on the official statistics of states that report illegitimacy, show 3.4 per cent of white births and 24.5 per cent of nonwhite births to be illegitimate in 1964.[5] The true figures are doubtless higher. A reasonable inference is that mating outside of marriage, as well as marriage itself, has increased.

Thus the potential reservoir of parents has been increasing, and this was in part responsible for the "baby boom." What it means genetically is that a larger percentage than previously of each cohort is exposed to the "risk" of childbearing, especially through earlier and almost universal marriage. We are in this way "reverting" to the situation prevailing in premodern societies, where marriage and mating are almost universal. In this country, age at marriage is lower than in most other Western countries, but of course higher than in most premodern societies.

(2) *Childlessness:* In our society there are three reasons why adults fail to have children: biological infecundity, failure to mate or marry, and choice, i.e., the practice of birth control. The second and third are unimportant in most premodern societies, including our own at an earlier stage. In the United States 100 years ago only a tenth of American women living through the childbearing period were childless. With less marriage, later marriage, and more practice of birth control, 23 per cent of women born in 1909 were childless. Now, again with more and earlier marriage, and probably greater success in the medical treatment of sterility, only 7 per cent of married women and 12 per cent of all women at age 35 are currently childless.

Let us consider the joint effect of mortality and childlessness, which are genetically the same: the persons concerned have no descendants. Combining these effects, about half of women born in 1840 did not participate in the reproduction of the next generation. Now the combined effect is only 15 per cent. Thus at current experience some 85 per cent of each female cohort born now not only survives but also produces offspring.

(3) *Number of offspring:* As noted earlier, the average number of children born per woman is much lower than in the earlier years of our history—three or less versus seven. Surviving children per woman are much more comparable in number, thanks to the great saving of life at earlier ages.

While mortality has declined continuously, natality has not followed a continuous decline since the 1930's. As result of the "baby boom" after World War II, the average number of children per woman rose from a low of 2.3 (for the female cohort born in 1909) to estimates of about 3 for those born in 1940, who have not yet completed their childbearing.

From the point of view of natural selection the variance is more important than the average (see Table 2). Contrary to "common sense," the possible selective intensity of the natality component did not decline with the long secular decline in family size. This anomaly is explained by the fact that reduction of the mean number of children does not necessarily reduce the variance. On the contrary, in the United States the opposite has been the case. Thus the index of selection intensity for the fertility component[6] was 0.710 for women born 1871-1875, who averaged 3.5 children. For women born in 1909 the index was 0.881, with an average of only 2.3 children, the lowest number for any U.S. cohort.

Table 2. Number of Offspring of Selected Cohorts of Women in the United States

	Per Cent Distribution by Parity			
	Women of Completed Fertility			Expectations of married white women under age 40 in 1962
Parity	Born in 1871–75	Born in 1909	Born in 1928 (at age 36)	
0	17	23	11	4
1	16	22	12	7
2	14	22	24	26
3	12	13	22	28
4	10	8	14	21
5	7	4	8	7
6 and over	24	8	9	7
Total	100	100	100	100

With the recovery of the birth rate since 1940 the index will certainly decline sharply, although the women concerned have not completed their childbearing. There will almost certainly not be much increase at higher parities that would contribute heavily to variance since birth rates at the higher reproductive ages are now very low. Finally, if the expectations of married white women under age 40 in 1962 are realized (with 75% in the range 2 to 4),[7] the index for this group will decline to 0.26.

Since 1957 the "baby boom" has receded, and the birth rate in 1966 was the lowest for any year since 1936. This is related to changes in the age structure of the population and may or may not portend a reduction in family size. The recent drop certainly is not just the result of the introduction of the *pill*; during the 1930's couples were just as successful in birth control with other methods. But the fact that birth control methods are becoming easier and more reliable makes family planning more effective. This factor will probably further restrict the number of high parities.

(4) *Parental age:* The intrinsic rate of natural increase is determined not only by the number of children, but also by the length of generation. A couple marrying early and having their children early have a higher rate of reproduction than one marrying and having their children late, *even though* they have the same number of children. Variance in age at marriage and length of generation is a source of differential fertility and of selection.

Recent demographic trends have also reduced this source of differential fertility. As noted earlier, marriage is earlier; also intervals between marriage and first and subsequent births have been reduced so that births are increasingly concentrated in the early childbearing years. The average length of female generation has declined from 28 in the 1930's to 26 at the present time, with less variance around the mean. The median American woman is married before she can vote, has her first child at age 21, and her third and last at age 27. The average childbearing period for women has been reduced to about seven years.

This concentration of childbearing in young ages and at low parities has direct genetic influence since a number of genetic disorders are correlated with age of mother, birth order, and number of children per family. These include new mutations, such as Down's syndrome, and genetic factors in combination with environmental factors, such as Rh-erythroblastosis, congenital malformations of the circulatory and nervous systems, cerebral palsy, etc. Newcombe's Canadian data suggest twice as high an incidence for the latter type of disorder for older mothers of high parity as compared with young mothers of low parity.[8] Reviewing Japanese experience between 1947 and 1960 (i.e., earlier childbearing and smaller families), Matsunaga estimates a reduction value of one third for mongolism, more than one half for Rh-erythroblastosis, and on the order of one tenth for the remaining defects.[9] While these lack specificity for the experience in this country, they suggest some order of magnitude for gains deriving from our growing concentration of childbearing at lower ages and lower parities.

Furthermore, reduction of the number of siblings limits the opportunity for consanguineous marriage, and more broadly inbreeding, and hence cuts the number of births with defects arising from such marriages.

Differential Fertility. In the preceding I have been discussing the situation in the U.S. population as a whole, often more specifically the white population.

Within the total population there have been very major differences in reproduction or differential fertility between social, economic, religious, and ethnic groups. Historically there were, for example, enormous differences between the urban dweller and rural farm resident, the uneducated and the college graduate, the professional and the laborer, etc. These differentials evoked fear that we have been "breeding from the bottom," the least capable producing the most children.

Whatever their earlier significance, socioeconomic differentials in fertility have contracted since World War II, and especially between 1950 and 1960. There is abundant evidence in the censuses of this narrowing of socioeconomic differentials between 1950 and 1960 for such characteristics as rural or urban residence, occupation, income, and education.[10] Most elements in the population have shared in the increase in fertility, but especially the better educated and upper income and occupational groups. This has narrowed the differentials.

To take a single, rather dramatic example: in 1950, white women at age 40-44 with less than eight years of schooling who had ever been married had had over twice as many children as women with four or more years of college. In 1960, this margin had been cut to 50 per cent more, and with current trends it will decline by 1970 to below 40 per cent more. Women of intermediate educational levels fall between these extremes, and because of general upgrading of education there is now less variance about the mean. Thus the completed fertility of married women with four or more years of college was only 59 per cent of the average for all married women born in 1901-5; it rose to 73 per cent for women born 1916-20; and will rise further to 89-93 per cent for women born 1926-30, the latter figure depending on assumptions about future fertility of this younger cohort. The gap between the most educated and the average is much less than before.

This reduction is not so true of two other fertility differentials—those by religion and by race.

In this country Catholic fertility has been consistently higher than non-Catholic, and this difference has probably not been narrowed in recent years. The genotypic significance of this difference is obscure. In any case, there is now rapid change in Catholic attitudes about family planning and there may well be a convergence in birth rates by religion.

The position of the nonwhite population (chiefly Negro) is quite different from the white. The opportunity for natural selection remains much higher both as regards mortality and natality. Twice as high a proportion of Negroes die before reaching age 15 and age 45. Negro women have more children, more widely scattered through the childbearing period, but more Negro women are childless in marriage. The result is greater variance in number of offspring and more opportunity for natural selection.

Differential fertility is also much greater among Negroes than among whites. Reproduction of urban, educated, and middle-class Negores is similar to that of whites, whereas fertility of low-income and rural Negroes is much higher than that of whites of comparable socioeconomic status. At present the Negro birth rate is declining as the low-fertility elements become a larger part of the Negro socioeconomic structure. The same forces are at work among Negroes as among non-Negroes—reduction of mortality, reduction of family size, reduction of fertility differentials—but in each case the process has not gone so far as in the remainder of the population. In the population as a whole there is and will be for some time a growing proportion of Negroes—the per cent of nonwhite births increased from 14.4 in 1950 to 15.8 in 1960 and to about 16.9 in 1965.[11]

Summary and Conclusions. Demographic trends in the United States are strongly in the direction of reducing the opportunity for natural selection. This is the result of several convergent elements.

(1) The mortality component in selection intensity has been dramatically reduced. The genetic load is doubtless being increased by the survival of deleterious mutations but environmental changes have made some hereditary defects irrelevant, such as susceptibility to now-curable diseases.

(2) A larger proportion of persons in the reproductive ages are married, and hence exposed to the "risk" of childbearing, than a generation ago. There is also a retreat from childlessness and perhaps some decrease in physiological sterility. About 85 per cent of each white female cohort born now lives to adulthood and participates in the reproduction of the next generation.

(3) Since 1940 the average number of children per couple has risen somewhat after the long secular decline in fertility ending in the 1930's. There is a growing concentration of couples with two to four, and especially two and three, children. The variance and hence the opportunity for natural selection through variability in numbers of offspring is much reduced.

(4) The average length of generation has fallen and the variance has diminished, owing to the concentration of births to parents in their twenties and early thirties.

(5) Differential fertility between socioeconomic groups is generally declining. Potential selectivity in terms of residence, occupation, income, and education is thereby diminished. An important exception is the fertility differential by race. Negro fertility and mortality are declining but in both cases are still substantially higher than those for whites. The opportunity for natural selection is higher in the Negro population.

There is a growing homogeneity in reproduction. The white population of the United States appears to be moving toward a situation in which the great majority live to adulthood, marry, and have two or three children. To the extent that this occurs, each generation will be close to a genetic carbon copy of its predecessor, aside from mutations that are absorbed by the genotype.

At the same time it must be emphasized that present fertility trends could change rapidly, as they have in the past. Furthermore, their genetic implications could be modified, for example, by an increase in assortative mating, bringing a rearrangement of genes without necessarily changing their frequency in the total population.

The levels and trends of mortality have been more stable than those of fertility. Barring catastrophe, very low mortality at the younger ages will be a continuing feature of our civilization. A relaxation of selection intensity of the degree and durability now existing among American and Western peoples has surely never before been experienced by man.

The potential results are not clear and will require far more sophisticated analysis. In the short run present demographic trends are reducing the incidence of serious congenital anomalies because of the younger average age at childbearing, the lower average parities, and the reduction of consanguineous marriages. And surely in the foreseeable future the possibility of medical and environmental correction of genetic effects will far outrun the effects of the growing genetic load. The longer-run effects depend upon the increase in mutational loads and how serious these may prove to be in a rapidly changing environment more and more created by man himself.

Footnotes

[1] A good summary of changes in causes of death is given by Spiegelman, Mortimer, *Significant Mortality and Morbidity Trends in the United States since 1900* (Bryn Mawr, Pa.: American College of Life Underwriters, 1964).

[2] Computed from U.S. Public Health Service, *Vital Statistics of the United States, 1965* (Washington, D.C.: Government Printing Office, 1967), vol. 2, *Mortality*, pt. B, Tables 1-4 and 1-9.

[3] Spuhler, J.N., "Empirical studies on human genetics," *Proceedings of the UN/WHO Seminar on the Use of Vital and Health Statistics for Genetic and Radiation Studies, September 5-9, Geneva, 1960* (New York: United Nations, 1962).

[4] These and following illustrative data are from the appropriate U.S. censuses and from U.S. Department of Commerce, Bureau of the Census, *Current Population Reports*, Series P-20, 159 (Washington, D.C.: Government Printing Office, Jan. 25, 1967).

[5] U.S. Public Health Service, *Natality Statistics Analysis*, 1964, Series 21, 11 (Washington, D.C.: Government Printing Office, February 1967).

[6] After Crow, J. F., *Human Biology*, 30, 1-13 (1958). The index of selection intensity of the fertility component (I_f) is the variance in the number of children born divided by the square of the average number born per parent.

[7] Freedman, R., D. Goldberg, and L. Bumpass, *Population Index*, 31, 3-20 (1965).

[8] Newcombe, H. B., *Ann. Human Genetics*, 27, 367-382 (1964).

[9] Matsunaga, E., *J. Am. Med. Assoc.*, 198(5), 120 (1966).

[10] U.S. Department of Commerce, Bureau of the Census, *U.S. Census of Population, 1950* (Washington, D.C.: Government Printing Office, 1955), Special Report P-E No. 5C, *Fertility*; and *U.S. Census of Population, 1960* (Washington, D.C.: Government Printing Office, 1964), subject reports, final report PC(2), pt. 3A, *Women by Number of Children Ever Born.*

[11] Figures for 1950 and 1960 from U.S. Department of Health, Education and Welfare, "White-nonwhite differential," in *Indicators*, September 1965; data for 1965 from U.S. Public Health Service, *Vital Statistics of the United States*, 1965 (Washington, D.C.: Government Printing Office, 1967), vol. 1, *Natality*, Tables 1-3. Data for the two earlier years had been corrected for under-registration of births.

3. The Study of Mutation and Selection in Human Populations*

HOWARD B. NEWCOMBE

Interest in the genetics of human populations may be either of an *historical* or of a *predictive* kind. That is to say, most population genetic studies relate fairly directly to one or other of two sorts of question, namely *"How did we get this way?"* and *"Where are we heading?"* Both emphases imply a certain amount of interpretation and speculation, but both have served to stimulate the gathering of relevant empirical data of quite non-speculative kinds. In view of the sponsorship of the present lecture by the Eugenics Society, and the special association with the name of Galton, it will deal primarily with the second of these two questions.

To those investigators who are concerned with current genetic trends and their long-term consequences, the large amount of information about people which is gathered routinely for a variety of purposes in modern societies may serve either as a special spur or, because of the immensity and relative inaccessibility of the resulting accumulation of facts, as a major source of discouragement. The difference depends on the view taken of the respective magnitudes of the difficulties as compared with the opportunities for extracting from the existing documentation genetic facts of the kinds needed for the more important sorts of human population studies.

The opportunities are of course substantial, even if as yet largely unrealized, because for no organism other than man have the facts of procreation and family composition, and of mortality and morbidity, been so minutely documented over such large populations. Furthermore, in no experiment with laboratory mammals is it even remotely conceivable that so much money and effort could ever be spent identifying the natures and causes of deviations from physical and mental well being, and social adjustment, as is done in any advanced human society.

SOURCE: Howard B. Newcombe, "The Study of Mutation and Selection in Human Populations," *Eugenics Review*, 57: 109-125 (London, England: The Eugenics Society, 1965).

*The Galton Lecture, delivered in London on 5th May 1965.

A major part of the resulting documentation is, in fact, relevant to studies of those influences which operate to alter the collective germ plasm, or pool of genetic material, and which therefore determine the hereditary basis of health and well being of future generations. However, genetic studies are, virtually by definition, family studies; and the biggest barrier to the use of much of the enormous body of recorded facts about people for any sort of genetic research lies in the difficulty of relating, or linking, these facts to the family histories of the persons concerned.

The present account will deal with the nature of the information needed for studies of current genetic trends, and with the ways that may be used to extract and apply such information. As we proceed, the difficulties will seem to loom larger. And yet, to admit to the possibility that we may never know in any substantial detail what changes are in fact taking place in the gene pool, and what factors are potent in influencing these changes for better or worse, would seem quite improper until we get much closer than we are now to exhausting the information that is already potentially at our disposal.

The Nature of Population Genetics

The kinds of information that are required may best be understood by considering for a moment the changes that take place, over a single generation cycle, in the collective germ plasm of which each generation is, in turn, the temporary custodian.

The diversity of hereditary characteristics represented at the outset owes its presence to past events of *mutation*, that is of *de novo* hereditary changes, and also, if one thinks of a population less than that of the world as a whole, of migration into the population of genetically diverse people. Such occurrences, in each generation, contribute further to the genetic diversity that has been already accumulated over the past.

The outward expressions of the resulting heterogeneity within the collective pool of genes are only in part determined by the external environment; just as important are the so-called patterns of mating among the immediate ancestors of the generation under observation. If marriages were to occur in a strictly random fashion between members of various population subgroups, the effect would be a blending of many widely different inheritances, so as to obscure much of the hereditary diversity present in the gene pool. In practice however, matings tend, in varying degrees, to occur preferentially within the various population subgroups, as based on family relationships, religious affiliations, ethnic origins and economic circumstances. There is in addition a measurable tendency for persons of similar physical and mental characteristics to marry one another. The effect of these inbreedings and assortative matings, when they occur, is to provide increased opportunity for outward expression of the genetic diversity inherent in the collected inheritance. An extended range of phenotypes is thus subjected to the tests of survival and procreation.

The composition of the gene pool of the next generation depends, of course, upon the selective effects of the differences in mortality and fertility or, in other words, upon what kinds of people reproduce, and how much. Selection does not operate directly upon the genes, to favour or suppress the propagation of one as compared with an alternative gene. Rather, it is the phenotypes or outward expressions which are in part determined by the genes but which are also frequently influenced by the environments as well, that are "selected" on the basis of their abilities to survive and reproduce. Genes that show a tendency to be associated with phenotypes of high selective value will, as a result, be transmitted preferentially to the gene pool of the next generation.

From all this it is apparent that human population genetics is concerned with a relatively small number of major phenomena, notably mutation, migration, mating patterns and selection. For the study of any of these, family histories or pedigrees are an essential requirement and the more of such histories that can be marshalled the better.

The point I wish to emphasize at this juncture is that the basic pedigree information is accurately recorded on a continuing basis as part of the vital statistics systems of most countries, and is essentially complete for all who remain within the registration areas. These systems also record our geographic and racial origins, religious affiliations and occupations; and if other routine records such as the census enumeration forms, income tax returns and so on are regarded as further potential sources, the documentation of our social characteristics is more complete still. Similarly, our physical and mental attributes are recorded in even greater detail, partly because of our demands for medical and health services, but in other ways as well, as in the case of records of school performance, enlistment in the armed services, and of unemployment, crime and delinquency.

Such documentation is in most instances tied to the names of the individuals concerned so that, in principle at least, virtually all of it could be integrated with the family pedigrees. Of all the recorded information, however, the most fundamentally important is that which serves to identify immediate relatives, and which is contained in the vital records systems in a form that makes possible the compilation on a statistical scale of histories of marriage, procreation and death in individual families. Thus we already possess much of what is needed for detailed and precise studies of fertility differences, mating patterns and so on, and the scope and reliability of the documentation is increasing continuously.

Access to such information is not always simple. Use of some of the records is restricted by law, and in most instances it is limited by the sheer magnitude of the files. In addition, there are organizational and technical problems of interrelating, on any substantial scale, the facts contained in two or more independently derived records pertaining to the same person or family.

For such reasons, major advances in the use for research purposes of any substantial fraction of the potentially available information are not to be expected in a hurry; but modest developments are in progress now, and some early products of one of these, relating particularly to mutation and selection will be described.

Pedigrees from Record Linkage

The study from which these products have been derived relates to the Canadian province of British Columbia, with a current population of 1.7 million, in which about 10,000 marriages and 40,000 births occur annually. A ten-year marriage file (1946-55) and a six-year birth file (1952-58) have been used. In addition, there are records of stillbirths over the same six-year period, and of child deaths, together with registrations of handicaps among children born in this period, provided by the exceedingly well run British Columbia Registry of Handicapped Children and Adults.

By methods that have been described fully elsewhere,[4,5,9,10,11] these registrations have been integrated into the form of family histories, starting in each instance with the marriage record (where this is present in the files), followed by those of any procreations arising out of the marriage, and with the records of handicaps and deaths among offspring of the married couples interpolated behind the birth records of the children to which they relate. Somewhat over 200,000 records are currently in use, and the integrated files will in due course be extended backward to 1946 and forward to include current records as they become available.

The technology is now well developed, and the actual merging and linking operation in which current records are used to update the family groupings on an existing master file can be carried out by an electronic computer exceedingly rapidly and economically.

Mutations

One of the applications of the record linkage technology which I propose to discuss in detail has to do with the possibility of detecting and measuring the consequences of recurrent mutations in the human germ plasm. Most of the genes which arise from these mutational events are believed, with considerable reason, to be harmful; and in recent years much thought has been given to the possible importance of such artificially induced increases in mutation rate as must almost certainly result from exposures of human reproductive tissues to man-made radiation.

The problem has substantial financial overtones, because upper limits to the exposures of whole populations from the future peaceful uses of atomic energy are being suggested now by an authoritative body.[15] There is, as a result, considerable pressure on individual geneticists to assess the likely consequences of the mutations that would arise in human populations from given radiation exposures, even where the amount of the harm cannot be judged to within an order of magnitude or so in either direction.

Such estimates of genetic damage from artificially induced mutations are unlikely to attain increased precision until more is known of the importance of the mutations that occur anyway as a result of natural causes. Some fraction of the current burden of hereditary ill health and handicap, and possibly a substantial fraction of it, is believed to be maintained in the population by repeated mutations of natural origin.

There has been a tendency on the part of geneticists in the past to concentrate on studies of mutation frequencies at particular gene loci which control striking and readily scorable characteristics, such as achondroplastic dwarfism, the heritable eye cancer known as retinoblastoma and the conspicuous nerve tumours of von Recklinghausen's disease or multiple neurofibromatosis, to take a few examples. Although the results of such studies are often quoted, they are essentially disappointing for present purposes, not because of any of the inherent imprecisions, but rather because they relate to a small and quite unrepresentative group of mutant traits and indicate little indeed concerning the possible importance of mutations throughout the total complement of gene loci.

There is at least one other approach to the detection of the consequences of mutation throughout the genome as a whole. If mutations or mutagenic influences accumulate during the reproductive life spans of individuals, genetically determined harm might be detectably increased in offspring from older parents. This suggestion, which was made originally by Penrose [14] and Sonneborn, [16] was in fact later borne out in the case of one quite striking hereditary disease, namely mongolian idiocy or Down's syndrome, the risk of which is greatly elevated among offspring from mothers who are approaching the end of the reproductive period. The mutation that determines this condition, however, involves not just an alteration of a single gene, but rather a gross chromosomal change consisting usually of the incorporation of an additional chromosome over and above the normal number. Nevertheless, other similar but less striking increases in the risks of various hereditary conditions with advancing maternal or paternal ages have been interpreted as probably due to increased incidences of particular gene mutations in gametes from older mothers and fathers.

I should like at this stage to mention certain difficulties in the use of the approach that may perhaps make the early applications of it seem, at first sight, light-hearted indeed, but which should not be taken to imply that we must forever forgo knowledge of the consequences of mutation, or that the approach should be abandoned in favour of some superior alternative when in fact there may be none.

The first difficulty relates to the availability of suitable empirical data. The early observations were based on quite limited ascertainments of the parental age distributions for children with various genetic disorders, and the control data related to populations that were not necessarily representative of those from which the cases were drawn, and which could have differed substantially in the distributions of parental ages at birth. Such limitations of the data arise from essentially artificial causes, in that very substantial numbers of affected children are known to exist and the necessary parental age information is almost universally recorded in the birth registrations. The difficulty of securing the necessary data arises only because the two sorts of information relating to a particular individual tend to be recorded independently on separate documents of quite different kinds, both of which may at some later date be effectively buried in massive files of accumulated records. As will be seen later, other facts needed for the study of mutation by means of this approach are likewise recorded but tend to be inaccessible for the same reason, and may remain so unless economical means are used to interrelate, on a substantial scale, specific by individual, the information contained in various large, independently derived files.

Our own studies of parental age correlations have been based on registrations of live births, stillbirths, handicaps and deaths among just over 200,000 children born in British Columbia during 1953-58, of whom more than 12,000 were registered either as stillborn, handicapped or as having died by the end of 1959. Record linkage techniques were employed which make use of computers to merge the parental ages and birth orders from the birth records, with the diagnoses from the handicap and death records of the same individuals. Although larger amounts of data were derived in this manner than have been analyzed in previous studies, the results must be regarded as modest in comparison with those that may be obtained through applications of similar methods on a continuing basis to these and other files of vital and health records. Furthermore, the most important objective of the study is judged to be the continuing development and testing of methods by which such information may be more fully utilized in the future.

The results serve, however, to illustrate the kinds of effect on risks to the offspring, that are not causually related to the advancing age of the father or mother, but are only secondarily associated with parental age, and which therefore must be carefully distinguished if the approach is to be used at all to study the consequences of mutation. Some of these secondary associations will tend to obscure the effects of mutations; others will tend to simulate or mimic them.

In the former category (i.e., those which tend to obscure) are the special risks to firstborn offspring and to offspring of very young mothers, which fortunately can be distinguished from one another by suitable statistical methods. A whole group of diseases peculiar to early infancy, and in particular the intracranial and spinal injuries at birth, have about twice the frequency among firstborn as compared with secondborn children, when effects of maternal age differences are removed (Table 1). Similarly, children of very young mothers, age nineteen and under, are about twice as prone to a wide variety of conditions, which include intracranial and spinal injuries at birth and postnatal asphyxia, as are children of mothers in the next five-year age group (Table 1); this is true even when effects of birth order differences

Table 1. Summary of Special Risks Associated with Birth Order and Parental Ages.
(From B.C. Handicap and Death Records Linked with Birth Records, and From Stillbirth Records)

Disease Category	Code	Total Cases	Relative Risk	χ^2 (D.F. = 1)	Reference
Firstborn children (vs. 2nd) (excluding maternal age effects)					
Certain diseases of early infancy	760–776	58	2·16	10·6	7
Intracranial and spinal injury at birth	760	123*	1·80	9·1	12
Higher birth orders (3rd and over vs. 1st and 2nd) (excluding maternal age effects)					
Infective and parasitic diseases	001–138	146	1·76	9·9	7
Strabismus	384	285	1·58	11·7	12
Other c.m. of nervous system and sense organs	753	175*	1·84	12·1	12
Postnatal asphyxia and atelectasis	762	773*	1·70	39·2	12
Children of very young mothers (0–19 vs. 20–24) (excluding birth orders effects)					
Categories VI–XII	400–716	58	2·44	8·6	7
Intracranial and spinal injury at birth	760	72*	2·19	9·6	12
Postnatal asphyxia and atelectasis	762	413*	1·66	15·4	12
Children of older mothers (35 and over vs. 0–34) (excluding birth order effects)					
Mongolism	325·4	191	7·68	172·6	12
Cerebral palsy	351	215	1·84	11·6	12
C.m. of circulatory system	754	868*	1·66	29·1	12
Children of older fathers (40–99 vs. 0–39) (excluding maternal age effects)					
Diseases of the respiratory system	470–527	955*	1·61	17·8	13
Congenital malformations	750–759	1727*	1·28	9·1	13
Other c.m. of nervous systems and sense organs	753	135*	2·07	7·8	13

*Includes cases ascertained through death and/or stillbirth records.

are removed (for an example of the method for removing the effect of one variable or the other, see Table 2).

Table 2. Example of Special Risks to Children of Older Mothers, when Effects of Birth Order Differences Are Removed. Congenital Malformations of the Circulatory System (Comparing Mothers Aged 35 and Over with Those Under 35)

	Cases		Controls			
Birth Order	Older Mothers a	Young Mothers b	Older Mothers A	Young Mothers B	Relative Risk aB/Ab	χ^2
1	8	236	2,611	59,675	0·77	...
2	24	189	4,381	53,411	1·55	...
3	43	136	6,039	36,239	1·90	...
4	37	71	5,372	18,755	1·82	...
5	25	28	3,533	8,598	2·17	...
6–9	31	333	5,161	7,519 }	1·42	...
10	7	0	1,432	627 }		...
Weighted mean relative risk (D.F. = 1)					1·66	29·1*

*P < 10^{-7}.
Data from Newcombe and Tavendale, 1964.[12]

More serious for present purposes are the effects that can mimic those of increased mutation frequencies with advancing parental ages. Not surprisingly, the infective and parasitic diseases (notably tuberculosis and polio) tend to be more common among children of birth orders three and over, the same being true of postnatal asphyxia (Table 1). As a result, these conditions exhibit a parental age effect that is in reality secondary to the birth order effect. An environmental factor is almost certainly involved, since for both sorts of condition the effect of birth order is most striking when the mother is young (for an example, see Table 3). Possibly this is because the pregnancies have been closely spaced, or because of crowding in the home, or perhaps because the mothers who have had many children while still very young tend to be from a socio-economic group in which the risks of such conditions are high. To distinguish between these alternatives, further information about the social circumstances and the reproductive histories of the families would be needed, much of which is already documented in the vital registrations, census enumerations and elsewhere. There is no certainty, of course, that genetic factors play any substantial role in the causation of these particular conditions, but the examples serve nevertheless to illustrate the ways in which biases can arise. Similar interactions appear to be quite common (Table 4).

Of less obvious origins are the increased risks of strabismus, or "squint," and of certain "other" congenital malformations of the nervous system and sense organs

Table 3. Example of an Interaction Between Birth Order and Mother's Age as
Influencing the Risk to the Child. Postnatal Asphyxia and Atelectasis
(Comparing Birth Orders 4 and Over with 1 to 3)

Mother's Age Group	Cases		Controls		Relative Risk aB/Ab	χ^2
	Higher Birth Order a	Lower Birth Order b	Higher Birth Order A	Lower Birth Order B		
0–19	3	86	95	18,584	6·82	...
20–24	42	181	4,715	56,495	2·78	...
25–29	63	86	14,084	47,584	2·48	...
30–34	64	83	16,605	26,662	1·24	...
35–39	41	51	11,474	10,643	0·75	...
40–44	17	12	3,748	2,282 ⎫	1·06	...
45–49	2	0	276	106 ⎭		
Weighted mean relative risk (D.F. = 1)					1·70	39·2*
Heterogeneity (D.F. = 5)						39·9†

*P $<$ 10^{-9}.
†P $<$ 10^{-7}.
Data from Newcombe and Tavendale, 1964.[12]

Table 4. Summary of Interactions Between Birth Order and Mother's Age
as Influencing the Risk to the Child. (Chi Squares for Heterogeneity;
For Details of Derivation see Table 3)

Code	Cause	Groups Compared	χ^2	D.F.	P
325	Mental deficiency	mother's age 0–19/20–24	8·2	2	0·02
330–398	Nervous system	mother's age 0–19/20–24	8·6	3	0·04
351	Cerebral palsy	mother's age 0–19/20–24	14·0	3	0·003
353	Epilepsy	birth order $>$ 2/1–2	10·8	4	0·03
751, Y38·2	Spina bifida	birth order $>$ 2/1–2	15·7	5	0·008
758	C.m. bone and joint	birth order $>$ 2/1–2	10·2	4	0·04
762	Postnatal asphyxia	birth order $>$ 2/1–2	39·9	5	0·001

Data from Newcombe and Tavendale, 1964.[12]

among children of higher birth orders (Table 1) which might, if birth orders were
not taken into consideration in the analysis, appear as maternal age effects of kinds
that would mimic the consequences of mutations accumulating in the female germ
cells. Genetic causes of another sort might conceivably be involved, since maternal/
foetal incompatibilities could very well be operating to increase the likelihood of
sensitization of mothers of higher parities, thereby increasing the risks to offspring
of higher birth orders. Such effects would not, of course, indicate any influence of
parental ageing as such.

Of the increased risks shown to be associated primarily with advancing age of mother, that of mongolian idiocy is known already to be mutational in origin, while that of cerebral palsy could perhaps be attributed to the changes in intra-uterine environment that occur in older mothers. Evidence to rule out such an effect of the prenatal environment is not wholly unobtainable, but would require another generation of pedigree information to permit comparison of the risks in the current generation of children for different ages of the maternal grandmother where the mother's age is not a variable. No one has seriously attempted to carry out such multi-generation comparisons on any substantial scale as relating to particular diseases or groups of diseases, although data have been obtained on correlations of the grandmaternal age with the sex ratio, of a kind indicating possible differences in prenatal mortality due to lethal mutation in the sex chromosomes.[2] Extensions of the approach to include mortality and morbidity of children, specific by disease or broad category of disease, would require application of pedigree information of a kind that is already contained in the accumulated vital records of most countries, but on a scale that would heretofore have seemed impracticable in the extreme.

Observed correlations of risks with advancing age of the father at the time of the birth, independent of the mother's age, cannot, of course, be attributed in the above manner to a changing intra-uterine environment, but other difficulties of interpretation present themselves. One such correlation, involving the risk of congenital malformation, and more particularly the group of such conditions known in the International Classification as other congenital malformations of the nervous system and sense organs, might perhaps be genuinely indicative of increased mutant frequencies in gametes from ageing fathers (Table 1). However, a similarly increased risk of diseases of the respiratory system (mainly deaths from pneumonia and bronchitis) which has been observed among offspring from older fathers seems quite unlikely to be mutational in origin (Tables 1 and 5). A relatively simple explanation is in this instance apparent from examination of the data; most of the deaths from respiratory diseases among British Columbia children occur in North American Indians, and Indian fathers tend to be older than is usual for the rest of the population. The correlation disappears when the data are broken down by Indian versus non-Indian parentage.

Findings of this sort emphasize the importance of distinguishing those correlations with paternal or with maternal age that are secondary in nature, and that depend on differences in the parental age distribution for population subgroups within which the risks may be higher or lower than for the population as a whole. The principal limiting factor in the identification of such non-mutational effects is not the amount of recorded knowledge of social characteristics, which is really quite extensive. Rather, it is the mechanical difficulty of bringing together such recorded information, as relating to particular individuals, and of doing so on a large enough scale to permit multiple breakdown of the data without spreading them too thinly between the resulting pigeon-holes or matrix cells.

Table 5. Example of a Special Risk to Children of Older Fathers, when Effects of
Maternal Age Differences are Removed, Which is Known to be Secondary to
Racial Inhomogeneity in the Population. Respiratory Diseases
(Comparing Fathers 40 Years and Older with Those of 39 and Younger)

Mother's Age Group	Cases		Controls		Relative Risk aB/Ab	χ^2
	Older Fathers a	Young Fathers b	Older Fathers A	Young Fathers B		
0–19	1	108	8	2,238	3·67	...
20–24	4	306	94	8,922	1·36	...
25–29	22	239	427	8,988	2·14	...
30–34	37	124	1,203	5,707	1·59	...
35–39	45	42	1,638	1,774	1·28	...
40–44	21	3	819	176 ⎫	1·65	...
45–99	3	0	70	1 ⎬		
Weighted mean relative risk (D.F. = 1)					1·61	17·8*

*$P < 10^{-4}$.
Data from Newcombe and Tavendale, 1965.[13]

To complicate matters still further, the extent of the difference between the age
of the father and that of the mother, is itself correlated with the risk of infant
death (Table 6), an effect which is presumably likewise associated with population
heterogeneities involving simultaneously the ages of the parents and the risks to the
children.

Table 6. Effects of Parental Age Differences, Per Se, on the Risk of Infant Death

Difference in 5-Year Age Group of Husband and Wife		Relative Risk*	χ^2 (D.F. = 1)	P
(Wife younger)	−4 to −6	1·31	7·6	0·006
	−3	1·29	19·8	0·000,01
	−2	1·11	10·9	0·001
	−1	1·11	0·2	...
(Same age group)	0	0·93	11·6	0·000,6
(Wife older)	+1	0·91	3·6	...
	+2	0·87	0·8	...
	+3 to +4	1·62	1·3	...

*I.e., the risk for a given difference in parental age group as compared with that for all other
age differences.
The values are weighted means of those derived separately for each 5-year group of father.
From analysis of Newcombe and Tavendale, 1965.[13]

This relatively long list of the difficulties that besets attempts to derive information about the importance of mutation from studies of parental age correlations may at first sight seem discouraging in the extreme. And yet, it is doubtful if there are any better approaches, and we are still in the happy position of having potentially available to us a vastly greater body of recorded personal information, of strictly relevant kinds, than anyone has so far attempted to analyze.

Selection

Some part of the hereditary causes of disease is, of course, maintained in the population, not by repeated mutation but by selection; that is by differences in fertility and mortality that operate in some circumstances to favour the perpetuation of genes which in other settings are the cause of ill health or handicap. To those investigators who are especially concerned with estimation of the consequences of artificially induced increases in the mutation rate, it is important to make full use of any information which will serve to identify those conditions that are maintained in the population by selective forces, and to distinguish them from those that are maintained by repeated mutations. To others who are concerned with all of the various influences that may affect the future quality of the gene pool, it is important to be able to identify the social and other factors that alter the directions in which the selective forces operate.

Precision in detecting differences of fertility and mortality depends to a major degree on the quality and accuracy of the available pedigree information. The identifications of family relationships contained in the vital registration systems are almost universal in their extent, covering as they do whole populations, and they are substantially more reliable than similar information which has been derived by personal interview or questionnaire. Because of this unique precision of the family information from the vital records, the most interesting early products of genetic studies employing record linkage methods relate to the detection of deviations from normal fertility and mortality displayed by families in which certain diseases and handicaps of possible genetic interest have occurred; or, in other words, to the study of selection pressures as associated with major causes of ill health.

The investigations of this kind that have been carried out using the vital records have been modest in size and have not as yet made full use of electronic data processing methods. They serve two purposes at this stage, to demonstrate that information on selection can be obtained from the vital registration system, and to give some hint of the kinds of empirical findings that are likely to come out of future applications of the approach to larger files and to a wider range of diseases.

The particular selection studies which will be described were prompted by a current belief that the Rh-negative gene is kept in the population by a tendency on the part of mothers of infants who die of erythroblastosis to "compensate" for their losses by having another child sooner than would otherwise be the case. If there is such a tendency, it is altogether likely that mothers also compensate in the same fashion for early infant deaths of other kinds, and that the phenomenon may be quite widespread.

Reproductive performances of mothers were therefore studied following still-births, infant deaths from a number of different causes and also following the births of mongoloid idiots who did not die. Comparisons have in each case been made with the reproductive histories of mothers of similar ages and parities who were not so selected (Figure 1)

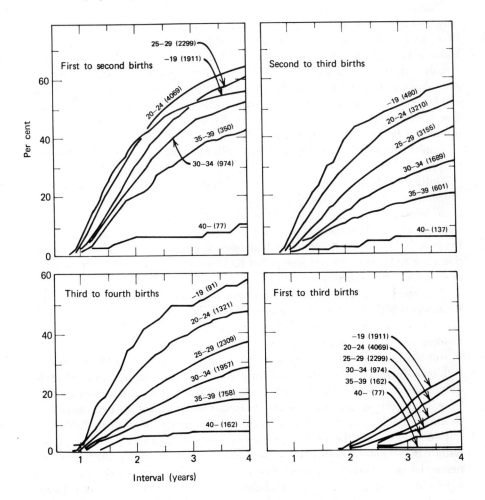

FIGURE 1. Percentage of mothers of liveborn children (born in 1954) having offspring throughout a four-year period following these births.

There are distinct advantages in being able to use the whole of a birth population as a control in a study of this kind. Approximately 170,000 birth records were available on magnetic tape representing the five-year period 1954-58. Each record of a birth in 1954, of which there were about 35,000, was used to initiate searches by computer for other births to the same mother in that year and in each of the subsequent years. This was a major undertaking, even with the help of a computer, and the total of over 170,000 searches (i.e., 35,000 per year over the five-year period) took eighty hours on a now obsolete machine (Burroughs 205). However, with the more recent machines now in use, two to three hours would be sufficient.

Mothers of stillborn children, in fact exhibited reproductive compensation, but only over a brief period. During the first twelve months following the stillbirth, between two and three times as many subsequent births took place to these mothers as would be expected in view of their ages and the numbers of children which they already had. [11] Over the second year their collective reproductive performances were about average, while in the third and fourth years these declined to about 70 and 60 per cent of normal. Taking the reproductive histories over the whole four-year period mothers of stillborn children tended to be less fertile than other mothers of the same ages and parities.

Similar studies of conditions other than stillbirth have been carried out more recently.

Mothers of children registered as dying of erythroblastosis, for example, have shown no tendency to compensate for their losses, the observed number of later born children being almost exactly the number expected of them (Table 7). This does not necessarily mean that compensation has not taken place at some period in the past, and perhaps even up to the time when the hereditary nature of the condition came to be generally recognized, but there is certainly no evidence from the present study that it is still occurring.

The same may be said of mothers of children dying of haemorrhagic disease of the newborn although, for this condition, knowledge of the cause would be less of a deterrent to further reproduction (Table 8). The observed number of later born children was again almost exactly what would be expected.

Not all of the conditions studied, however, failed in this manner to be associated with reproductive compensation, and for deaths from postnatal asphyxia, which is often associated with failure of the lungs to expand at birth, the effect is striking. From the ninety-seven mothers of affected children, instead of the expected sixty-six later born children, there were 111, an excess of nearly 70 per cent (Table 9).

Consideration of some of the possible causes of such an association serves to illustrate the complexities that are introduced into selection studies whenever environmental factors may be involved. From the standpoint of the mother, the loss of a child from postnatal asphyxia is probably not so very different from having a stillborn child, or from the death of a child shortly after birth from erythroblastosis or haemorrhagic disease. It seems unlikely, therefore, that the effects on her reproductive behaviour will be strikingly different in the case of postnatal asphyxia. The true reason for the association with an elevated fertility might therefore be that this latter condition tends to occur preferentially in an unusually fertile subgroup of the population.

Table 7. Reproduction Following Deaths of Infants From Erythroblastosis (Born 1953-55)

Birth Order	Number of Deaths	Mothers' Ages	Subsequent Births Over 4 Years*	
			Exp.	Obs.
1	5	20 20 20 21 28	4·10	4
2	15	20 21 23 23 24 25 26 27 28 29 30 30 31 33 38	7·54	9
3	8	23 26 28 31 33 33 34 37	2·63	2
4	8	24 24 26 28 29 29 31 35	3·22	2
5	5	21 27 32 33 36	1·69	1
6—8	7	33 33 33 33 36 41 41	1·75	3
9—	4	30 30 33 34	1·14	1
Combined	52		22·07	22†

*I.e., to end of 4-year period following the birth of the dead infant, but not beyond 1958.
†Includes 3 stillbirths and 2 infants dying before the end of the 4-year period.
Av. age = 1,514/52 = 29·12.

This view is supported in the case of postnatal asphyxia by evidence from the parental age and birth order studies discussed earlier. The risk was shown to be substantially elevated among children of very young mothers when contributions from birth order effects were excluded, and also among children of higher birth orders when contributions from maternal age effects were removed (Tables 1 and 3). This seemingly paradoxical relationship is easier to visualize when one considers that maternal age and birth order interact in such a way as to exaggerate the risk to a child of high birth order when it is born to a very young mother (as, for example, a third child to a nineteen-year-old girl), and also the risk to a child of low birth order when it is born to an older mother (as, for example, a first or second child to a woman of thirty-five or over). Stated in another way, it is risky for the child if its birth order is strikingly unusual in view of the mother's age.

All this implies that there are strong intra-uterine environmental influences operating on the foetus, and that a breakdown of the data by social factors might

Table 8. Reproduction Following the Deaths of Infants From
Haemorrhagic Diseases (Born 1953–55)

Birth Order	Number of Deaths	Mothers' Ages	Subsequent Births Over 4 Years	
			Exp.	Obs.
1	5	18 20 20 26 27	4·44	4
2	7	18 22 25 26 28 28 29	4·63	5
3	6	27 28 30 34 38 40	1·86	2
4	4	25 28 29 39	1·29	1
5	4	25 34 36 36	1·44	2
6–8	3	26 35 37	0·37	0
9–	0	...	...	...
Combined	29	...	14·03	14*

*Includes 1 stillbirth and 1 infant dying before the end of the 4-year period.
Av. age = 834/29 = 28·76.

conceivably serve to indicate effects of the environments to which the mothers themselves are exposed. Thus, analyses in terms of social variables will eventually be needed if genetic conclusions are to be drawn from selection studies of hereditary conditions whenever these are influenced in their expressions by environmental factors. Ease of interpretation is not to be expected, but to follow the superficially simple alternative of studying only those traits that are determined wholly by genetic factors, and not at all by the environment, would exclude from consideration the bulk of the important hereditary and partially hereditary conditions of man.

Curiously, for one well-known genetic condition, namely mongolism, that is believed to be determined entirely by inherited factors, there is an elevated maternal fertility following the birth of an affected child. The 118 mothers of mongol children in the present study produced a total of forty-five subsequent offspring in the period under consideration, as against an expected thirty-five, an excess of 27 per cent (Table 10).

Just why this should be so is not clear. There is reasonably good ascertainment of cases of mongolism in British Columbia so that more nearly complete reporting of affected children from the lower, and more fertile, socio-economic subgroups is not a likely explanation. No one has seriously proposed that the risk of mongolism, for a given age of mother, is greater in these subgroups than for the rest of the population, but this is a possibility.

Such a suggestion should not be taken to imply that mongolian idiocy may be caused by an environmental effect upon the foetus in the absence of the gross chromosomal anomalies usually responsible. But effects of the external environment

Table 9. Reproduction Following Deaths of Infants From
Asphyxia (Born 1953—55)

Birth Order	Number of Deaths	Mothers' Ages	Subsequent Births Over 4 Years	
			Exp.	Obs.
1	27	16 18 18 18 19 20 20 21 21 21 22 22 22 23 23 23 23 24 24 26 28 29 32 32 32 39 40	22·76	28
2	20	18 20 21 22 23 23 24 24 26 26 27 27 28 29 30 33 35 35 38 41	12·71	22
3	20	18 21 22 22 22 23 23 23 25 26 27 27 27 27 29 30 30 31 36 43	13·22	25
4	15	20 23 24 26 26 27 27 27 28 28 29 32 33 34 35	8·90	17
5	8	22 22 24 25 26 27 28 31	5·18	10
6—8	7	21 26 30 31 34 38 38	3·19	9
9—	0	...	...	...
Combined	97	...	65·96	111*

*Includes 2 stillbirths; no liveborn infants died before the end of the 4-year period.
Av. age = 2,580/97 = 26·60.

upon the mother could perhaps serve to increase the frequency of chromosomal non-disjunction in her reproductive cells, in much the same way that ageing does. Such an effect would tend to escape detection because of a lower average age of motherhood in lower socio-economic groups, and its demonstration would depend upon comparisons being made of the age specific risks in different population subgroups.

Current findings from studies of selection thus serve to emphasize that increased availability of empirical information on fertility differences brings with it unexpected problems of interpretation and that, for the solution of these, further knowledge of the social characteristics of the individuals is required. Fortunately, our religious, ethnic and economic particulars are fairly well documented, even if only the vital records are considered, and an immediate task is the fuller utilization of the information that is already potentially available, on a scale appropriate to permit multiple breakdowns of the data.

Table 10. Reproduction Following the Births of Mongoloid Infants
(Born 1953–55)

Birth Order	Number of Cases	Mothers' Ages	Subsequent Births Over 4 Years	
			Exp.	Obs.
1	21	17 17 18 19 19 21 21 22 22 23 24 25 25 27 29 34 39 39 41 41 42	15·08	20
2	18	22 25 27 28 28 30 33 36 37 38 38 39 40 40 41 41 41 42	5·04	4
3	35	19 26 28 28 30 31 31 32 32 33 33 33 34 34 34 34 34 35 36 36 37 38 38 39 40 40 40 41 41 41 41 41 42 43 47	8·14	10
4	23	23 25 30 30 31 31 34 34 35 36 37 37 37 38 38 39 40 40 41 42 43 45 48	4·81	7
5	7	29 33 34 39 40 40 46	1·08	0
6–8	7	35 37 38 39 40 43 46	0·77	0
9–	7	39 40 41 41 43 43 46	0·46	4
Combined	118	...	35·38	45*

*Includes 2 infants dying before the end of the 4-year period; no stillbirths occurred in this period.

As a minor refinement in the present study, the effective fertilities of the mothers in terms of the numbers of offspring surviving to the end of the period under consideration have been derived for the four conditions, erythroblastosis, haemorrhagic disease, postnatal asphyxia and mongolism (Table 11). When this is done, a slight reduction of fertility is observed in the case of the first two conditions, and a slight increase in the case of the other two.

Further attempts to interpret the observed effects would be unprofitable at this stage when we have only just started to describe them. The study so far has considered only five conditions, only 500 to 600 affected infants, and the reproductive performances of the mothers over a period of only four years maximum. The methods are applicable equally to conditions of other kinds and to much larger numbers of families followed over much longer time spans.

Table 11. Under- and Over-Fertility Following Infant Mortality and Morbidity
(Measured Over a 4-Year Period)

Condition	Number of Cases*	Subsequent Births			Surviving to End of 4 Years		
		Obs.	Exp.	Ratio	Obs.	Exp.	Ratio
Stillbirths†	269	136	146·0	0·93	...	...	...
Erythroblastosis (deaths)	52	22	22·1	0·99	17	21·3	0·80
Haemorrhagic disease (deaths)	29	14	14·0	1·00	12	13·5	0·88
Mongolism	118	45	35·4	1·27	43	33·6	1·28
Asphyxia (deaths)	97	111	66·0	1·68	109	63·0	1·73

*Stillbirths in 1953, and births in 1953—55 of children with other conditions.
†Previous data (Newcombe[11]) cases are limited to birth orders 1 to 4 inclusive.

Disease Conditions for which Studies of Mutation and Selection are Profitable

The empirical data yielded by studies of mutation and selection are, of course, most simply interpreted where the traits under investigation are readily recognizable, are determined by single gene differences and are always expressed when the causal gene is present. For this reason, such studies have in the past tended to make use of a few quite rare genetic traits that come reasonably close to meeting the ideal specifications. Achondroplastic dwarfism and Huntington's chorea are examples in point. While appropriate for establishing genetic prinicples, this sort of approach concentrates, as it were, on a small and unrepresentative part of the total genetic variability affecting man's well-being.

At the other end of the spectrum, attempts to estimate the likely total magnitudes of the consequences of mutation and of selection have sometimes been based on gross mortality data, not subdivided at all by disease cause.[2,5]

In between the two extremes there is a wide range of possible lines of study awaiting exploration, information on which is abundantly available even if a little tricky to get at. While routine records such as those of deaths and of hospital discharges are correctly regarded as inefficient sources of ascertainments of cases of the rarer genetic diseases, they do contain diagnoses of other sorts of conditions that are appropriate for studies either of mutation or of selection. Very large numbers of blood groupings, for example, are recorded annually in hospitals and as a part of the running of blood banks and transfusion services. Cases of important diseases that are partially genetic in origin and inherited in an irregular fashion are likewise reasonably well documented, although not necessarily in any single set of files; diabetes, idiopathic epilepsy and certain of the congenital malformations being notable examples. A number of the metrical, or quantitative, traits of man, of kinds that are suitable for studies of selection pressures, are likewise widely recorded; examples include birth weight which is stated on virtually all birth registrations, intelligence which is measured routinely in many school systems, and arterial pressure which is commonly taken in the course of medical examinations.

The possible applications of linkages with family pedigree information, such as may be obtained from the vital registrations, varies somewhat with the nature of the trait under study. In the case of the blood groups an important aim is the detection of correlations with fertility, mortality and morbidity, which will help to account for the continued prevalence of a diversity of gene alleles. In the case of birth weight variations there is a rather special unsolved problem, in that the modal value differs unexpectedly from the selective optimum as judged from the abilities of the infants to survive the first year of life.[3] There are also special problems associated with the study of selection as relating to intelligence and school performance, in that most investigations have in the past been concerned mainly with a single generation of test scores. The refinements that might derive from use on a substantial scale of two or more generations of scores within the same pedigrees, to help disentangle a transmissible component in mental ability from the more obviously environmental components for the purposes of selection studies, has hardly been explored at all even in theoretical discussions of the degree of precision that might eventually be possible.

For the diseases that display irregular inheritance another sort of problem arises, that bears on the manner in which tests for selective pressures may be designed, and that can in part be investigated by the use of pedigree information. Because of the nature of these conditions, many of the carriers of the causal genes may not be affected, and still others may have diseases that do not especially resemble the usual expressions of the genes. A few aetiological associations between outwardly diverse traits of an hereditary or partially hereditary sort have already been established, as in the case of spina bifida and anencephalus, and where these exist it is desirable to study the effects of selection on all of the known expressions of the underlying causal factors.

That such aetiological groupings may perhaps be more common than has been recognized in the past is suggested by a recent study of the risks of death and disease among the brothers and sisters of children with strabismus. This irregularly inherited trait is particularly well reported in British Columbia, so that the cases appearing in the Register of Handicapped Children and Adults represent a substantial fraction of the total. In the 207 affected families which were studied using the linked files of births, deaths and handicaps, the risk of strabismus among the later born brothers and sisters was between forty and fifty times that for the birth population as a whole (i.e., 6.3 per cent as compared with 0.13 per cent; see Table 12). This however, was not the only risk to which the families were exposed.

Table 12. Handicaps, Stillbirths and Deaths Among Sibs of Children with Strabismus. (Based on Files Relating to Children Born in British Columbia During 1953–58 and Registered as Stillborn, and Registered as Handicapped Before the End of 1960 or as Dying Before the End of 1959)

Conditions	Cases Obs.	Cases Exp.	Ratio Obs. Exp.	Excess Obs. Minus Exp.
Among 207 liveborn plus stillborn sibs, born after the first child with strabismus				
Strabismus	13	0·27	47·62	12·73
Other registered handicaps	7	4·32	1·62	2·68 ⎫
Stillbirths	9	2·31	3·90	6·69 ⎬15·20
Deaths	12	6·18	1·94	5·83 ⎭
Combined	41	13·08	3·14	27·93
Among a total of 310 liveborn plus stillborn sibs, born before and after the first child with strabismus				
Other registered handicaps	11	6·48	1·70	4·52
Stillbirths	10	3·45	2·90	6·55
Deaths	17	9·25	1·84	7·75
Combined	38	19·18	1·98	18·82

Note: The expected numbers are based on the numbers of registered cases of strabismus (286 cases representing 273 families), other handicaps (4,509), stillbirths (2,403) and deaths to the end of 1959 (6,437) out of the total birth population, i.e., live plus still, for the 6-year period 1953–58 (215,795).

An unexpected finding of the study was that of an approximate doubling of the risks of stillbirth, death and of handicapping conditions of other kinds. In fact, the excess of these other sorts of casualty, taken together, equalled in number the strabismus cases, each being represented by about six affected children per 100 brothers and sisters. Thus strabismus would appear to be aetiologically related in some way with a wide range of conditions, and about half of the special risk to the later born siblings of a child who has strabismus come from the aetiologically related conditions that may not even remotely resemble strabismus (see Table 13).

How then do we define in this instance the genetic entity which we wish to study? If selection is shown to operate in a certain way when individuals or families with strabismus are investigated, might it not perhaps operate in a different way when the genes responsible, or some part of them, are expressed differently? The interpretation of empirical data on selection pressures is in such circumstances complicated by the possibility that there may be still further expressions of the genes that have not so far been identified; but this is preferable to the possible oversimplifications that might arise from recognition of only a single kind of expression of the genes.

Table 13. Causes of Stillbirth, Handicap and Death Among 310 Sibs of
Children with Strabismus (Excluding Cases of Strabismus)

Code	Cause	Cases
Stillbirths		
Y35·3	Ill-defined causes in the mother	1
Y36·1	Placenta praevia	2
Y36·2	Prematurity	2
Y36·6	Abnormality of the placenta	1
Y38·1	Hydrocephalus	1
Y39·6	Unspecified	2
Y38·6, Y38·2	Monster, spins bifida	1
Handicaps		
004·0	Tuberculosis	1
326·2	Behaviour disorder	1
351·0	Cerebral palsy	1
389·4	Retrolental fibroplasia	1
748·0	Clubfoot	1
754·4	Fibroelastosis cordis	1
757·2	C.m. of external genitalia*	1
325·4, 748·0	Mongolism, clubfoot	1
325·5, 351·0	Mental deficiency, cerebral palsy	2
759·3, 758·6	Multiple malformations, including skeletal	1
Deaths		
057·0	Meningococcal infections	1
330·0	Subarachnoid haemorrhage	1
490·0	Lobar pneumonia	1
493·0	Pneumonia	1
754·2	Interventricular septal defect	2
754·4	Fibroelastosis cordis	1
762·5	Postnatal asphyxia, with immaturity	2
763·0	Pneumonia of the newborn	1
771·5	Haemorrhagic disease, with immaturity	1
776·0	Immaturity	4
910·0	Accidental injury	1
924·0	Accidental suffocation	1

*C.m. = congenital malformation.

The exploration of aetiological associations has in the present instance been
aided by two circumstances: (*a*) the use of systematically collected files of all live
births, stillbirths, child handicaps and child deaths in a large defined population,
regardless of any preconceived notions concerning the genetic or non-genetic origins
of the various events of morbidity and mortality; and, (*b*) the integration of these

files into the form of family histories of vital and health events using rapid modern methods of data processing. Under no other circumstances could the relationships with strabismus have been demonstrated with the same degree of precision, or in the relatively short period of time (about two days) which the study took. Current work is being directed to eliminating some remaining manual steps in such operations so that a large number of conditions can be treated similarly in a single study using larger files of records.

Costs and Administrative Applications

I will not at this stage discuss in any detail the methods used for this kind of work, since these have already been fully described elsewhere.[4,5,9,10,11] The present status of the undertaking has also been reported earlier.[8]

However, since descrpitions of this kind of work may give it the appearance of being excessively elaborate and costly in relation to the likely yield of genetic information, some clarification may be in order. The record linkage approach makes use only of information that is already being collected and recorded. In our own case, it has also employed mechanically readable punchcards that are already being prepared routinely from these original records for the extraction of annual statistics, and for the compilation of alphabetic name indexes of the kinds used for searching out birth registrations and such. The only innovation in relation to these existing procedures has been a minor modification of the vital statistics name index cards to include additional family identification such as the mother's maiden surname and the parental initials, ages and provinces of birth. The cost involved is slight, and the modified index cards, suitable for purposes of family linkage, are now used routinely for all events of marriage, live birth, stillbirth and death over the whole of Canada.

The most demanding part of such an undertaking is the development of increasingly versatile and rapid computer methods for inter-filing and linking the successive annual crops of records into family packets, and for extracting the resulting pedigree information in the required form after this has been done. Using the most recent programmes developed for a fast, modern computer, the cost of the actual merging and linking of the marriage, birth, death and handicap records, when suitably arranged in magnetic tape form, is very much less than that of preparing the punchcard records in the first place, being only about a quarter of a cent per updating record; and the current speeds of the updating and linking operation are in the vicinity of 1,000 to 2,000 incoming records per minute, so that the 40,000 annual birth records for British Columbia can be incorporated into the master family file in less than an hour.

What may not be generally recognized is the extent to which methods for the automatic linkage of records may serve existing administrative needs, relating to the uses of these same records. One example will serve to illustrate the point. The bringing together of information from all birth registrations into family groupings is, in fact, already being carried out over the whole of Canada by laborious manual methods as a necessary part of the operation of a government scheme of family allowances. To ensure that false claims will be detected, information from the mother's application form is in each instance compared visually with alphabetic lists of the names of newborn infants derived from the birth index cards, and with information from the existing family files wherever the family is already "in pay." These verifications of the mothers' statements account for 20 per cent of all family allowance file upkeep operations, which together cost in the vicinity of $1.25 per family per year. Substantial financial savings are thus potentially possible, using a rapid automatic record linkage procedure which could in addition yield family groupings of records for population research.

There are numerous other examples of uses to which the linkage technology might be put, and if family groupings of routine records are ever to be employed extensively as a source of information on the genetics of human populations, it will be because cognizance has been taken of these administrative needs in setting up stable, continuing systems which will permit scientific information to be derived as a by-product.

In the long run, it is doubtful whether precision in the study of selective forces, and perhaps in the study of mutation as well, is likely to be achieved in any other way, in view of the inaccuracies and the limited quantity of family pedigree information that can be derived from *ad hoc* surveys of conventional design based on personal interviews and questionnaires. Precision in the study of mutation and selection will undoubtedly be needed if we are to understand, and eventually control as Galton suggested, the influences that determine our genetic future.

Summary

Studies of mutation and selection, as influencing the genetically simple traits of man, have been popular in the past but they tell little about the overall magnitudes of the effects of these two forces throughout the genome as a whole. Similar studies would be desirable, relating to the collectively more important characteristics of man that are less simply inherited, and that are determined by combinations of environmental and hereditary factors. But such work is beset with difficulties arising out of the numerous environmental inhomogeneities characteristic of human populations. To be interpretable, data from these more difficult sorts of investigations must be broken down by a multiplicity of social variables; and correspondingly larger amounts of data are needed to make possible the required multivariate approach.

An abundance of relevant information about people is in fact recorded, which we have only begun to tap, describing not only their biological characteristics and family relationships, but their social circumstances as well. Application of this body of knowledge depends largely upon bringing together, or "linking," the separately recorded facts about the same individuals and families. Such "record linkages" will have many uses if the process of linking can be made cheap enough and rapid enough. In the long run, understanding of the magnitudes of the effects of recurrent mutations, and of various selective forces, on the quality of the human gene pool may depend upon the use of such methodological developments to take maximum advantage of information we already possess.

References

1. Cavalli-Sforza, L. L. 1962. Demographic attacks on genetic problems: some possibilities and results. In *The Use of Vital and Health Statistics for Genetic and Radiation Studies.* Pp. 221-30. United Nations, New York.

2. Crow, J. F. 1962. Selection. In *Methodology in Human Genetics.* Pp. 53-75. Ed. W. J. Burdette. Holden-Day Inc., San Francisco.

3. Karn, M. N., and Penrose, L. S. 1952. Birthweight and gestation time in relation to maternal age, parity and infant survival. *Ann. Eugen. Lond.* 16, 147.

4. Kennedy, J. M. 1961. *Linkage of Birth and Marriage Records Using a Digital Computer.* Document no. A.E.C.L.-1258, Atomic Energy of Canada Limited, Chalk River, Ontario.

5. Kennedy, J. M. 1962. The use of a digital computer for record linkage. In *The Use of Vital and Health Statistics for Genetic and Radiation Studies.* Pp. 155-60. United Nations, New York.

6. Morton, N. E., Crow, J. F., and Muller, H. J. 1956. An estimate of the mutational damage in man from data on consanguineous marriage. *Proc. nat. Acad. Sci., Wash.* 42, 855.

7. Newcombe, H. B. 1964. Screening for effects of maternal age and birth order in a register of handicapped children. *Ann. hum. Genet. Lond.* 27, 367.

8. Newcombe, H. B. 1964. Pedigrees for population studies. A progress report. *Cold Spr. Harb. Symp. quant. Biol.* 29, 21.

9. Newcombe, H. B., and Kennedy, J. M. 1962. Record linkage: Making maximum use of the discriminating power of identifying information. *Communications of the Association for Computing Machinery,* 5, 563.

10. Newcombe, H. B., Kennedy, J. M., Axford, S. J., and James, A. P. 1959. Automatic linkage of vital records. *Science,* 130, 954.

11. Newcombe, H. B., and Rhynas, P. O. W. 1962. Child spacing following stillbirth and infant death. *Eugen. Quart.* 9, 25-35.

12. Newcombe, H. B., and Tavendale, O. G. 1964. Maternal age and birth order correlations. Problems of distinguishing mutational from environmental components. *Mutation Research,* 1, 446.

13. Newcombe, H. B., and Tavendale, O. G. 1965. Effects of father's age on the risk of child handicap and death. *Amer. J. hum. Genet.,* 17, 163.

14. Penrose, L. S. 1955. Parental age and mutation. *Lancet,* ii, 212.

15. *Recommendations of the International Commission on Radiological Protection.* 1959. Pergamon Press, New York.

16. Sonneborn, T. M. 1956. Paternal age and stillbirth rate in man. *Records of the Genetics Society of America,* Number 25, p. 661.

Selected Bibliography Including References Cited

Part I

Allison, A. 1969. Natural Selection and Population Diversity. Pp. 1-15 in Harrison, G. and Peel, J. (eds). 1969. *Biosocial Aspects of Race.* Blackwell.

Bajema, C. 1967. Human Population Genetics and Demography: A Selected Bibliography. *Eugenics Quarterly,* 14: 205-237.

Berry, A., Berry, R., and Ucko, P. 1967. Genetical Change in Ancient Eygpt. *Man,* 2: 551-568.

Bilsborough, A. 1969. Rates of Evolutionary Change in the Hominid Dentition. *Nature,* 223: 146-149.

Bodmer, W. 1968. Demographic Approaches to the Measurement of Differential Selection in Human Populations. *Proceedings National Academy of Sciences,* 59: 690-699.

Brues, A. 1964. The Cost of Evolution vs. the Cost of Not Evolving. *Evolution,* 18: 379-383.

Brues, A. 1968. Mutation and Selection—A Quantitative Evaluation. *American Journal of Physical Anthropology,* 29: 437-440.

Brues, A. 1969. Genetic Load and Its Varieties. *Science,* 164: 1130-1136.

Campbell, B. 1963. Quantitative Taxonomy and Human Evolution. Pp. 50-74 in Washburn, S. (ed.). 1963. *Classification and Human Evolution.* Aldine, Chicago.

Campbell, B. 1966. *Human Evolution.* Aldine, Chicago.

Cavalli-Sforza, L. 1969. "Genetic Drift" in an Italian Population. *Scientific American,* 221 (Aug): 30-37.

Coon, C. 1962. *The Origin of Races.* Knopf, New York. 724 pp.

Coon, C. 1965. *The Living Races of Man.* Knopf, New York. 344 pp.

Crow, J. 1968. Rates of Genetic Change Under Selection. *Proceedings National Academy of Sciences,* 59: 655-661.

Darlington, C. 1969. *The Evolution of Man and Society.* Allen and Unwin, London.

Darwin, C. 1871. *The Descent of Man and Selection in Relation to Sex.* Murray, London

Dobzhansky, T. 1962. *Mankind Evolving.* Yale University Press, New Haven.

Fisher, R. 1930. *The Genetical Theory of Natural Selection.* Oxford Univ. Press, Oxford.

Gajdusek, D. 1964. Factors Governing the Genetics of Primitive Human Populations. *Cold Spring Harbor Symposium on Quantitative Biology,* 29: 121-135.

Haldane, J. 1932. *The Causes of Evolution.* Harper, New York.

Haldane, J. 1964. Defense of Beanbag Genetics. *Perspectives in Biology and Medicine,* 7: 343-360.

Harrison, G. 1967. Human Evolution and Ecology. Pp. 351-359 in Crow, J. and Neel, J. (eds.). 1967. *Proceedings of the Third International Congress of Human Genetics.* Johns Hopkins Press, Baltimore, 578 pp.

Hockett, C., and Ascher, R. 1964. The Human Revolution. *Current Anthropology,* 5:135-168.

Larsson, T. 1956/1957. The Interaction of Population Changes and Heredity. *Acta Genetica,* 6: 333-348.

Livingstone, F. 1969. The Founder Effect and Deleterious Genes. *American Journal of Physical Anthropology,* 30: 55-59.

Morton, N. 1968. Problems and Methods in the Genetics of Primitive Groups. *American Journal of Physical Anthropology,* 28: 191-202.

Morton, N. 1969. Human Population Structure. Pp. 53-74 in Roman, L., Campbell, A. and Sandler, L. (eds.). 1969. *Annual Review of Genetics.* Vol. 3. Annual Reviews, Palo Alto. 585 pp.

Neel, J., and Schull, W. 1968. On Some Trends in Understanding the Genetics of Man. *Perspectives in Biology and Medicine,* 11: 565-602.

Newcombe, H., and Tavendale, O. 1964. Maternal Age and Birth Order Correlations. Problems of Distinguishing Mutational from Environmental Components. *Mutation Research,* 1: 446-467.

Newcombe, H. 1967. Use of Vital Statistics. Pp. 494-497 in United Nations. 1967. *Proceedings of the World Population Conference, Belgrade, 1965, Vol. II.* United Nations, New York. 510 pp.

Newcombe, H. 1967. Present State and Long-Term Objectives of the British Columbia Population Study. Pp. 291-313 in Crow, J. and Neel, J. (eds.) 1967. *Proceedings of the Third International Congress of Human Genetics.* Johns Hopkins Press, Baltimore.

Newcombe, H. 1968. Risks to Siblings of Stillborn Children. *The Canadian Medical Association Journal,* 98: 189-193.

Roberts, D. 1968. Genetic Effects of Population Size Reduction. *Nature,* 220: 1084-1088.

Roberts, D. 1968. Genetic Fitness in a Colonizing Human Population. *Human Biology,* 40: 494-507.

Salzano, F. (ed.) 1971. *The Ongoing Evolution of Latin American Populations.* Charles Thomas, Springfield.

Sutter, J. 1969. The Effect of Birth Limitation on Genetic Composition of Populations. Pp. 213-251 in Behrman, S., Corsa, L., and Freedman, R. (eds.). 1969. *Fertility and Family Planning: A World View.* University of Michigan Press, Ann Arbor.

Sved, J. 1968. Possible Rates of Gene Substitution in Evolution. *American Naturalist,* 102: 283-293.

United Nations Scientific Committee. 1967. *The Genetic Risk From Radiation.* United Nations, New York. 153 pp.

Weyl, N. 1968. Some Possible Genetic Implications of Carthaginian Child Sacrifice. *Perspectives in Biology and Medicine,* 12: 69-78.

Wolpoff, M. 1969. The effect of Mutations Under Conditions of Reduced Selection. *Social Biology,* 16: 11-23.

Workman, P. 1968. Gene Flow and the Search for Natural Selection in Man. *Human Biology,* 40: 260-279.

Wright, S. 1931. Evolution in Mendelian Populations. *Genetics,* 16: 97-159.

Wright, S. 1932. The Roles of Mutations, Inbreeding, Crossbreeding and Selection in Evolution. *Proceedings 6th International Congress Genetics Vol. 1,* 356-366.

Wright, S. 1968. *Evolution and the Genetics of Populations. Vol. 1. Genetic and Biometric Foundations.* University of Chicago Press, Chicago. 469 pp.

Wright, S. 1969. *Evolution and the Genetics of Populations. Vol. 2: The Theory of Gene Frequencies.* University of Chicago Press, Chicago.

PART II

Natural Selection in Relation to Physical Characteristics

Biologists have utilized human skeletal remains to estimate the rate of evolutionary change with respect to such human skeletal characteristics as dentition (Bilsborough, 1969), cranial capacity (Haldane, 1949; Campbell, 1963; and Kurten, 1959) and shape of head (Shapiro, 1943; Weidenreich, 1945; and Hunt, 1959). Much of the evidence concerning the past operation of natural selection on such characteristics as weight, height, and pigmentation is based on the correlation of climatic factors with the geographical distribution of the physical characteristic (for example, Schreider, 1964; Loomis, 1967).

The physical characteristics of a race (pigmentation, body build, etc.) are primarily adaptations to the physical conditions of the environment. By controlling the climates within his dwellings via artificial heating, lighting, and air conditioning systems, by wearing appropriate clothing, and by improving and supplementing his diet, man is altering the direction and intensity of the very selective forces that led to the formation of human racial groups.

The secular increase in human height that has taken place during the past century provides a good example of how a physical pattern that has a strong genetic basis can change over a relatively short period of time. This increase in adult height that has taken place in many populations of the human species over the past several generations does not appear to have been due to an increase in the frequency of genes favoring the development of taller humans. Damon's paper in this section provides little support for the hypothesis that natural selection is favoring an increase in the frequency of genes for tallness in the human species. Damon (1970, personal communication) has also analyzed the relationship between height and fertility in three hitherto unreported populations (a new group of Harvard alumni, a group of U.S. veterans aged 40–50, and a group of British Solomon Islands natives). He found that natural selection was not directional with respect to height in these populations but rather was stabilizing, that is, natural selection did not favor tallness but rather the individuals who were in the average range with respect to height.

What, then, accounts for the observed increase in human height over the past several generations if an increase in the genetic factors favoring tallness in human populations has not occurred during this period of time? Although better nutrition and physical health have been important factors, there is evidence that at least half and possibly more than half of the increase in adult height that has taken place is because of a change in mating patterns, due to heterosis resulting from outbreeding. Tanner (1968) points out that a number of studies have shown that "the height of adults is significantly and inversely correlated with the degree of inbreeding in the region" under study (Hulse, 1957; Schreider, 1967). Thus the increase in adult height that has taken place in some human populations over the past several generations appears to be largely due to the genetic reconstitution of populations as the result of changes in mating patterns rather than being primarily due to improved nutrition or the genetic consequences of directional natural selection.

The papers reprinted in this section include original reports on the operation of natural selection with respect to birth weight, congenital malformations, body dimensions (height, weight) and shape of the head. The paper by Hulse, which discusses the differential population growth rates of different racial groups in the past, is very important because it points out that evolution can occur in the human species via competition between populations as well as competition between individuals of the same population. Hulse very correctly points out that the differential population growth rates of the various races is primarily a function of cultural and technological differences rather than a function of racial differences in physical or immunological characteristics.

4. Selection in Natural Populations. New York Babies (Fetal Life Study)*

LEIGH VAN VALEN and GILBERT W. MELLIN

Introduction

Birth, and the many new environmental challenges that soon follow it, are well known to provide the most critical period, with respect to survival, that can be readily studied in a human life. Prenatal mortality rates, as manifested by visible and invisible abortions and some stillbirths, are undoubtedly higher but are more difficult to investigate systematically. A number of studies, reviewed below, have been made on the relation between birth weight and probability of survival. In the present study two additional characters are used: presence or absence of congenital malformations, and rate of tooth eruption during the first year after birth.

The following aspects are examined in the present paper: the relationships between birth weight and abnormality, between abnormality and survival, between birth weight and survival, and between tooth eruption and abnormality. Terminology is as follows: foetal deaths are deaths before or during delivery, neonates are suvivors of delivery, neonatal deaths are deaths from delivery to 28 days after delivery, perinatal deaths are foetal deaths plus neonatal deaths, and survivors are survivors past 27 days. Modifications of this terminology in comparisons with other studies will be obvious.

Some of the present results appeared in an abstract (Van Valen & Mellin, 1966).

SOURCE: Leigh Van Valen and G. Mellin, "Selection in Natural Populations. 7. New York Babies," Annals of Human Genetics, 31: 109-127 (London, England, 1967).

*Supported in part by Grant GM-12549 from the National Institutes of Health, to Van Valen. The Fetal Life Study has been aided by grants from the Life Insurance Medical Research Fund, the Rockefeller Foundation, the New York State Department of Health, the Association for the Aid of Crippled Children, the New York Foundation, the National Institute of Child Health and Human Development, and the National Foundation.

Material

The 6053 births used were obtained through first trimester registration of the mothers at the ante partum clinic of the Columbia-Presbyterian Medical Center in Manhattan, New York City. 3378 were from mothers who described themselves as "white," 2646 were "Negro," and 29 were "other." Approximately 10% of this population were of the Puerto Rican minority, but exact data on this point are unavailable. For this reason, and because of the relatively small sample size, racial and ethnic classification has been ignored in the present study. The sample is unified primarily on an economic and geographical basis, and the undoubted lack of panmixia would affect the statistics only if there is a sampling bias between the subgroups of the economic-geographical population from which the sample is drawn. Such a bias has been assumed to be absent, but there is no evidence on this point.

The births occurred from 1947 to 1963 inclusive, and have been described in more detail elsewhere (McIntosh et al. 1954; Mellin, 1963, 1966). The mothers came from a predominantly lower class and lower middle class area of uptown Manhattan. The 318 abortions, defined as concepti weighing less than 501 grams, were excluded from the present study, leaving a total sample of 5735.

Procedure

The data on birth weight were recorded to gram, 5 g., or 10 g. in different cases. For analysis these data were grouped at intervals of 100 g.; e.g., the lowest group was 501-600 g. inclusive. There were 54 weight groups in the observed range. All the raw data in the form of a magnetic tape were prepared by Mellin; the analysis and interpretation are by Van Valen.

The malformations have been discussed in detail by McIntosh et al. (1954) and Mellin (1963, 1966). The malformations studied are of course only those that could be detected by the relatively extensive examinations given, and the degree of abnormality of a character necessary for it to be classified as a malformation is in some cases arbitrary although repeatable. Undoubtedly some malformations with little external or clinical manifestation were missed, especially in the survivors. Only 50% of the known malformations of the survivors were diagnosed at birth. About 95% of the survivors were re-examined at 6 months, 1 year, or rarely somewhat later. Structures, such as teeth, not evident at birth had few abnormalities. For the perinatal deaths 54% of the malformations were diagnosed clinically and 87% of the perinatal deaths had autopsies, which accounted for the remaining 46% of the malformations diagnosed. In the present analysis each individual was classified simply as normal or abnormal.

Rate of tooth eruption was measured by the number of erupted teeth at 6 and 12 months. These two numbers were treated as different characters in the analysis. All individuals over 500 g. at birth were classified as either male or female. The classification of abnormalities gives no indication of the distribution of any intersexes, as none were observed by external examination or on autopsy.

Results and Discussion

Tooth Eruption

The main point of interest with regard to the teeth is whether there is any difference between the malformed and normal survivors. Teeth are ordinarily all unerupted at birth. As shown by the statistics in Table 1, there is no detectable difference between normal and abnormal individuals in the mean or variation of the rate of tooth eruption. This is true both for statistics calculated from the untransformed data and for statistics calculated from a logarithmic transformation of the data; the latter statistics are omitted for brevity.

Table 1. Statistics for Birth Weight,* Tooth Number, Malformation, and Survival, Fetal Life Study

	Normal				Malformed			
	$\overline{X}$	s^2	CV	N	$\overline{X}$	s^2	CV	N
			Birth Weight					
Survivors								
Male	3.201 ± 0.011	0.286	16.7	2558	3.169 ± 0.037	0.324	18.0	236
Female	3.113 ± 0.010	0.265	16.5	2586	3.040 ± 0.051	0.399	20.8	151
Foetal deaths								
Male	2.17 ± 0.19	1.22	51	33	2.35 ± 0.22	0.80	38	16
Female	1.74 ± 0.16	0.97	56	36	2.45 ± 0.27	0.82	37	11
Neonatal deaths								
Male	2.02 ± 0.17	1.35	58	49	2.58 ± 0.48	1.80	52	8
Female	1.84 ± 0.15	0.89	51	39	1.57 ± 0.30	0.82	58	9
			Tooth Number					
6 months								
Male	0.616 ± 0.023	1.23	180	2336	0.537 ± 0.071	1.07	192	214
Female	0.559 ± 0.022	1.15	192	2349	0.602 ± 0.082	0.89	157	133
12 months								
Male	6.48 ± 0.05	5.55	36	2004	6.43 ± 0.16	4.77	34	194
Female	6.15 ± 0.06	6.32	41	2009	6.08 ± 0.25	7.59	45	121

*In kilograms. CV is coefficient of variation.

The correlations between tooth number at 6 and 12 months, and between these numbers and birth weight, are given for each sex in Table 2. The several pairs of similar correlation coefficients are not significantly different from each other. Because about 0.029 of the variance of the rate of tooth eruption is predictable from knowledge of birth weight (and vice versa), a selection intensity of about 0.001 on the rate of tooth eruption would be expected from the selection on birth weight discussed below if there is no interaction. The data are far too inadequate to discriminate this value from the absence of any selection.

Table 2. Correlations of Rate of Tooth Eruption, Fetal Life Study

	r	N
Birth weight—teeth at 6 months		
Male	0.13	2550
Female	0.18	2482
Birth weight—teeth at 12 months		
Male	0.15	2198
Female	0.21	2130
Teeth at 6 months—teeth at 12 months		
Male	0.49	2138
Female	0.50	2077

Malformations

The relations of malformation to the probability of survival can also be found in Table 1. Grouping the foetal deaths and neonatal deaths, X^2 tests on 2 x 2 tables give $P < 0.001$ for each sex separately. There is thus a lower probability of survival for abnormal infants than for normal ones unless abnormalities are relatively more often detected in perinatal deaths. However, the occurrence of such lethal abnormalities as anencephaly supports, but does not establish, the existence of a greater mean mortality for abnormal infants.

The probability of survival for normal males is 0.969 and for normal females is 0.972, while for abnormal males it is 0.91 and for abnormal females it is 0.88. By equation (8) of Van Valen (1965), this difference corresponds to a perinatal selection on the population of about 0.0060 for males and about 0.0055 for females, with respect to the presence of an abnormality. The true selection intensity throughout the life span (including reproduction) is unknown.

There appear to be only two other sets of data that can be used in estimating the perinatal selection on congenital abnormalities (cf. Green, 1964). Neither was classified as to sex. For 29,024 births in Boston from 1930 through 1941, Stevenson, Worcester & Rice (1951) found the probability of survival to ten days for normal infants to be 0.968 and for malformed infants to be only 0.691. This difference is considerably greater than that in the Fetal Life Study, apparently in part because of a more restricted definition of malformation.

The methods of diagnosis appear comparable but the frequency of malformation is considerably less. The amount of perinatal selection is 0.0064, about the same as for the Fetal Life Study although the shorter follow-up period would give a prediction of a lower amount of selection in that time. The greater selection intensity therefore present in the Boston sample proves that part of the observed greater probability of mortality for malformed infants in that sample is real and not due to a restriction of the definition of malformation.

Böök (1951) gave data for 44,109 births in Lund, Sweden, from 1927 through to 1946. They were classified by clinical examination only, and sufficient data are given only for foetal deaths. The probability of surviving birth for a normal infant was 0.978 and for a malformed infant was 0.922. The selection intensity estimated from these data is only 0.0008, but because of the inadequacy of the data it is not trustworthy.

Obviously not all malformations are equivalent with respect to perinatal selection. Some are invariably lethal and some give no detectable disadvantage. This heterogeneity is well shown by McIntosh et al. (1954), who discuss the Fetal Life Study data from a more clinical viewpoint and give survival data for each malformation. The character "congenital malformation" is presumably at least as complexly determined as such characters as "mental deficiency" or "tall stature," but, like them, it can usefully be considered as an arbitrarily delimited pragmatic class. The data could be grouped by estimated or observed degree of severity of the malformation but for such groupings the selection intensities could be anywhere from 1 to almost 0, the values depending on how many groups were used and what boundaries were chosen. Such a procedure would seem to add little information to the study of McIntosh et al. (1954).

Birth Weight: Fetal Life Study

The data on birth weight with respect to survival and malformation are given in the appendix, and statistics for various groupings are given in Table 1.

The probabilities, from t and F tests, for identity of mean and variance of birth weight between various pairs of groups are given in Table 3. For the means as well as the variances the results of a one-tailed test are presented, although neither this nor a two-tailed test is strictly appropriate. The expectation from previous studies is that, if there is any real difference, the deleterious classes will have a lower mean birth weight as well as a larger variance. However, a higher mean would not be utterly implausible (this would be accepted at a lower significance level; cf. Van Valen, 1963), so the true P values should be somewhere between those for one-tailed and two-tailed tests although closer to the former. The two-tailed P values are twice the one-tailed values.

As in other studies, the survivors are less variable and average heavier than the

Table 3. One-Tailed Probabilities for Difference in Birth Weight Between
Classes (Fetal Life Study)

	Mean	Variance
Survivors–perinatal deaths		
Male	<0.001	<0.001
Female	<0.001	<0.001
Neonates–neonatal deaths		
Male	<0.001	<0.001
Female	<0.001	<0.001
Normal–malformed all combined		
Male	0.07	0.05–0.10
Female	0.005	<0.001
Survivors		
Male	0.2	0.1–0.25
Female	0.09	<0.01
Neonates		
Male	0.2	0.1–0.25
Female	0.01	<0.0005

non-survivors. In addition, normal individuals are less variable and average heavier than those with malformations (for the mean, the two-tailed P is about 0.01 for all individuals when the probabilities are combined for the two sexes), although the effect here is less striking.

The optimal birth weight can be estimated in various related ways. The initial assumption for the present analysis was that the probability of mortality is some quadratic function of the deviation from the optimal birth weight, i.e., if m is the probability of perinatal mortaility, X is birth weight, and a, b, and c are constants,

$$m = aX^2 + bX + c. \tag{1}$$

This assumption is one of the simplest consistent with the observed general pattern of selection, and because of this simplicity and its statistical properties it has been used in theoretical studies by various writers since Fisher (1930), p. 105), Haldane (1932), and Wright (1935).

The data were grouped by 200 g. intervals and quadratic regressions of the form of equation (1) were fitted to the data of each sex separately. Because of the greatly unequal numbers in the different weight classes, the data in each class were weighted by the square root of the number of individuals in that class. This weighting is approximately proportional to the reciprocal of the width of a confidence interval about the respective datum. This procedure weights each extreme datum more than each central datum, which is reasonable if the regression is to be fairly close to the extreme points as well as to the much more numerous central ones.

Weight classes with 100% observed mortaility were ignored. These classes would bias the result because the scale of mortality does not go beyond 1 even though the mean disability of the infant (in units of, e.g., lethal equivalents) might still increase. The equation for males is

$$m = 0.118\,X^2 - 0.855\,X + 1.519, \tag{2}$$

while that for females is

$$m = 0.134\,X^2 - 1.030\,X + 1.921. \tag{3}$$

From differentiation of these equations, the optimal birth weight for males is about 3.62 kg. and that for females is about 3.84 kg. Other regression models give slightly different estimates of the optima. The parabolic regressions of the kind used by Karn & Penrose (1951) are

$$m = -0.890\,Y^2 + 6.22\,Y - 6.32 \tag{4}$$

for males and

$$m = -0.745\,Y^2 + 5.81\,Y - 6.38 \tag{5}$$

for females, where Y is the natural logarithm of the ratio of survivors to non-survivors. The optimal weights derived from these equations are 3.49 kg. for males and 3.90 kg. for females. Additional aspects of the regression curves will be discussed in a later section.

The unselected distribution of birth weights is not normal, but rather is left-skewed and somewhat leptokurtic, as shown by graphs against the normal distribution expected for each sex with the given mean and variance. Much of this deviation from normality (on an arithmetic scale) is due to an excess of infants with very low birth weights. Because most of these infants die, the distribution of the survivors is more nearly normal. Graphs of the expected and observed distributions of the normal survivors of each sex, however, also show some excess at the lowest birth weights, although the rest of the distribution is close to normality. There is some bias towards low weight among the foetal deaths because many of them die before birth, and the time from death to their expected time of delivery would have permitted additional growth if the foetus were alive. This bias cannot be estimated and has been ignored. Its effect is to make the estimated selection intensities somewhat lower than the real ones. The time of birth is in any event to some degree arbitrary from the viewpoint of the foetus: even though the size of the foetus may influence the time of its birth, this is certainly not the only important cause.

It is possible to estimate the total number of selective deaths. The expected death rate for those individuals with the optimal birth weight, by equations (2) and (3), is about 0.009 for males and about 0.005 for females. The expected mortality at the optimal birth weight by equations (4) and (5) is about 0.011 for males and about 0.007 for females. The statistics derived from the latter values will be given in parentheses in the next paragraph and are probably the more accurate.

The excess of the mean death rate for the entire sample or for any subset of it, over the mortality at the optimal birth weight, defines the selective mortality on birth weight for that sample. The mean death rate for males is 0.037 and that for females is 0.034; thus the total selective mortality on birth weight, in the perinatal period, is about 0.028 (0.026) for males and about 0.029 (0.027) for females. The effective mortality (the proportion of the total mortality that is selective for birth weight) is relatively high, being about 0.75 (0.7) for males and about 0.85 (0.8) for females. The effective mortality and the number of selective deaths are shown for weight classes in Figs. 1 and 2, using the optimal mortalities calculated from equations (4) and (5).

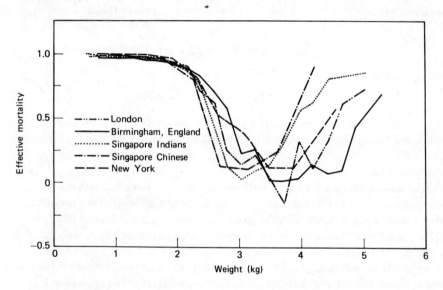

FIGURE 1. Effective mortality as a function of birth weight. The negative value is due to sampling error, which is greatest in this region. The mean of the effective mortalities calculated separately for the two sexes is used except for the Birmingham sample, in which the sexes were not separated in the data.

Birth Weight: Other Studies

Data given by previous workers provide other opportunities for measurement of selective mortalities. None of these studies is entirely comparable with the above results because of one or more important differences in procedure, but the general pattern or results can be compared and shows some differences.

Karn & Penrose (1951) estimated the optimal birth weights in a London hospital

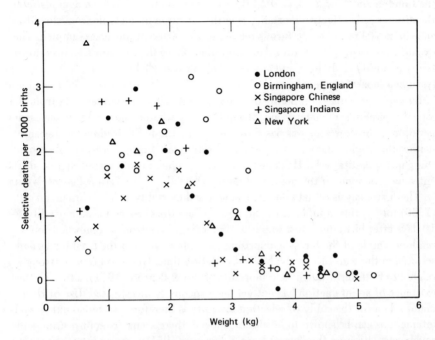

FIGURE 2. Number of selective deaths as a function of birth weight. The negative value is
due to sampling error. The sum of the selective deaths calculated separately for the two
sexes is used except for the Birmingham sample, in which the sexes are not separated in
the data. The ordinate is scaled so that different intervals of birth weight are comparable; as given it is really the number of selective deaths per 1000 births for a 100 g.
interval on each side of each datum. The numbers for New York and London are
to 28 days, for Birmingham they are for 1 year, and for Singapore
they are for foetal deaths only.

from 1935 through to 1946 to be 3.60 and 3.63 kg. for males and females respectively, and the perinatal selective mortality at these weights can be estimated from
their equations as about 0.019 and 0.018 respectively. The sample size is 13,730.
The total perinatal selective mortality on birth weight can then be calculated to be
about 0.029 for males and about 0.023 for females. These values are very close to
those for the New York sample. The effective mortality is, however, ostensibly lower in London, being about 0.6 for males and about 0.55 for females. The effective
mortality and the number of selective deaths are given for weight classes in Figs. 1
and 2. The differences from the New York (Fetal Life Study) data are due at least
in part to sampling error. Part of this sampling error cannot be evaluated, because in
Britain only the prospectively or actually more difficult births occur in hospitals,
while almost all births do in the United States. No difference is selection between

the London and New York (Fetal Life Study) populations has been demonstrated; the difference in effective mortality is in the direction predictable from the difference in mortality, which in turn is related to the different universes sampled. The exclusion of twins from Karn's data somewhat biases the London selection down; twins are individually light and are less viable than single births. The twins were analysed in a separate paper (Karn, 1952), but the inclusion of other twins as well in the latter paper precludes combining the data. Jayant (1966b) analysed data from the same hospital in some later years. She did not say whether she excluded twins, and her optimal birth weight was based on the "z" method criticized below. Because neither the original data nor the variances were given, I have not been able to use this paper quantitatively. Haldane (1954) and Spuhler (1963) found slightly different values for some of the numerical results given above for Karn & Penrose's data.

The first analysis of data on birth weight and mortality was by Hosemann (1950) for Göttingen infants. He did not separate the sexes and used mortality to 10 days after birth, but most importantly his data were from a somewhat non-random sample of the births, comprising somewhat over half the total births and selected in three groups on the basis of gestation time. Gestation time is strongly related to both birth weight and mortality (Karn & Penrose, 1951), although the existence of an interaction has not yet been adequatly investigated. Use of Hosemann's data for the study of selection intensity is therefore precluded unless such interactions can be shown to be absent. Indeed, there seems from inspection and rough calculation on the data of Karn & Penrose (1951) to be an interaction in that a longer than optimal gestation time does not obviously affect the position of or the mortality at the optimal birth weight, but increases the mortality at birth weights above and below the optimum.

Gibson & McKeown (1951) gave data for 22,454 births in Birmingham, England, in 1947, a random and 93% complete sample. Multiple births were excluded and the sexes were combined. The distribution by sex is not adequately given in a later paper that deals with this subject (Gibson & McKeown, 1952). The lumping of the sexes slightly biases up the expected mortality at the optimal birth weight because the optimal weight for the combined sample is expected to be slightly suboptimal for both sexes. This effect slightly biases down the expected selective mortality, the same direction as from the exclusion of multiple births. A parabolic regression of the Karn and Penrose variety for the sample gives an optimal weight of 3.93 kg. and an expected mortality at this weight of 0.027. These values coincide closely with the data points in this region. The total selective mortality is 0.028, for deaths to 1 year, and the effective mortality is about 0.5. The effective mortality and number of selective deaths to 1 year are plotted by weight classes in Figs. 1 and 2.

For Italy, there is the study of Fraccaro (1956) and 5486 hospital births in Pavia from 1942 through to 1951. In this paper abnormal infants as well as twins were excluded, and neonatal deaths beyond a period from 7 to 20 days after birth were unrecorded and therefore classed as survivors. All three factors slightly reduce both the expected selective mortality and the expected effective mortality. The total selective mortality from these data is 0.036 for each sex, and (as noted by Fraccaro) the effective mortality is about 0.55 for each sex.

Data from Ghana have been given by Hollingsworth (1965) for 4756 hospital births in Accra from 1954 through to 1956. Neonatal deaths are classed as survivors and multiple births are excluded. Despite heavy mortality (about 12% foetal deaths), the selective mortality is only 0.028 for males and 0.30 for females. The effective mortality is about 0.25 in each sex. An estimated selective mortality of 0.06 for the perinatal period would perhaps not be far wrong. Selection may well be even more intense in groups without hospital care, particulary the hunting-gathering tribes of our ancestry and a few corners of the present world. The effective mortality, however, would probably be relatively small in such groups because of a relatively greater mortality non-selective for birth weight.

Two subgroups are present in a sample from Singapore discussed by Millis (1959), namely, infants from Indian and from Chinese parents. The data are for ward births from 1950 through 1953. Neonatal deaths are classed as survivors. 48,070 Chinese and 5361 Indian infants were studied. For the Chinese, the optimal birth weight can be calculated to be 3.24 kg. for males and 3.47 kg. for females. The expected mortalities at these weights are 0.011 and 0.010 respectively. The selective mortality is 0.019 for males and 0.018 for females, and the effective mortality is about 0.65 for each sex. For the Indians, the expected mortality at the optimal birth weight is 0.009 for males and 0.008 for females. The selective mortalities are 0.025 and 0.031 respectively, and the effective mortalities are about 0.75 and 0.8 respectively. The effective mortality and the number of selective deaths for both Chinese and Indians are plotted by weight classes in Figs. 1 and 2. There is a detectable shift in the pattern toward lower birth weight as compared with the English and American samples.

The final available data are for two socioeconomic groups of Bombay, analysed by Jayant (1964). Multiple births were excluded and neonatal deaths beyond 7 days were classed as survivors. The data for the upper class are from 1947 through to 1961 and are based on 3822 infants. Using Jayant's equations for the parabolic regressions, the expected mortality at the optimal birth weight is 0.017 for males and 0.019 for females. The latter is inconsistent both with the data in this region and with the lower female mortality plotted in Jayant's Fig. 1. 0.012 is a better estimate for the expected female mortality at the optimal birth weight. The selective mortalities are about 0.029 for males and 0.028 for females, and the effective mortalities are about 0.65 and 0.7 respectively. The data for the lower class are for 1962 and are based on 2279 infants. Again using Jayant's equations, the expected mortalities for males and females at the optimal birth weights are 0.015 and 0.017 respectively. At least the former is also unquestionably wrong; our estimates are 0.033 and 0.020 respectively. Using the latter estimates, the selective mortalities are

0.064 for males and 0.057 for females, and the effective mortalities are about 0.65 and 0.75 respectively.

The effective mortality and the selective mortality of each population are plotted against its total mortality in Fig. 3. The trend lines have the slope d defined

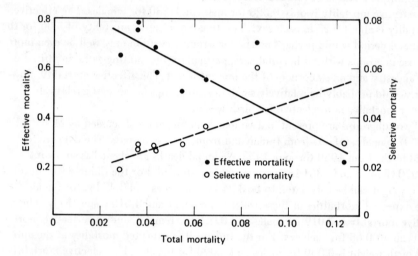

FIGURE 3. Effective mortality and selective mortality plotted against total mortality. One point represents each population and is the mean of the separate values for the two sexes.

by Van Valen (unpublished) which is a slight modification of the reduced major axis. The time after birth at which the statistics were computed differs among the populations. The effective mortality is constant as far as is known, but the total mortality and selective mortality increase with age. Therefore the relation shown between the latter two quantities is approximately correct, and by making an approximate adjustment of total mortality it can be shown that the age differences somewhat reduce the slope of the trend line of effective mortality on total mortality but have little effect on the correlation. The correlation between the effective and total mortalities (as plotted) is −0.76 and that between the selective and total mortalities is 0.51. These are both significant at the 0.01 level, while the correlation between the effective and selective mortalities is non-significant ($r = 0.10$) and remains non-significant when corrected for the age differences. Therefore, in populations with greater mortality a greater proportion of the population die because of selection on birth weight but a smaller proportion of the deaths are from this cause.

Birth Weight: Selection Intensities

Only in the Birmingham study is it possible to compare estimates of selection intensity (the selective mortalities) made from deaths at several different times. Assuming the same effective mortality at each period, the estimates of selection intensity made from different periods are given in Table 4. The assumption of equal

Table 4. Increase in Estimated Selection Intensity on Birth Weight From Birth to One Year (Birmingham Data)

	Birth	1 Day	1 Week	1 Month	6 Months	1 Year
Estimate	0.011	0.014	0.017	0.019	0.024	0.028

effective mortality is valid at least in the comparison of the situation at birth with that at 1 year; i.e., the postnatal deaths are about as selective for birth weight as the foetal deaths. The increase in the estimates of selection intensity with time is of course due to the fact that more infants have died with a greater passage of time. The true total selection intensity on birth weight would require life tables and fertility schedules to the end of reproduction.

In the Fetal Life Study data, comparison is possible between the weight distributions of foetal deaths and neonatal deaths, and also between the distributions of normal and abnormal infants. Several kinds of estimates of components of the selection intensity have been made. The estimate made from the distribution of all perinatal deaths, given above, is 0.026 or 0.028 for males and 0.027 or 0.029 for females, the effective mortalities being about 0.7 or 0.75 and 0.8 or 0.85 respectively. Considering only foetal deaths, the effective mortality is about the same and the estimates of the intensity are 0.012 for males and 0.014 for females. There were only eighteen known deaths between 28 days and 1 year in the Fetal Life Study sample, and only 528 of the 5732 were untraced. Assuming even several times the proportion of deaths in the untraced subset as in the known subset, and assuming the same effective mortality as for the perinatal deaths, the estimated selection intensity at 1 year would be increased by less than 0.004 over its value at 28 days, considerably less than the increase for the Birmingham sample.

If the distributions of birth weights are assumed to be normal before selection, which they are not, estimates of the selection intensity on the mean and variance can be obtained from the observed changes in mean and variance with selection by the method of Van Valen (1965). The graphs of Van Valen (1967) provide the necessary relations. For all perinatal deaths, the estimate for these parameters together is 0.043 for males and 0.047 for females, considerably greater than the values obtained without a normality assumption, but for foetal deaths alone the values are comparable; 0.012 for males and 0.016 for females when normality is assumed. The fact that these values are for selection on the mean and variance alone, while the values without the normality assumption are for total selection, is relatively unimportant, because almost all the selection is (as usual) on the mean and variance.

Although the sample size is too small to make an accurate estimate, inspection of the distributions in the appendix indicates that the optimal birth weight for malformed infants is about the same as that for normal ones. (As noted above, there is no obvious effect of gestation time on optimal birth weight in the data of Karn & Penrose, 1951). The causal conclusion does not follow, however, that part of the selection against malformed infants is due to the fact that they have a lower mean birth weight than the normal infants and are less clustered around their mean. There is undoubtedly some casual relationship in the reverse direction; e.g., anencephalous infants are lighter than most.

Some selection undoubtedly occurs throughout life against the class of individuals with congenital malformations (cf. Grüneberg, 1965). No estimate is available for this selection with respect to the entire class of congenital malformations. Several values of relative fitness were therefore used in estimating the effect of the distribution of congenitally malformed infants on selection for birth weight. Some proportion W of the malformed survivors was combined with the normal survivors, and the effect of this exclusion of some malformed survivors was then determined. If the subscripts 1 and 2 refer to normal and malformed respectively and N_i is the number in the respective class, then the resulting mean birth weight for the adjusted classes of survivors is

$$\bar{X} = \frac{N_1\bar{X}_1 + WN_2\bar{X}_2}{N_1 + WN_2}$$

and the variance of birth weight is

$$s^2 = \frac{N_1 s_1^2 + WN_2 [s_2^2 + (\bar{X}_1 - \bar{X}_2)^2]}{N_1 + WN_2}. \tag{6}$$

W, the relative fitness of the malformed survivors, was taken at various values from 0.95 to 0.5. Although the number eliminated by selection on both birth weight and malformation more than doubles when W is as low as 0.5, the distribution of birth weight among the malformed infants is so similar to that of the normal infants that the estimate of selection intensity on birth weight alone increases by only 0.001 for males and 0.003 for females. This calculation was made for selection on the mean and variance assuming normality of the unselected distribution, but the correct values should be similar, giving estimates of the total selection intensity on birth weight close to 0.030 for each sex. Considering the likelihood of other selection on birth weight after 28 days, the total selection intensity on birth weight could easily be as high as 0.04 or even 0.05.

The selection intensity on the mean is the minimum proportion of the unselected population that must be removed per generation in order to give a population with the same mean as the selected population, and analogously for the selection intensity on the variance. These intensities are more specifically the potential selection intensities (not to be confused with the maximum limit on selection developed by Crow, 1958). The realized selection intensity on, e.g., the mean, is the same as the potential intensity, except that the proportion removed must come only from the deaths themselves, and is therefore at least as large as the potential intensity. These concepts are defined and discussed more fully by Van Valen (1965). The potential and realized selection intensities on the mean and variance separately (more properly the selective perinatal mortalities, because only this aspect is considered) are given in Table 5 for the Fetal Life Study data and the data of Karn & Penrose (1951). The estimates were made by iteration from the distributions themselves; normality was nowhere assumed. The sum of the intensities on the mean and variance is greater than the total perinatal selection intensity, because some of the same individuals are removed in the estimation of the two intensities.

Table 5. Selection Intensities on Mean and Variance Separately (Birth Weight)

	Potential Intensity		Realized Intensity	
	Mean	Variance	Mean	Variance
Fetal Life Study				
Male	0.017	0.014	0.019	0.019
Female	0.019	0.015	0.024	0.021
London				
Male	0.018	0.018	0.020	0.020
Female	0.018	0.016	0.020	0.021

The means and variances for the selected and unselected samples of the London and New York (Fetal Life Study) samples are given in Table 6. In both a much greater proportional reduction is found for the variance (or the standard deviation) than for the mean even though the selection intensities on these two parameters are the same. This property of more efficient phenotypic selection on the variance comes from the shape of the unselected distribution itself and is true for almost all quantitative characters.

On the basis of data of Robson (1955) and others, Penrose (1954a) partitioned the variance of human birth weight (survivors only) into nine casual components. Because of the simplifying assumptions involved in this kind of analysis, and because the correlations on which the estimates were based are not as accurately known as is desirable, the values he obtained (without saying how they were derived) must be regarded with some scepticism even for the general population from

Table 6. Birth Weight Statistics*† for Selection,
Fetal Life Study and London Sample

$\overline{X}$	s^2	$\dfrac{\overline{X}_2-\overline{X}_1}{s_1}$	$\dfrac{s_1^2-s_2^2}{s_1^2}$	$\dfrac{s_1-s_2}{s_1}$	$\dfrac{CV_1-CV_2}{CV_1}$
		Fetal Life Study			
Total sample					
Male 3.160 ± 0.011	0.363	–	–	–	–
Female 3.067 ± 0.011	0.345	–	–	–	–
Survivors					
Male 3.198 ± 0.010	0.290	0.063	0.20	0.11	0.12
Female 3.109 ± 0.010	0.272	0.072	0.21	0.11	0.12
		London			
Total sample					
Male 3.310 ± 0.007	0.374	–	–	–	–
Female 3.212 ± 0.007	0.295	–	–	–	–
Survivors					
Male 3.347 ± 0.006	0.272	0.061	0.27	0.15	0.16
Female 3.246 ± 0.006	0.239	0.063	0.19	0.10	0.11

*In kilograms.
†Subscripts: 1 designates total (unselected) sample; 2 designates survivors. CV is coefficient of variation.

which the statistics were determined. Their relative values are, nevertheless, probably more or less correct. The foetal genotype, within each sex, was estimated to account for only 0.16 of the variance of birth weight. A low value for this component was confirmed by the study of Morton (1955). Jayant (1966a) found that both the optimal and mean birth weights increase with parity, the former apparently more rapidly. Therefore selection on birth weight is presumably for the most part on effects of various aspects of the foetal environment rather than on the foetal genotype.

A major source of variance in birth weight not estimated by Penrose (1954a) is that of gestation time. Karn & Penrose (1951) found the correlation between birth weight and gestation time, corrected the maternal age and parity, to be 0.43 for males and 0.39 for females. Although not so stated, the numbers of individuals involved indicate that these correlations are only for survivors. Because both birth weight and gestation time are positively related to survival, somewhat greater correlations would be expected if all births were included. A further small increase in the estimated correlations is mandated by the fact that the relation between gestation time and birth weight is not linear. The proportion of the variance of birth weight due to gestation time is therefore between 0.16 and about 0.20 (the latter value is the more likely estimate), making it a substantial component. Presumably the effect of gestation time cuts across several of Penrose's categories, for it itself is a character that must be determined.

Fitness and Deviation from Optimal Birth Weight

The conventional model of a quadratic relation between deviation from the optimum and probability of survival does not hold for birth weight. An equation of the form

$$S = aD^2 + bD + c, \tag{7}$$

where S is the probability of survival and D is the deviation from optimal birth weight, is readily obtainable as the parabolic regression of survival on weight (cf. equation (1)). This equation, although the closest fit of any quadratic model, deviates markedly from the Fetal Life Study data. The same is true for other simple polynomial expressions. Although a moderately close fit can probably be obtained by juggling a general fourth degree equation, this sort of curve-fitting by multiplication of terms has little to recommend it from the theoretical viewpoint.

If the probability of death is assumed to increase exponentially from the optimal birth weight, a more promising picture emerges. Figs. 4 and 5 show the relation of the natural logarithm of the probability of death, to birth weight, for the Fetal Life Study data and the data of Karn & Penrose (1951). The regressions were calculated after eliminating from consideration the classes with complete mortality in the sample. The regressions are of the form

$$-\ln M = bD + a, \tag{8}$$

where M is the probability of mortality and D is the absolute deviation from optimal birth weight. This equation is equivalent to

$$M = ce^{-bD} \tag{9}$$

a very simple form. The equations, which are the only simple ones to approximate the distribution of M, are as follows:

Fetal Life Study data

$$\text{Males:} \quad M = 0.0091e^{-1.64D},$$
$$\text{Females:} \quad M = 0.0052e^{-1.77D}.$$

Karn & Penrose data

$$\text{Males:} \quad M = 0.0155e^{-1.64D},$$
$$\text{Females:} \quad M = 0.0095e^{-1.82D}.$$

e can be eliminated from these equations if desired; e.g., the first is equivalent to $M = 0.0091 \ (0.193^D)$. The fit of the quadratic deviation model has not previously been tested for any data.

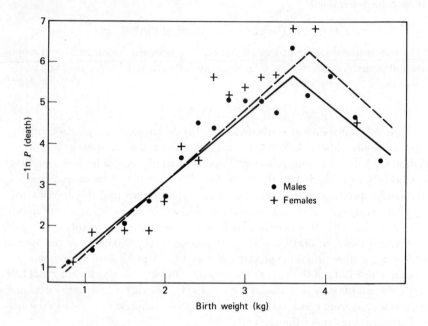

FIGURE 4. Relationship to birth weight of the negative logarithm of the probability of death, Fetal Life Study. The solid line is the male regression; the broken line is the female regression.

Although the fit of the exponential model is reasonably good, there is still some systematic deviation of a sigmoid kind. In fact the deviations are such as to suggest a truncate normal distribution. In this case the distribution would be of the double exponential form

$$-\ln M = fe^{-gx^2}, \tag{10}$$

where X is the birth weight. (D can be used in place of X and the distribution then truncated at its apex.) If R denotes the ratio of survivors to dead, expression (10) is equivalent to

$$\ln (1 + R) = fe^{-gx^2}. \tag{11}$$

With a change in variance, equation (11) is equivalent to the condition that R itself is approximately normally distributed over the range of birth weights. As noted by

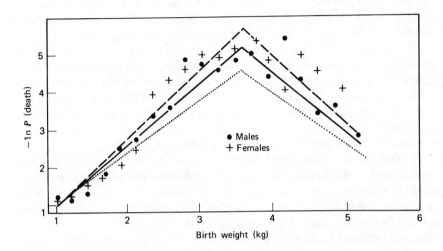

FIGURE 5.   Relationship to birth weight of the negative logarithm of the probability of
death, London data. The solid line is the male regression, the broken line is the female
regression, and the dotted line is the regression derived from the
extreme classes only. See text.

Haldane (1954), it is this normal distribution of R that is equivalent to the parabolic relationship of $\ln R$ to birth weight that was first found by Karn & Penrose
(1951) and that is present also in at least New York, Italian, and Ghanaian infants.
Karn & Penrose had assumed that the weights of both survivors and non-survivors
are normally distributed, a condition which practically implies the preceding but
which happens to be seriously in error for the Fetal Life Study data. The distribution of the weights of nonsurvivors is sometimes not far from rectangular, i.e., the
fact that an infant is dead may give little information about its birth weight. The
adequate fit of the parabolic relationship of $\ln R$ to X in several diverse sets of data
demonstrates the approximate normality of R and $-\ln M$. Direct tests of fit for
these distributions are possible but superfluous. The non-normality of the distributions of non-survivors to some degree invalidates the "z" method for estimating the
optimal birth weight (Penrose, 1947, 1954b), because this method is relatively sensitive to deviations from a symmetrical distribution.

Because of the incremental effect of small main effects and interactions, it is
easy to understand why most quantitative biological characters (such as birth
weight) are approximately normally or lognormally distributed. But why, of all
things, should the ratio of survivors to non-survivors, or the logarithm of the probability of mortality, be more or less normally distributed with respect to birth

weight? A possible answer is provided by the operation of natural selection over a large number of generations. We assume that in some fictitious period of human ancestry, "before an effect of selection," there was some kind of regular decrease of viability as birth weight departed from an optimum, such as a quadratic or exponential decrease. As usual, natural selection would operate to increase viability or prevent its deterioration. If the mechanisms available for increasing viability differ in their effects on infants of different birth weight, there would be greater selection for those mechanisms that improved most the viability of the most numerous classes, i.e., those around the mean. This is the pattern that is seen in all studies: the most numerous classes have a greater viability than that predicted by a single relationship with the viability of the less numerous classes.

The fact that the optimal birth weight is somewhat above the mean is not a serious drawback to this hypothesis, because the difference is small and the optimum is always in a numerous size-group. An optimum different from the mean can be stable (James, 1962), but it is also possible that the mechanism or set of mechanisms that happened to be selected because it most reduced mortality, simply operates best at a birth weight somewhat above the mean. These alternatives are not mutually exclusive.

For the operation of the above mechanism it is not necessary that selection act directly on birth weight itself. Selection would operate most strongly on the most numerous classes of any phenotype and, if the phenotype is developmentally related to birth weight, these classes would tend to include the same individuals as the most numerous classes of birth weight. The period "before an effect of selection" should not be taken literally. It is simply a convenience for expressing the action of natural selection, which acts continuously but in the same way as if it had started from an unselected population.

An aspect of the above hypothesis is that there has been selection for an increased canalization of the mortality pattern, i.e., that birth weights near the optimum now have proportionally less difference in mortality than previously. The same difference between two birth weights near the optimum does in fact produce much less difference in the probability of death than does the same difference in weight far from the optimum, even on the exponential scale of Figs. 4 and 5. The effect of this selection for canalization can be roughly quantified if a relationship between mortality and birth weight without canalization can be specified. Such a relationship is more or less arbitrary; for convenience we use an exponential one. As with any reasonably simple initial relationship in the present case, the effect of selection is maximal at some distance from the optimum (Fig. 6). The figure is meant only to illustrate the general pattern of response expected. The smaller improvement at the initial optimum is to be expected if it becomes progressively more difficult to reduce the probability of mortaility as this probability decreases. In that case the initially optimal birth weight would have a relatively smaller increase in fitness than the values near it that still have a sufficiently large proportion of the population to influence the course of selection.

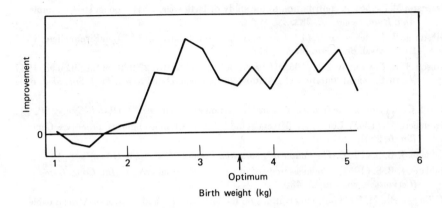

FIGURE 6. Amount of relative improvement of the logarithm of the probability of survival as a result of a hypothetical response to selection. Sexes combined, London sample. The ordinates are proportional to the distance above the dotted regression of Fig. 4.

Summary

There is no detectable selection on rate of tooth eruption. Some selection occurs against congenital malformations; the selective mortality on this character is about 0.006 in the perinatal period. Data from several studies give perinatal selective mortalities on birth weight from about 0.03 to perhaps 0.06 and effective mortalities of 0.25 to 0.8. Populations with greater mortality tend to have more intense selection on birth weight but a smaller effective mortality. Selection intensities on the mean and variance separately are each about 0.02 for two samples although the proportional change in mean by selection is considerably less than that in variation. The probability of mortality increases about exponentially with deviation from the optimal birth weight. A simple model of the action of natural selection accounts for the residual inaccuracy of this relationship by inferring the evolution of a more canalized pattern of birth weight and mortality.

We are indebted to Dr. N. E. Morton for comments.

References

Böök, J. A. (1951). The incidence of congenital diseases and defects in a south Swedish population. *Acta Genet. Stat. Med.* **2**, 289.

Crow, J. F. (1958). Some possibilities for measuring selection intensities in man. *Human Biol.* **30**, 1.

Fisher, R. A. (1930). *The Genetical Theory of Natural Selection.* Oxford: Clarendon Press.

Fraccaro, M. (1956). A contribution to the study of birth weight based on an Italian sample. *Ann. Hum. Genet.* **20,** 282.

Gibson, J. R. & McKeown, T, (1951). Observations on all births (23,970) in Birmingham, 1947. III. Survival. *Br. J. Soc. Med.* **5,** 177.

Gibson, J. R. & McKeown, T. (1952). Observations on all births (23,970) in Birmingham, 1947. VI. Birth weight, duration of gestation, and survival related to sex. *Br. J. Soc. Med.* **6,** 152.

Green, C. R. (1964). The frequency of maldevelopment in man. *Am. J. Obstet. Gynec.* **90,** 994.

Grüneberg, H. (1965). Congenital defects as indicators of lifelong abnormal processes. *Genet. Res.* **6,** 263.

Haldane, J. B. S. (1932). *The Causes of Evolution.* London: Longmans, Green.

Haldane, J. B. S. (1954). The measurement of natural selection. *Proc. IX Int. Cong. Genet. (Caryologia,* vol. suppl.), 480.

Hollingsworth, M. J. (1965). Observations on the birth weights and survival of African babies: Single births. *Ann. Hum. Genet.* **28,** 291.

Hosemann, H. (1950). Kindliche Masse and Neugeborenensterblichkeit. *Naturwissenschaften* **37,** 409.

James, J. W. (1962). Conflict between directional and centripetal selection. *Heredity* **17,** 487.

Jayant, K. (1964). Birth weight and some other factors in relation to infant survival. A study of an Indian sample. *Ann. Hum. Genet.* **27,** 261.

Jayant, K. (1966a). Effect of parity on optimal and critical birth weights. *Ann. Hum. Genet.* **29,** 363.

Jayant, K. (1966b). Birth weight and survival: A hospital survey repeated after 15 years. *Ann. Hum. Genet.* **29,** 367.

Karn, M. N. (1952). Birth weight and length of gestation of twins, together with maternal age, parity and survival rate. *Ann. Eugen.* **16,** 365.

Karn, M. N. & Penrose, L. S. (1951). Birth weight and gestation time in relation to maternal age, parity and infant survival. *Ann. Eugen.* **16,** 147.

McIntosh, R., Merritt, K. K., Richards, M. R., Samuels, M. H., & Bellows, M. T. (1954). The incidence of congenital malformations: A study of 5,964 pregnancies. *Pediatrics* **7,** 505.

Mellin, G. W. (1963). The frequency of birth defects. In Fishbein, M. (ed.), *Birth Defects,* p. 1. Philadelphia: L. B. Lippincott, for National Foundation.

Mellin, G. W. (1966). The Fetal Life Study of the Columbia-Presbyterian Medical Center. A prospective epidemiological study of prenatal influences on fetal development and survival. In Chipman, S. S. et al. *Research Methodology and Needs in Perinatal Studies,* p. 88. Springfield, Illinois: C. C. Thomas.

Millis, J. (1959). Distribution of birth weights of Chinese and Indian infants born in Singapore: birth weight as an index of maturity. *Ann. Hum. Genet.* **23,** 164.

Morton, N. E. (1955). The inheritance of human birth weight. *Ann. Hum. Genet.* **20,** 125.

Penrose, L. S. (1947). Some notes on discrimination. *Ann. Eugen.* **13,** 228.

Penrose, L. S. (1954a). Some recent trends in human genetics, *Proc. IX Int. Cong. Genet. (Caryologia,* vol. suppl.), 521.

Penrose, L. S. (1954b). Genetical factors affecting the growth of the foetus, *La prophylaxie en gynecologie et obstetrique (Cong. Int. Gynec. Obstet.),* p. 638. Geneva: Libraire de l'Université, Georg et Cie, S. A.

Robson, E. B. (1955). Birth weight in cousins. *Ann. Hum. Genet.* 19, 262.

Spuhler, J. N. (1963). The scope for natural selection in man. In Schull, W. J. (ed.), *Genetic Selection in Man*, p. 1. Ann Arbor: University of Michigan Press.

Stevenson, S. S., Worcester, J. & Rice, R. G. (1951). Congenitally malformed infants and gestation characteristics. *Pediatrics* 6, 37, 208.

Van Valen, L. (1963). Selection in natural populations: *Merychippus primus*, a fossil horse. *Nature, Lond.* 197, 1181.

Van Valen, L. (1965). Selection in natural populations. III. Measurement and estimation. *Evolution* 19, 514.

Van Valen, L. (1967). Selection in natural populations. 6. Variation genetics, and more graphs for estimation. *Evolution.* 21, 402.

Van Valen, L. & Mellin, G. W. (1966). Natural selection on New York babies (abstract). *Third Int. Cong. Hum. Genet., Abst. Contrib. Pap.*, p. 101.

Wright, S. (1935). The analysis of variance and the correlations between relatives with respect to deviations from an optimum. *J. Genet.* 30, 243.

APPENDIX. Birth Weight Data (Fetal Life Study)

Weight* (kg.)	Males Normal S†	F†	N†	Males Malformed S	F	N	Females Normal S	F	N	Females Malformed S	F	N
0.5–0.6	0	0	4	0	0	0	0	2	0	0	0	2
0.6–0.7	0	0	5	0	0	1	0	4	4	0	0	0
0.7–0.8	0	2	2	1	0	0	0	1	1	0	0	0
0.8–0.9	0	1	2	0	1	0	1	4	1	0	1	1
0.9–1.0	1	2	2	1	0	1	0	0	2	0	0	0
1.0–1.1	0	3	0	0	1	0	1	0	2	0	0	0
1.1–1.2	2	1	1	1	0	0	0	2	1	1	0	0
1.2–1.3	2	1	2	0	0	0	8	1	1	2	0	0
1.3–1.4	4	0	1	0	2	0	2	2	1	1	1	1
1.4–1.5	0	0	1	0	0	0	7	0	1	0	0	0
1.5–1.6	3	0	1	0	0	0	6	1	2	0	0	1
1.6–1.7	5	1	1	1	0	0	3	2	2	0	0	0
1.7–1.8	11	1	1	1	0	0	13	1	3	0	0	0
1.8–1.9	9	3	0	1	0	1	12	1	1	2	1	0
1.9–2.0	13	2	0	1	1	0	15	1	1	0	0	1
2.0–2.1	19	3	1	0	0	0	12	2	2	2	0	0
2.1–2.2	23	0	1	2	0	0	18	0	1	2	0	1
2.2–2.3	25	0	1	3	2	0	30	0	1	2	0	0
2.3–2.4	38	0	0	4	1	0	35	1	3	6	1	1
2.4–2.5	52	0	1	2	1	0	48	1	0	10	1	0
2.5–2.6	58	0	2	6	0	0	71	0	0	3	1	0
2.6–2.7	68	0	2	12	1	0	117	1	0	10	1	0
2.7–2.8	117	1	2	11	0	0	132	2	1	4	1	0
2.8–2.9	126	1	1	13	0	0	158	0	1	5	0	0
2.9–3.0	172	2	1	17	0	0	216	0	2	6	1	0
3.0–3.1	178	1	1	8	1	1	207	1	1	15	1	0
3.1–3.2	210	0	2	18	1	0	236	1	0	11	1	0
3.2–3.3	197	1	2	19	1	1	217	2	0	15	0	1
3.3–3.4	196	2	2	24	1	2	190	1	1	6	0	0
3.4–3.5	215	1	2	17	1	0	201	1	1	12	0	0
3.5–3.6	184	0	0	13	0	0	163	1	0	7	0	0
3.6–3.7	168	1	0	19	1	0	141	0	0	8	0	0
3.7–3.8	128	1	1	11	0	0	100	0	0	4	0	0
3.8–3.9	99	0	2	11	0	0	74	0	1	4	0	0
3.9–4.0	56	0	2	5	0	0	58	0	0	4	0	0
4.0–4.1	52	0	0	6	0	0	28	0	0	6	0	0
4.1–4.2	41	0	0	4	0	0	21	0	0	0	1	0
4.2–4.3	30	0	0	0	0	0	16	0	1	1	0	0
4.3–4.4	19	0	0	2	0	0	7	0	0	0	0	0
4.4–4.5	14	1	0	1	0	0	6	0	0	0	0	0

(continued)

APPENDIX. Birth Weight Data (Fetal Life Study)

Weight* (kg.)	Males Normal			Males Malformed			Females Normal			Females Malformed		
	S†	F†	N†	S	F	N	S	F	N	S	F	N
4.5–4.6	8	1	0	0	0	1	3	0	0	0	0	0
4.6–4.7	8	0	0	0	0	0	6	0	0	0	0	0
4.7–4.8	1	0	0	0	0	0	3	0	0	1	0	0
4.8–4.9	2	0	0	1	0	0	1	0	0	0	0	0
4.9–5.0	2	0	0	0	0	0	1	0	0	0	0	0
5.0–5.1	1	0	0	0	0	0	0	0	0	1	0	0
5.1–5.2	0	0	0	0	0	0	1	0	0	0	0	0
5.2–5.3	0	0	0	0	0	0	0	0	0	0	0	0
5.3–5.4	1	0	0	0	0	0	0	0	0	0	0	0
.												
.												
.												
5.8–5.9	0	0	0	0	0	0	1	0	0	0	0	0

*Weight classes are, e.g., from 0.501 kg. through 0.600 kg.
†S, survivors; F, foetal deaths; N, neonatal deaths.

5. Fertility and Physique–Height, Weight, and Ponderal Index

ALBERT DAMON[1] and R. BROOKE THOMAS

Introduction

All populations are exposed to selective forces which, through differential actions upon individuals, tend to bring the population as a whole into equilibrium with the stresses upon it. Differential reproductive performance among members of a population is the ultimate mechanism effecting this homeostatic process. Investigations of differential fertility can therefore indicate the intensity and direction of such selective forces as well as evolutionary trends within a population. The present study examines, among males, the relationship to fertility of the ponderal index (height/$\sqrt[3]{\text{weight}}$) and of its components, height and weight.

Several studies have dealt with associations between physique and fertility. Davenport (1923), working with 506 American couples, assigned each mate to one of five body builds (weight/stature^2) ranging from very slender to very fleshy. Based on their builds, couples were classified as to mating types, and the mean number of progeny was calculated for each type. In comparing these means, Davenport found that fleshy couples (fleshy x fleshy) produced 1.3 and 2.3 times as many children as medium and slender couples, respectively. He therefore concluded that stockier parents produce larger families than slender parents. Unfortunately there was no information as to the age distribution of each mating type. Weight increases among men and women during adult life have been well documented, both cross-sectionally (Stoudt, et al., 1965) and longitudinally (Damon, 1965). American men gain on the average at least 6-8 kg beyond their late teens or early twenties. An association between fleshy build and high fertility may therefore merely reflect the fact that fleshier (older) couples are closer to completing their families than are

SOURCE: Albert Damon and R. Thomas, "Fertility and Physique: Height, Weight, and Ponderal Index," *Human Biology,* **39**: 5-13 (Detroit, Michigan: The Wayne State University Press, 1967).

[1]This work was done during the tenure of an Established Investigatorship of the American Heart Association and with the support of a Research Grant from the same source. Computations were aided by NSF Grant GP-2723 to the Harvard Computing Center.

younger, more slender couples. This age effect is enhanced by the secular trend toward smaller families.

Among 1,450 Italians with at least seven children, Frassetto (1933, 1934), using methods like Davenport's (1923), reached similar conclusions, unsupported by his data. Besides Frassetto, twenty other investigators made anthropometric studies among prolific Italians, under the sponsorship of the Italian Committee for the Study of Population Problems (Gini, ed., 1934). Unfortunately the highly selected samples, general failure to control for age, analysis by irrelevant categories, and above all the lack of proper controls vitiate most of this monumental effort.

Working with a sample of 97 single females, 227 married females, 43 single males, and 187 married males from Ann Arbor, Michigan, Clark and Spuhler (1959) have investigated differential fertility in relation to body dimensions. Their sampling frame was random, but "inasmuch as about a quarter of the individuals selected for study refused to participate, the sample cannot be regarded as wholly random" (*ibid.*, p. 123). The subjects were divided into 5-year age groups, and males and females were analyzed separately. Within each age-sex category, individuals were further classified as to whether they reported above or below median fertility for that group. Relationships between 41 anthropometric traits and fertility were analyzed separately using a series of 82 r x 2 analyses of variance, r being the number of age groups and 2 the fertility categories. This method eliminates the effect of age on fertility.

For males, the following body dimensions were significantly larger in the above-average-fertility group than in the below-average group: weight; head length; bizygomatic, biacromial, and bicondylar knee breadths; neck, midarm, forearm, wrist, and calf circumferences. Except for head length, which differed by 2.33 mm, all of the differences in body breadths and circumferences could reflect the weight difference. Several other breadths, stature and all length measurements (except head length), and 8 out of 9 indices showed no significant difference between the two fertility groups. The stature/span index was lower in the above-average-fertility group by 0.54 percentage points ($.05 > p > .025$).

While the shorter stature (3.66 mm) of the more fertile men was not significant ($.75 > p > .50$), there was a significant interaction between age and fertility for stature ($.005 > p > .001$). Hence, the authors argued, stature might be significantly related to fertility, "although the nature of the relationship would vary from age group to age group" (*ibid.*, p. 131). To interpret such a relationship, the complete table would have to be examined, and even then no meaningful biological pattern might be discernible. Moreover, no weight-stature ratio, the customary measure of stockiness, was included among the 41 anthropometric traits. The closest approximation, shoulder breadth/stature, was not significantly associated with fertility.

From the above results, Clark and Spuhler (1959, p. 136) concluded: "The analyses show . . . that for both males and females of the Ann Arbor population, there is a tendency for individuals of above average fertility to be more stockily built than individuals of less than average fertility. However, it is unclear from these analyses whether a stocky build tends to result in above average fertility or whether the factors associated with above average fertility—such as marital status and socio-economic level—tend to result in a stocky build." Even if one accepts the authors' conclusion for men, about which our reservations will be discussed later, the mixing of single and married men and the definition of "single" as currently unmarried make it indeed impossible to decide whether or to what extent physique is associated with marital status or with fertility within marriage.

Subjects and Methods

The present study conforms to Clark and Spuhler's recommendation (1959, p. 136) for prospective anthropometric research on young adults before marriage, to be correlated with later reproductive performance. Described previously in this Journal (Damon and Nuttall, 1965), the sample consists of 2,616 Harvard men measured nude, mostly in college, between 1880 and 1912 by the late Dudley A. Sargent and two assistants. For the present purposes, only Caucasian subjects were retained, 15 Negroes or Orientals being omitted from Sargent's total of 2,631. The group, with a mean age of 20.25 years, was homogeneous ethnically and ultimately approached socio-economic homogeneity. Approximately 85% were Old Americans of British ancestry, and 10% were of other Northern European descent. Ponderal index (height/$\sqrt[3]{\text{weight}}$) for the 2,616 men was calculated and coded on a 7-point scale, ranging from stocky to lean, by Mr. F. L. Stagg and the late Dr. E. A. Hooton. Mr. Stagg has kindly made these data available to us.

From their published Class Reports and from correspondence, marital status and number of children (excluding step- and adopted-children) were obtained; 578 men had remained single, 527 married but were childless, and 1,511 became fathers. Their completed reproductive histories were available—an important but often neglected point in fertility studies. "Married" as used here means ever-married, since whatever the varying circumstances of marriage (age at marriage, separation, death of spouse, etc.), the individual has generally had a chance to produce offspring. "Fertility" refers to the number of reported live-born offspring per man.

Results

The total sample has been divided into three groups: 578 single men, 527 childless married men, and 1,511 married fathers. Table 1 shows that in height, weight, and ponderal index while at Harvard, the future single and the married but childless men did not differ at all, nor did the married fathers differ from the single men or

Table 1. Stature, Weight, and Ponderal Index of Single Men, Childless
Married Men, and Married Fathers—Harvard Series

Group	N	Stature (cm.)		Weight (kg.)		Stature/$\sqrt[3]{\text{Weight}}$	
		Mean	S. D.	Mean	S. D.	Mean	S. D.
Single men	578	162.7	7.9	72.9	6.0	43.63	15.22
Married, childless	527	162.5	7.5	73.1	6.8	43.66	15.29
Married fathers	1511	163.5	7.6	73.7	6.2	43.62	15.54

the married but childless men in weight or ponderal index. In stature, however, the
three groups differed significantly, analysis of variance yielding a p-value below .01.
The married fathers were 0.8 cm taller $(.05 > p > .02)$ than the single men and 1.0
cm taller $(p < 0.01)$ than the married, childless men. For the ponderal index, not
only the means and standard deviations (Table 1) but also the distributions (not
given here) were virtually identical in the three groups of subjects.

Table 2 presents, for married fathers, the number of children by ponderal index
category. No association is discernible from examination of all seven categories, the
greatest fertility appearing in the middle categories, 4 and 5, and in the small category 7. The excess mean fertility, 0.16 child per father, of the 4 leaner categories
(4-7) over the 3 stockier ones (1-3) was not significant $(.25 > p > .20)$.

Table 2. Fertility of Married Fathers by Ponderal Index Category—Harvard Series

	Ponderal Index Category*							
	Stocky						Lean	
Interval	<38.3	38.4-40.0	40.1-41.7	41.8-43.4	43.5-45.1	45.2-46.8	46.9+	
Code	1	2	3	4	5	6	7	Total
No. Fathers	2	16	117	560	593	198	25	1,511
Children/ Father:								
Mean	4.00	2.50	2.50	2.79	2.64	2.55	2.64	2.68
S. D.	0	1.10	1.50	1.49	1.40	1.32	1.58	1.43
Combined categories		1, 2, 3			4, 5, 6, 7			
Children/ Father:								
Mean		2.53			2.69			
S. D.		1.43			1.43			

*Ponderal index = height (cm) divided by cube root of weight (kg).

Possible relationships between ponderal index and fertility among married fathers were further investigated by single-classification analysis of variance and by the coefficient of correlation, neither of which showed significant association. The between-category/within-category F-value was 1.48, with 3.84 needed for 5% significance; the r was $-0.03 \pm$ s.e. 0.03. The correlation ratio showed no non-linear association ($eta = 0.03 \pm$ s.e. 0.03).

Fertility of fathers was then examined in relation to height and weight (Table 3). Stature alone ("Total" row) showed no consistent trend, borne out by the r of 0.002 between stature and number of children. The apparent positive association between fertility and weight ("Total" column) did not reach statistical significance, as shown by single-classification analysis of variance between weight classes and mean fertility ($F = 0.48$) and also by the negligible coefficient of correlation between weight and number of children ($r = 0.023 \pm$ s.e. 0.03). The individual cells of Table 3 show variation without any trend.

Discussion

The investigation yields the following results:

1. As young men in college, those who never married, those who later married but remained childless, and those who married and produced at least one child were similarly distributed in all ponderal index categories—stocky, medium, and lean. The three groups of men did not differ in mean ponderal index or in weight, but married fathers were taller than the two other groups.

2. Among married men who produced offspring, fertility and ponderal index were independent.

3. Looked at separately, the variables used to calculate the ponderal index— namely, stature and weight—were not significantly related to fertility.

Two trends were observed which did not reach statistical significance and which were contradictory in direction; hence, they are best regarded as chance variations. They were greater fertility of lean fathers (height/$\sqrt[3]{\text{weight}}$—Table 2) but also of heavy fathers (weight alone, over all stature categories—Table 3).

Our findings differ from those of Davenport (1923) and of Clark and Spuhler (1959), who reported stockier individuals to be more prolific. Davenport neglected the effect of age on fertility, as we have noted. Clark and Spuhler reported fertility to be independent of stature, as we found among fertile men, but Harvard men who produced any children at all were taller than those who produced none. Other differences between the two groups concern weight and the definition of stockiness of build: unlike the Ann Arbor men, the Harvard men showed no association between fertility and weight or build.

The discrepancy between the two sets of findings could arise from differences in

Table 3. Mean Number of Children Per Father, by Stature and Weight—Harvard Series

Weight (kg)		<159	160–1	162–3	164–5	166–7	168–9	170–1	172–3	174–5	176–7	178–9	180–1	182–3	184–5	186+	Total
<49	N	5	7		4	5	4	2	1	2	1						31
	Mean	3.00	2.00		2.25	2.40	1.50	5.00	3.00	4.50	1.00						2.55
	S.D.	1.67	1.07		1.09	0.49	0.87	2.00	0.0	0.50	0.0						1.50
50–59	N	8	15	26	35	61	66	68	47	42	26	17	8	6	1		426
	Mean	2.25	2.80	2.36	2.54	2.74	2.53	2.78	2.68	2.57	2.65	2.70	2.12	2.17	2.00		2.62
	S.D.	1.20	1.94	1.21	1.34	1.43	1.18	1.70	1.26	1.24	1.21	1.02	1.36	1.34	0.0		1.38
60–69	N	1	6	10	16	33	64	92	107	108	140	72	52	39	15	12	767
	Mean	3.00	2.00	2.90	3.25	2.54	2.69	2.77	2.68	2.72	2.75	2.56	2.48	2.92	2.07	2.75	2.69
	S.D.	0.0	1.00	1.58	2.16	1.35	1.21	1.53	1.40	1.26	1.54	1.43	1.34	1.83	1.29	1.01	1.45
70–79	N			1		1	9	10	11	27	42	37	40	34	13	18	243
	Mean			2.00		2.00	3.00	3.20	2.64	2.56	2.48	3.38	2.58	2.76	2.46	2.83	2.76
	S.D.			0.0		0.0	1.56	2.04	1.72	1.31	1.40	1.53	1.22	1.46	1.01	1.50	1.47
80+	N						2		4	1	7	3	4	5	6	12	44
	Mean						3.00		3.00	1.00	2.71	2.00	3.50	2.80	3.00	2.08	2.61
	S.D.						1.00		1.00	0.0	0.88	0.82	0.87	1.17	1.41	1.04	1.17
Total	N	14	28	37	55	100	145	172	170	180	216	129	104	84	35	42	1,511
	Mean	2.57	2.43	2.51	2.73	2.65	2.61	2.82	2.69	2.67	2.67	2.80	2.53	2.80	2.37	2.60	2.68
	S.D.	1.40	1.64	1.33	1.64	1.36	1.23	1.66	1.37	1.28	1.46	1.45	1.30	1.63	1.24	1.29	1.43

Stature (cm.)

the sample studied, in the method of analysis, or in interpretation. Taking the customary definition of stockiness as the relationship of weight to height, and gauging stockiness directly by the ponderal index, height/$\sqrt[3]{\text{weight}}$, we could not demonstrate a relationship to fertility. Neither could Clark and Spuhler for shoulder breadth/stature, a less direct measure of body build but the only one they used.

It is possible that Clark and Spuhler's finding of a positive association between weight and fertility reflects in part the lesser weight of single than of married men, who for the most part in their study were those with children. It is general observation, which we can unfortunately not document, that men gain weight on marriage and while married. They may also lose weight as widowers ("single" by the Clark-Spuhler definition). The pattern of physical differences reported by Clark and Spuhler suggests endomorphy (fat) rather than mesomorphy (bone and muscle), since most of the body dimensions and indices which when increased characterize mesomorphy were not associated with fertility. Such traits include hand length and breadth, foot length, ankle circumference, bicondylar breadth of elbow, and most head and face dimensions, as well as the sitting height/stature, hand breadth/hand length, and bigonial breadth/minimum frontal breadth indices. Our proposed explanation—which cannot be tested from the published data—would probably not account for all of the discrepancy between the two studies, since there were relatively few "single" men in the Michigan series.

Other possible explanations for the divergent findings of the two studies include differences in definition of terms (e.g., single and married); in the composition of samples in respect to age when measured, age when fertility was recorded, ethnic origin, socio-economic status, or the period under study (1880-1930 vs. 1950-1955); or in study design (partial versus completed reproduction). If, for example, heavy men marry and reproduce earlier than light men, the former would appear to be more fertile at any age before both groups had completed reproduction—even if fertility and physique were quite independent. The present study, by its prospective design, its use of completed reproductive histories from a homogeneous sample, and its distinction between single, married but childless, and married men who produced children, reduces the range of variables which confound the biological relationship between physique and fertility.

In brief, a sizeable, homogeneous group of men measured before marriage showed no association between weight or body build and the number of children they later produced. Men who produced at least one child were taller than those who produced none, but among the fertile men, stature and fertility were independent. Until associations between fertility and physique are firmly established, speculation on their underlying mechanisms or their genetic and evolutionary significance is premature.

Summary and Abstract

Among 2,616 men measured at Harvard College between 1880 and 1912, the 1,511 who subsequently married and had children were 0.8 cm taller than the 578 men who remained single, and 1.0 cm taller than the 527 men who married but had no children. The three groups of men did not differ in weight or in ponderal index, height/$\sqrt[3]{\text{weight}}$.

The 1,511 fathers showed no association between ultimate fertility and height, weight, or ponderal index.

Acknowledgment

We are grateful to Mr. F. L. Stagg for access to data which he collected.

Literature Cited

Clark, P. J. and J. N. Spuhler. 1959. Differential fertility in relation to body dimensions. *Human Biol.*, 31: 121-137.

Damon, A. 1965. Notes on anthropometric technique. III. Adult weight gain, accuracy of stated weight, and their implications for constitutional anthropology. *Am. J. Phys. Anthrop.*, 23: 306-311.

Damon, A. and R. L. Nuttall. 1965. Ponderal index of fathers and sex ratio of children. *Human Biol.*, 37: 23-28 and 216.

Davenport, C. B. 1923. *Body Build and Its Inheritance.* Carnegie Institution of Washington, Washington, D.C.

Frassetto, F. 1933. Prolificité et constitution. Congrès International d'Anthropologie et d'Archéologie Préhistoriques, 15e Session, Paris, France, 1931.

—————.1934. I principali caratteri antropologici e costituzionalistici studiati i 1,450 genitore prolifici della regione emiliana. Pp. 145-220 in Gini, C. (ed.), *infra.*

Gini, C. (ed.). 1934. *Proceedings of the International Congress for Studies on Population, Rome, 1931. Vol. IV, Anthropology and Geography.* Instituto Poligrafico dello Stato, Rome.

Stoudt, H. W., A. Damon, R. A. McFarland and J. Roberts. 1965. Weight, height and selected body dimensions of adults. United States 1960-62. Publication No. 1000—series 11—No. 8, Vital and Health Statistics, U.S. Public Health Service, Gov't. Print. Off., Washington, D.C.

6. The Operation of Natural Selection on Human Head Form in an East European Population

TADEUSZ BIELICKI and ZYGMUNT WELON

Introduction

Perhaps the most conspicuous morphological alteration undergone by man since the end of the Pleistocene has been brachycephalization, a marked roundening (shortening and widening) of the braincase. The proportion of brachycephalic individuals, and consequently the mean of the cephalic index, have during the last millennia increased appreciably in many populations. Evidence of this trend has been reported from areas as widely apart as Europe, India, North America, and Polynesia (cf. Shapiro, 1943; Weidenreich, 1945; Kocka, 1958; Hunt, 1959).

The causes of brachycephalization have long been a subject for speculation in anthropological literature. Broadly speaking, two main groups of hypotheses have been advanced:

1. Brachycephalization is a non-genetic, developmental response to some changes in those elements of the environment which affect the pattern of growth of the human braincase in infancy and childhood. It is, then, a phenomenon of an ontogenic rather than phylogenic (evolutionary) nature.

2. Brachycephalization has been produced by natural selection favouring round-headedness, i.e., by the fact that in certain populations roundheads have been leaving, on the average, more viable offspring than longheads have. It involves, then, changes in the human gene pool, and hence has to be regarded as true evolution.

The first of these hypotheses assumes that human head form is environmentally conditioned, the second—that it is under genic control. Twin studies have shown that both these assumptions are valid, i.e., that the variance of head form contains an environmental as well as a genetic component (Clark, 1956; Osborn and DeGeorge,

SOURCE: Tadeusz Bielicki and Z. Welon, "The Operation of Natural Selection on Human Head Form in an East European Population," *Homo*, 15: 22-30 (Goettingen, West Germany, 1966).

1959); hence, both hypothesis 1 and 2 are plausible. It may also be noted that they are mutually not exclusive, as the possibility must be reckoned with that brachycephalization is a joint effect of hereditary *and* environmentally induced changes pointing in the same direction (a situation known as the "Baldwin effect"— cf. Thoma, 1959, p. 252).

The present study is an attempt to test the hypothesis of selection by an anthropometric and demographic investigation of a modern population from East-Central Europe.

Brachycephalization in East-Central Europe

For the area between the Carpathian Mountains and the Baltic Sea the trend toward roundheadedness is well documented by abundant craniological materials. Table 1 gives some of the pertinent data. As can be seen, since ca. 1300 A.D., i.e.,

Table 1. Changes of the Cephalic Index in Time

Site or Region	Century	Mean Cephalic Index	Reference
Radom	XII	73.5	Rosiński 1951
Końskie	XI—XII	74.0	Dambski 1955
Płońsk	XI—XII	74.0	Kočka 1958, p. 219
Czersk	XVII	81.0	Miszkiewicz 1954 a
Warsaw	XVII	81.0	Miszkiewicz 1954 b
Radom	XX	83.0	Mydlarski 1928
Końskie	XX	84.0	Mydlarski 1928
Płońsk	XX	83.0	Mydlarski 1928

in the course of less than 700 years (ca. 30 generations) the mean of the cephalic index has in various parts of Poland increased by nearly 10 units; this is a very great shift, almost equal to three times the present within-group standard deviation of this trait. It should be added that during the above indicated period the area in question, notably Poland, has exhibited a relatively high degree of ethnic continuity and stability, which makes it practically certain that the change in head form must have been largely due to *some factor other than population mixture* (Czekanowski, 1930, 1935; Kočka, 1958).

Now, if selection is really the agency responsible for the phenomenon in question—it seems logical to expect that (1) it may still be at work in some modern populations, and that (2) it is likely to be most intense, and hence most easily detectable, in those groups in which brachycephalization has been particularly recent and rapid. From this point of view, then, the populations of Central and East-Central Europe seem to constitute an exceptionally promising object for study.

Material and Method

An ideal test of intra-group selection in head form would, of course, be a direct comparison of the average reproductive performance of persons belonging to different categories of the cephalic index and drawn from one panmictic population. Such a test is infeasible, since materials containing that sort of data, and large enough to permit statistically valid inferences, are lacking.[1] The present writers decided therefore to utilize what was available, namely some of the data collected during the late 1920's in the course of the Military Anthropological Survey of Poland.[2] This material consists of army recruits, all roughly of the same age (21–25, i.e., born in the first decade of the century). The present analysis was limited to individuals born in the district of Nowogródeke (6 counties) and in the adjoining part of the district of Wilno (3 counties). The total area involved is ca. 35,000 square km, and constituted the North-Eastern corner of pre-war Poland; it now belongs to the White-Russian Republic, U.S.S.R. Only those men whose civilian occupation was listed as "farmer" were taken into account. A sample totalling 6229 individuals was thus obtained.

The area in question was at that time one of the poorest and economically most backward regions in Europe. Among the rural population ca. 60 per cent of adults were illiterate. Farming was relatively primitive and characterized by low productivity; rye and potatoes were the basic crops. Of the infectuous diseases scarlet fever, diphtheria, and typhus were the main killers. Fertility was very high, with an estimated average of over 6 live births per family; infant mortality was of the order of 17–20 per cent in the years preceding World War I, soared up to 22–25 per cent during the war, and fell back to the pre-war level in the early 1920's. Social and spatial mobility was low. In 1926 more than 85 per cent of the people (of peasant stock) were residents of the same counties in which they had been born.

The material contains only 3 variables pertaining to the problem in question:

1. The subject's head form (cephalic index)
2. Total number of the subject's siblings born

---

[1] A project aimed at securing such materials has been only very recently (in April 1963) started in Poland.

[2] The survey covered ca. 120,000 individuals, with over 70 measurements taken on each, thus yielding what was probably the largest anthropometric material ever collected in the course of a single project. Unfortunately, the bulk of it was destroyed during the Warsaw Uprising in 1944.

3. Number of the subject's siblings who were alive at the time when the examination was being made.

Nothing is known of the head form of the subject's siblings or parents. Also, the age at which the deceased siblings died was not recorded (however, from the fact that the subjects are all in their twenties it follows that most of these deaths must have occurred before or during the reproductive period).

The analysis was made in the following manner: The frequency distribution of the cephalic index was divided into 5 categories: "ultradolichocephalic" (x—77.50), "dolichocephalic" (77.51–80.50), "mesocephalic" (80.51–83.50), "brachycephalic" (83.51–86.50), and "ultrabrachycephalic" (86.51–x). The number of individuals who fell in each of these categories was 544, 1570, 2177, 1398, and 540, respectively, from which it can be seen that the division was roughly symmetrical in relation to the population mean. Subsequently these 5 groups were compared with one another with regard to the mean size of sibship. Such comparisons were performed in several different ways; moreover, a smaller control material from another part of Eastern Poland was analyzed (details will be given below). The writers' idea was that significant differences in the mean size of sibship between persons from different categories of head form, if ascertained, would indicate that the respective genotypes differ in fertility or mortality or both. There are, of course, two assumptions underlying this reasoning: (1) There exists correlation between head form as expressed by the cephalic index, and a person's genetic constitution;[3] in particular, the extreme phenotypical classes contain more homozygous or nearly homozygous individuals than do the intermediate ones. (2) There exists between-sib correlation in head form, so that e.g. round heads have, on the average, more roundheads among their siblings than do longheads or mediumheads.[4]

Attention must be drawn to the fact that sibship of soldiers do not constitute a purely random sample of all sibships in the general population. There is a statistical bias inherent in this material and resulting from the fact that big sibships are more likely to contribute to a man to the army than are small ones. E.g., of all sibships consisting of one person only, ca. 50 per cent (namely all those consisting of 1 female) have no chance at all to be represented in such a sample; the likelihood of a two-person sibship to be included in this sample is 3 times smaller than such a likelihood for a six-person sibship; etc. It must be assumed, therefore, that there is an excess of big sibships in our material—a fact which may (though does not have to) distort the results of the analysis of differential survivorship. In order to eliminate this bias, in each class of sibship the following transformation was employed:

---

[3]At the loci controlling head form.

[4]The strength of this correlation was estimated on the basis of another pre-war Polish material, consisting of 180 sets of sibs from Mazowsze, Central Poland; an r = 0.45 was obtained (B. Rosiński, personal communication).

$$N_t = k \frac{N_o}{n}$$

where N_t is the transformed ("corrected") number of siblings, N_o is the total number of living siblings actually observed in a given class, n is the number of persons of whom each sibship belonging in the class consists, and k is a coefficient which brings the total number (sum) of living siblings in all classes back to the actually observed figure. E.g., if in the class *no* 8 (made up of all eight-person sibships) the total number of living siblings is 400 then N_t = k 400/8 = 50 k. The relative "weight" of big sibships in the sample is thus reduced.

Results

1. The mean number of persons *born*, per sibship, is:

> 6.316 in the ultradolichocephalic group
> 6.308 in the dolichocephalic group
> 6.262 in the mesocephalic group
> 6.201 in the brachycephalic group
> 6.192 in the ultrabrachycephalic group

The numbers decrease from the left to the right end of the scale. This might suggest that the genotype responsible for dolichocephaly is characterized by a slightly higher average *fertility* than that producing brachycephaly; the differences, however, are not significant.

2. Table 2 gives the frequencies of sibships of various size (14 classes: from 0 to 13 living siblings), and the mean size of sibship, in each of the 5 categories of the subjects' head form. For each category two means are given, one computed from untransformed, the other from transformed numbers. The two kinds of means are then plotted against the scale of the cephalic index [Figure 1 (a) and 1 (b)].

The table and the diagrams show that mesocephals have the highest average number of living brothers and sisters. The differences appear small. Yet in distribution 1 (b) (transformed data) two of them prove statistically significant, namely the differences between the mesocephalic group and either extreme group; in the other distribution (untransformed data) only the difference between the mesocephals and ultradolichocephals is significant, while that between the mesocephals and ultrabrachycephals falls slightly below the 5 per cent level of significance. But what seems even more remarkable than the degree to which the means differ from one another—is the distinctly trend-like fashion in which they decrease gradually toward both ends of the scale. The trend seems unmistakable, and shows up both in the transformed and untransformed data.

Table 2. Numbers of Sibships of Various Size in
Relation to the Subjects' Head Form

Size of Sibship (Number of the Subject's Living Siblings)	Subject's Head Form				
	x−77.5	77.5−80.5	80.5−83.5	83.5−86.5	86.5−x
0	5	4	13	7	8
1	33	76	82	67	19
2	62	205	226	157	61
3	111	286	386	254	113
4	127	321	427	269	95
5	84	304	453	244	106
6	59	181	305	213	71
7	29	126	178	112	37
8	21	48	66	48	19
9	9	17	22	19	8
10	4	1	15	7	3
11	−	1	2	1	−
12	−	−	1	−	−
13	−	−	1	−	−
Total number of sibships	544	1570	2177	1398	540
Mean size of sibship:					
a) Untransformed data	4.162	4.251	4.439	4.390	4.313
b) Transformed data	4.353	4.544	4.688	4.611	4.490

(Number of Sibships)

3. Another interesting tendency noticeable in the diagrams is the fact that the mean size of sibship is lowest in the ultradolichocephalic group, and in the dolichocephalic group it is lower than in the brachycephalic one. In other words, the means decrease more rapidly on the left ("longheaded") part of the scale: the distribution is skewed.

The following checks were made to find out whether this assymmetry is accidental or real.

4. Brachycephals and ultrabrachycephals were added together to form one group of roundheads (ceph. index 83.51−x); similarly, dolicho- and ultradolichocephals were considered jointly as one group of longheads (ceph. index x−80.50). Each of these two groups was then divided into 12 classes, according to the total number of the subject's sibs *born:* 0, 1, 2, ..., 10, 11−15. Subsequently the mean number of *living* sibs was calculated in each class (except class 0); thus, e.g., in the longheaded group there were 306 soldiers each of whom had had 6 brothers and sisters born, and on the average 4.26 of them per soldier survived. Results are

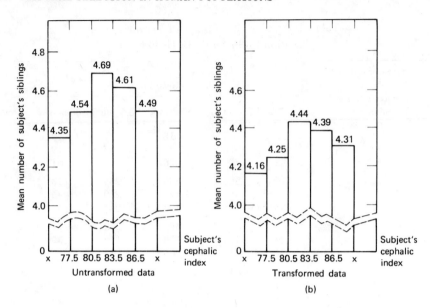

FIGURE 1. Cephalic index and size of sibship.

summarized in Table 3. It can be seen that in every class except one, roundheads have on the average more living sibs than do longheads. I.e., there are 10 differences whose sign is in accordance with the hypothesis that roundheads are more viable, and only 1 difference which is discordant. This ratio of 1:11 is, of course, a highly significant departure from randomness. It should be added that among these 10 concordant differences as many as 5 (those marked with an asterisk) are significant, whereas the only discordant is not.

5. Survivorship of longheads' and roundheads' siblings (the two groups delimited as above) was analyzed separately in each of the 9 counties of which the analyzed area was made up (Table 4). Again, it can be seen that a distinct tendency exists for the roundheads to have, on the average, slightly more living siblings. It must be stressed that this finding is fully corroborated by the results of a similar analysis of a control material consisting of soldiers of Ukrainian nationality born in the district of Wolyń (767 roundheads and 736 longheads); in all the five counties analyzed (Dubno, Horochów, Kowel, Krzemieniec, Równe) the longheads' sibships turned out to be on the average slightly smaller than those of roundheaded individuals.

6. Finally, a following four-fold association table, based on the material from the Nowogródzkie and Wileńskie districts, may be of interest:

	Roundheads	Longheads
Total number of sibs deceased	3305	4105
Total number of living sibs	8311	8690

Table 3. Size of Sibship in Relation to the Subject's Head Form*

Class of Sibship (Number of Sibs Born)	Brachycephals		Dolichocephals		Difference Between Means
	Number of Sibships	Mean Number of Living Sibs	Number of Sibships	Mean Number of Living Sibs	
0	2	–	3	–	–
1	34	0.97	49	0.96	+0.01
2	85	1.74	95	1.72	+0.02
3	137	2.63	153	2.42	+0.21
4	230	3.18	217	3.23	−0.05
5	270	3.96	274	3.86	+0.10
6	302	4.52	306	4.26	+0.26
7	276	5.08	285	4.80	+0.28
8	220	5.59	258	5.24	+0.35
9	145	6.04	142	5.80	+0.24
10	98	5.99	116	5.96	+0.03
11–15	78	6.33	128	6.16	+0.17
Total	1877	4.43	2026	4.29	+0.14

*Numbers of sibships used in this table are somewhat lower than those which can be obtained by summing up the respective numbers in Table 2. This discrepancy is due to the fact that some of the anthropometric cards which contained data on the number of living sibs lacked information on the number of sibs born and were therefore useless for the construction of Table 3.

Table 4. Mean Size of Brachycephals' and Dolichocephals' Sibships in Particular Counties

County	Brachycephals		Dolichocephals		Difference Between Means
	Number of Sibships	Mean Number of Living Sibs	Number of Sibships	Mean Number of Living Sibs	
Baranowicze	63	4.14	54	4.00	+0.14
Lida	311	4.44	366	4.35	+0.09
Nieświez	201	4.19	179	4.37	−0.18
Nowogródek	153	4.48	161	4.38	+0.10
Oszmiana	130	4.42	221	4.40	+0.02
Stolpce	356	4.37	236	4.11	+0.26
Wilno-Troki	137	4.58	185	4.32	+0.26
Wilno	186	4.62	205	4.53	+0.09
Wolkowysk	340	4.26	419	4.10	+0.16
	1877	4.43	2026	4.29	

The value of chi-square computed from this table is 37.8 which reveals a very strong (<0.001) association between the subject's head form and the viability of his siblings.

Discussion

Perhaps the most noteworthy of the above described results is the fact that individuals with intermediate values of the cephalic index—mesocephals—tend to have more living siblings than those exhibiting extreme values of this trait. It seems difficult to explain this finding in any other way than by assuming that the genotype (or class of genotypes) responsible for mesocephaly apparently endows its bearers with above-average viability. Since in polygenically transmitted metric traits the genotype which is most likely to produce an intermediate phenotype is the heterozygote (possessing approximately equal numbers of "plus" and "minus" genes)— the above finding strongly suggests that natural selection directed against both homozygotes is operative in the analyzed population. Such selection is known to have a stabilizing effect and lead to genetic equilibrium.

In a way, this finding comes as no surprise, since the idea has already been considered a number of times in anthropological literature (recently, e.g., Penrose, 1959; Garn, 1962) that many human metric traits may be subject to two-sided selection, and that it is this mechanism—rather than their seeming "adaptive neutrality"—that may well be responsible for the relative stability of the population means of such traits and for the maintenance of polymorphism. On the other hand, however, head form is exactly one of those few human morphological characteristics which have recently undergone dramatic secular changes, and thus has been conspicuously *not* stable. This, therefore, is a situation in which directional (one-sided) selection would at first glance seem the most likely explanation.

In this connection the results described above under paragraphs 3, 4, 5, and 6 merit special attention, since they do suggest that the dolichocephalic homozygote, apart from being selectively inferior to the mesocephalic heterozygote, is inferior to the brachycephalic homozygote, too. On the whole, the differences are not large enough to allow firm conclusions. This, however, may partly be due to the limitations imposed by the method of analysis which, of necessity, had to be "circuitous" and to make use of several highly simplifying assumptions; e.g., it is obviously not true that *all* siblings of dolichocephals must be dolichocephalic, too, or that *all* extreme dolichocephals are closer to homozygosity than are all mesocephals. Similarly, limitations inherent in the material itself are more likely to blur than to sharpen the actual picture of differential survivorship. Namely, it should be noted that the frequency distribution of the cephalic index in our material—as compared with the world-wide range of variability of this trait—is somewhat shifted toward the bracycephalic end of the scale; class x—77.5, regarded here as representing extreme dolichocephaly, cannot be regarded as such in the absolute sense. Now, it seems logical to suppose that it is at the extremes of the *world-wide* spectrum (rather than at the extremes of any *local* within-group distribution) that the really "pure" homozygotes tend to be concentrated. Thus, our dolichocephals are, on the whole, probably less "pure" homozygotes than are our brachycephals, i.e.—in terms of a simple model of polygenic inheritance—the first are likely to harbour a higher admixture of "plus" genes than the latter of "minus" genes. If in our material the right-hand limit of the longheaded class could be withdrawn further down the scale—without at the same time drastically reducing the number of individuals falling in this class (which is, of course, possible only considerably enlarging the whole material)—the differences in the mean size of sibships between the two extreme phenotypes might come out more clearly.

Conclusions

The mere fact that one homozygote has a higher fitness than the other does not, of course, exclude the possibility that the population is in a state of balanced polymorphism; on the other hand, the selective superiority of the heterozygote is not sufficient proof that such a state has already been reached. All this makes the general picture rather obscure, and the final interpretation difficult. For, even if one assumed that the above presented findings offer clear-cut evidence for the existence of a selective advantage of (a) roundheads over longheads, and (b) mediumheads over both extreme categories—this would still leave the question open whether the studied population is evolving (undergoing brachycephalization), or whether it has already achieved temporary genetic equilibrium. In order to find out which of these alternatives is the case one would have to know (1) the frequencies of genes controlling head form (which would require an exact knowledge of the mode of inheritance of this trait), and (2) the relative fitness of the various genotypes involved. Both information are not available.

Summary

(1) A statistical analysis of a demographic and anthropometric material collected in the 1920's among the rural population of the North-Eastern part of pre-war Poland strongly suggests that in the studied group head form is subject to selective pressures. The mechanism of this selection is differential survival rather than differential fertility. (2) There is reason to believe that selection acts with greatest intensity against longheadness, and that, as the result of this, brachycephalization is still in progress. However, (3) counter-selection against extreme roundheadness—seems to be at work, too. This suggests that the process of brachycephalization is

waning; but it is possible, too, that it has already come to an end, and stable genetic equilibrium reached.

Bibliography (References Cited)

Clark, P. J.: The heritability of certain anthropometric measurements as ascertained from the measurements of twins. *Am. J. Hum. Genetics* 8, 1, 1956.

Czekanowski, J.: *Zarys Antropologii Polski.* Lwów 1930.

Czekanowski, J.: *Clowiek w Czasie i Przestrzeni.* Warszawa 1935.

Dambski, J.: Wczesnosredniowieczne cmentarzysko w Konskich. *Materialy i Prace Antropol.,* 1955.

Garn, S. M.: Comment on the articles by A. Wiercinski and T. Bielicki. *Current Anthropology* 3, p. 28, 1962.

Hunt, E. E.: The continuing evolution of modern man. *Cold Spring Harbor Symp. on Quant. Biol.* 24, 245-54, 1959.

Kočka, W.: Zagadnienia Etnogenezy Ludów Europy. *Materialy i Prace Antropol.* 22, 1958.

Miszkiewicz, B.: Analiza antropologiczna serii czaszek z Czerska kolo Warszawy. *Przeglad Antropol.* 20, 1954a.

Miszkiewicz, B.: Fragmenta Craniologica, Cmentarzyska z Polski i ziem osciennych. *Materialy i Prace Antropol.* 4, 1954b.

Mydlarski, J.: Przyczynek do poznania struktury antropologicznej Polski i zagadnien doboru wojskowego. *Kosmos* 53, 1928.

Osborn, R. H. and F. V. DeGeorge: *Genetic basis of morphological variation.* Cambridge, 1959.

Penrose, L. S.: Natural selection in man, some basic problems. In *Natural Selection in Human Populations.* New York, 1959.

Rosinski, B.: Charakterystyka antropologiczna kostnych szczatków ludzkich cmentarzyska wczesnohistorycznego w Radomiu. *Wiadomosci Archeologiczne* 17, 327-340, 1951.

Shapiro, H. L.: Physical differentiation in Polynesia. *Peabody Museum Papers* 20, 1943.

Weidenreich, F.: The brachycephalization of recent mankind. *Southwestern J. Anthropology* 1, 1945.

Thoma, A.: Age at menarche, acceleration, and heritability. *Acta Biologica Academiae Scientiarum Hugaricae* 11, 3, 1960.

7. Factors in the Alteration of Reproductive Potential of Chondrodystrophics

WILLIAM B. COTTER

The reproductive potential of the chondrodystrophic dwarf as estimated for western European societies is based mainly on the data gathered and interpreted by Mørch[8] concerning the incidence of chondrodystrophics in the Danish population. The estimates of the rate of mutation to the chondrodystrophy-producing allele by the direct method and indirect method of Mørch[8] and Haldane[3] have achieved the status of classic genetic studies. As such, they have been utilized in the preparation of such textbooks of genetic principles as those by Stern,[14] Clarke,[1] Sinnott, Dunn and Dobzhansky,[11] Harrison, Weiner, Tanner and Barnicot,[4] Whittinghill,[16] Herskowitz,[5] and Neel.[9]

An obvious feature of utility of Mørch's data is the closeness of agreement in the estimates by the direct and indirect methods of the rate of mutation from the normal allele to the chondrodystrophy-inducing allele, yet these data are not completely accepted by all authors as the most meaningful ascertainment of the chondrodystrophic alleles in any population. Krooth[6] and Slatis[12] have discussed the bias of the truncated sibship, which featured the terminal sib as chondrodystrophic, and its effect on family size, which exaggerated the relative lack of fecundity of the chondrodystrophic when compared to a normal sib. Stevenson[15] has demonstrated an additional source of error in failure to ascertain chondrodystrophics at birth.

Of 37 nonfamilial cases, only three were recognized in the nursery, and the majority were diagnosed at about one year of age when an aberrant walk or failure of normal growth appeared. This group of individuals did not show a marked reduction in vitality. A second group of nine chondrodystrophics that were identified at birth were characterized by a pregnancy complicated with hydramnios, and by a very low viability, with eight of the nine still-born or dead within five days. This latter group was probably genetically distinct from the main chondrodystrophic population; Stevenson has suggested that they might represent a phenocopy.

SOURCE. William B. Cotter, "Factors in the Alteration of Reproductive Potential of Chondrodystrophics," *Journal of Heredity*, **58** (2): 59-63, (Washington, D.C., 1967).

Neel,[9] in his review of the general aspects of mutation-rate estimation for human populations, has reemphasized some of these problems. In addition, he suggested a careful reexamination of the assumption that the given population under consideration is in genetic equilibrium. If the population is in genetic equilibrium, the formula for the indirect method of mutation rate may be used when an acceptable value for the relative fitness (f) is determined. In the case of chondrodystrophy, the dominant allele has been assigned the value of 0.196 on the basis of Mørch's data on the differential reproductive capacity of chondrodystrophics and their sibs.

The concept of relative fitness (also referred to as Darwinian fitness, selective value, or adaptive value) of an allele is an expression describing the proportions of the alleles at one locus in the progeny in comparison with the allelic proportions at the same locus in the preceding population.[7] Selection for or against a given allele depends on several environmental parameters among which are the several biological and physical aspects of the environment. In human populations the additional aspect of a sociological selection parameter must be considered.

The hypothesis to be developed is that the value assigned to the relative fitness of the dominant gene producing chondrodystrophy is a function of the society or the general culture in which the study is made. It is important to note that the value for relative fitness may vary as the society changes in time, as the breeding population's attitudes concerning this genetic trait change in time. To test this hypothesis, the extensive pedigree of familial chondrodystrophy published by Stephens[13] was examined.

Pedigree Analysis

The family examined was early Mormon pioneer stock dating back to the opening of the Utah area to farmers in 1833. A spontaneously appearing chondrodystrophic male gave rise to two sibships polygamously. The descendants of these sibships had existed largely *in situ* to the date of the initial study.[13] Dedication to large sibships and genealogy has produced accurate records of 16 major sibships produced by chondrodystrophic x normal matings. Of the 16, only nine contained individuals who were old enough to have produced some progeny. In Table 1 those sibships are indicated and the affected parent numbered (not according to Stephens, but according to presently accepted practice).

In the nine sibships (A to I) the total number of sibs was 76 and the segregation of chondrodystrophics and normals (34:42) did not differ significantly from chance expectations. These 76 individuals produced 209 offspring; there were 70 children of chondrodystrophic x normal matings and 139 offspring of normal x normal matings (Table 2). Hence, the average sibship size for chondrodystrophics was 2.06 as compared with 3.31 for the normal sibs of the chondrodystrophics—when all adults are considered potential parents. The data of progeny size are shown in

Table 1. Segregation of Sibs in Nine Matings of
Chondrodystrophic X Normal

Sibship	Affected Parent	No. Progeny		
		Total	Normal	Chondrodystrophic
A	II-4	11	8	3
B	II-4	11	7	4
C	III-5	8	6	2
D	III-6	7	1	6
E	III-9	10	4	6
F	III-12	4	2	2
G	III-13	12	8	4
H	III-17	8	4	4
I	IV-70	5	2	3
Total		76	42	34

$x^2 = 0.842$ $.50 > P > .30$

Table 2. Analysis of the Nine Sibships (A-I) Showing the Total Sibs, the Number of
Sibs Who Were Progeny Producers or Non-Producers, and the
Total Number of Offspring Produced by Those Sibs

Sibship	Chondrodystrophic*				Normal			
	n	n_p	n_{n-p}	n_o	n	n_p	n_{n-p}	n_o
A	3	3	0	25	8	4	4	43
B	4	3	1	24	7	5	2	37
C	2	0	2	0	6	4	2	24
D	6	1	5	3	1	1	0	4
E	6	2	4	5	4	4	0	16
F	2	2	0	8	2	1	1	2
G	4	0	4	0	8	3	5	10
H	4	2	2	3	4	2	2	3
I	3	1	2	2	2	0	2	0
Totals	34	14	20	70	42	24	18	139

*n = number of phenotype in sibship
n_p = number of phenotype producing progeny
n_{n-p} = number of phenotype not producing progeny
n_o = total number of offspring produced

Figure 1, and when the mean sibship size of the chondrodystrophic is compared by
the t-test with the mean of the progeny produced by the normal sibs, the t-value

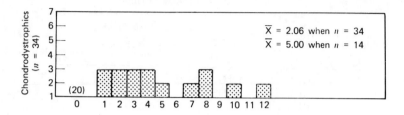

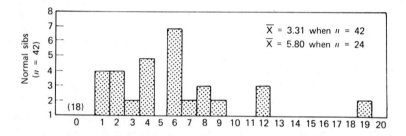

FIGURE 1. Distribution of sibship size produced by 76 chondrodystrophic and normal sibs.

lies outside the .975 confidence limits, indicating a highly significant difference. This result is consistent with a value for relative fitness, which is lower than unity, as indicated by other workers.

However, this average number of offspring per potential parent does not adequately express the reproductive potential of the chondrodystropic. When the data (of Figure 1 and Table 2) were examined for actual performance in progeny production, the sibs within each sibship were separated into two additional classes, progeny producers and progeny nonproducers. Of the 76 sibs, only 38 became parents (14 chondrodystrophics and 24 normals). The average number of offspring per actual parent then becomes 5.0 for the chondrodystrophic and 5.8 for the normal sib. The difference in mean sibship size for the two classes is found to be nonsignificant with the t-test (within the 80 percent confidence limits).

Of the 76 sibs, less than 50 percent of the chondrodystrophics became parents, while more than 50 percent of the normals did so; the actual numbers were tested by a 2 x 2 contingency table to determine if the differences were significant (see Table 3). The P value for one degree of freedom is between .20 and .10; hence the differences observed, while indicative of a trend, are not significant. But when the chondrodystrophics and their normal sibs are taken only from generation V and

Table 3. 2 x 2 Contingency Table Test for Significance of Difference Noted in Totals of Producers and Non-Producers in Chondrodystrophics and Their Normal Sibs

	Producers	Non-Producers	n
Chondrodystrophic	14	20	34
Normal	24	18	42
Totals	38	38	76

x^2 = 1.92 with df = 1.   .20 > P > .10

compared, the differences observed are significant (.05 > P > .02). Fewer chondrodystrophics became parents, and relatively more of their normal sibs produced some progeny (cf. Table 4).

Table 4. 2 x 2 Contingency Table Test for Significance of Differences Noted in Generation V Only in Relation to the Totals of Progeny Producers and Non-Producers in Chondrodystrophics and Their Normal Sibs

	Producers	Non-Producers	n
Chondrodystrophics	7	17	24
Normals	15	10	25
Totals	22	27	49

x^2 = 4.74 with df = 1, .05 > P > .02

In assessing the meaning of these statistics, several additional factors should be considered. First, the territory that became the state of Utah was sparsely populated by settlers during the nineteenth century. Second, the chondrodystrophic dwarf was able to function well in an agrarian society, some becoming community leaders and men of substance. Third, while the pelvis of the chondrodystrophic female is flattened dorso-ventrally (antero-posteriorly), childbearing is not a problem, and as Eastman[2] has pointed out, the female is "exceptionally fertile," in decided contrast to the relative sterility of other dwarf types.

Therefore, in the cultural framework of the Mormon religion that was supportive of large sibships and of inbreeding largely within the religious aggregate, in the economic framework of an agricultural community in a sparsely populated region that put a premium on vigor and physical strength, and in the framework of essentially normal fertility of the male and female chondrodystrophic dwarf, the estimate of the value of biological fitness for this population must be close to unity, in contrast to the f = 0.196 most often cited[3,10, et al.].

As the population structure, economic structure, and educational structure have shifted in that region, the ideas and attitudes toward the genetic condition have changed. Eugenic practices now tend to reduce the overall incidence of the trait in the family. Not only have fewer chondrodystrophics married and produced children, but their age at marriage was greater and sibships, if produced, tended to be smaller.[13]

In addition, Stephens was of the opinion that relatively fewer normal sibs of chondrodystrophics were progeny producers compared with normal individuals from families free of the condition because of the mistaken notion that their offspring could be afflicted with the condition. In Figure 2, the sibships produced by chondrodystrophics and their normal sibs are shown in a histogram that separates the progeny producers of generations IV and V. In addition, the sibships used as controls are those produced by normal individuals from families of normal sibs of chondrodystrophics in generation IV. According to Stephens' rationale, a comparison of the ratio of nonproducers to total adults in the two normal populations should be different; however, in both populations about 40 percent of the total were nonproducers. A t-test comparison of mean sibship size ($\bar{X}$) of the three populations (chondrodystrophic, normal, and control) showed that the differences observed were not statistically significant in this generation, with the result that the primary agent of selection against this allele is withdrawal from the breeding population.

Hence the selective value for the chondrodystrophy-producing allele is quite low in most societies as a function of psychological attitude toward the genetic condition, but not as a function of physiological capabilities of reproduction. But, as with the selective value for the albino allele in the San Blas Indians, the values of a given society must be estimated relative to the given genetic trait for a realistic approximation of an allele's evolutionary potential.

Summary

Published estimates of mutation rates for the chondrodystrophic allele in general have failed to recognize a potential (if not actual) genetic heterogeneity in the aggregate of chondrodystrophics. The allied concept of the allele's relative fitness is reviewed in the light of a previous extensive pedigree of familial chondrodystrophy, supporting the hypothesis that the value for relative fitness of a given allele may vary as a society or culture changes in time.

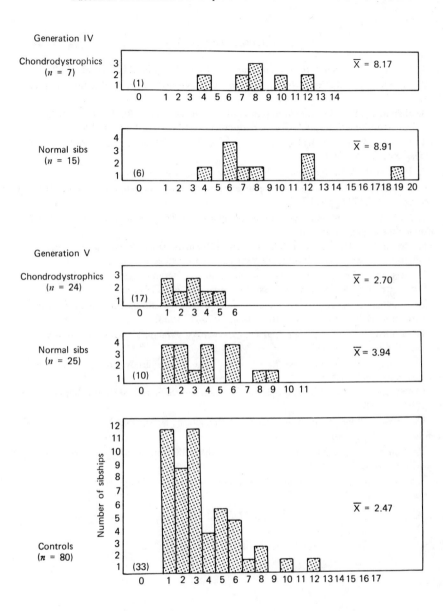

FIGURE 2. Distribution of sibship size according to generation and parental phenotype.

Literature Cited

1. Clarke, C. A. *Genetics for the Clinician.* Blackwell Sci. Pub. Ltd., Oxford. 1962.

2. Eastman, N. J. *Williams Obstetrics.* 11th ed., Appleton-Century Crofts, Inc., New York. 1956.

3. Haldane, J. B. S. The rate of mutation of human genes. *Proc. 8th Int. Cong. Genetics* (Supplement to *Hereditas*) pp. 267-273, Berlingska Boktrycheriat, Lund. 1949.

4. Harrison, G. A., J. S. Weiner, J. M. Tanner, and N. A. Barnicot. *Human Biology.* Oxford U. Press, New York. 1964.

5. Herskowitz, I. H. *Genetics.* Little, Brown and Co., Boston, 1962.

6. Krooth, R. S. Comments on the estimation of the mutation rate for achondroplasia. *Am. J. Hum. Gen.*, 5:373-376. 1953.

7. Li, C. C. The hemophilia gene in the population. In *The Hemophilias.* Univ. of North Carolina Press, Chapel Hill. 1964.

8. Mørch, E. T. Chondrodystrophic Dwarfs in Denmark. Opera tex Domo. *Biologiae Hereditariae Humanae*, Vol. 3:1-200, E. Munksgaard, Copenhagen, 1941.

9. Neel, J. V. Mutations in the human population. In *Methodology in Human Genetics*, W. J. Burdette, Ed. Holden-Day, Inc., San Francisco. pp. 203-219. 1962.

10. Popham, R. E. The calculation of reproductive fitness and the mutation rate of the gene for chondrodystrophy. *Am. J. Hum. Gen.*, 5:73-75. 1953.

11. Sinnott, E. W., L. C. Dunn, and Th. Dobzhansky. *Principles of Genetics.* 5th ed. McGraw-Hill, New York. 1958.

12. Slatis, H. M. Comments on the rate of mutation to chondrodystrophy in man. *Am. J. Hum. Gen.*, 7:76-79. 1955.

13. Stephens, F. E. An achondroplastic mutation and the nature of its inheritance. *J. Hered.*, 34:229-235. 1943.

14. Stern, C. *Principles of Human Genetics*, 2nd ed., W. H. Freeman & Co., San Francisco. 1960.

15. Stevenson, A. C. Achondroplasia: an account of the condition in Northern Ireland. *Am. J. Hum. Gen.* 9:81-91. 1957.

16. Whittinghill, M. *Human Genetics and its Foundations.* Reinhold Publ. Corp., New York. 1965.

8. Congenital Malformations in
Man and Natural Selection*

L. S. PENROSE, F.R.S.

In spite of the obvious variations in size, shape and colour which differentiate one person from another, the idea is still prevalent that the human species is at the top of an evolutionary ladder and thus represents an almost perfect animal. Even though few women are expected to have the proportions of Milo's Venus and few men resemble the Apollo Belvedere or Hermes of Praxiteles, the species as a whole is felt to be a biological masterpiece. Man's domination of terrestrial life is the result of his very large and efficient central nervous system. In respect of other traits, however, there is always some animal which can be found to do better. The egocentric view that man is the highest form of life is comparable to the Ptolemeic system of astronomy and, in modern times, to the assumption that, if life should develop in distant parts of the Universe, it must follow the terrestrial pattern culminating in human forms.

Belief in the perfection of the human species is completely shattered by considering the very high rate of malformation among the offspring in every generation. In earlier times a freakish event like the birth of a child with severe malformation was believed to be a divine portent—even a warning of disaster. Hence the name "monster" from the Latin "monstrum," a sign. Accordingly such births were viewed as being extremely exceptional.

Accurate observation, however, leads to a different evaluation. Darwin, in the *Origin of Species*, differentiated between "monstrosities" which suddenly appeared in a population and variations of possible usefulness. It is difficult sometimes to know where to draw this line of demarcation. Imperfections of the skin, such as pigmented patches, are present almost universally but these are not scored as abnormal unless they are very noticeable. The same applies to minor defects of hands and feet, eyes, ears, etc. Gross malformations occur in about 2 per cent of all infants born at term and careful observation tends to raise the proportion to about 5 per

SOURCE: L. Penrose, "Congenital Malformations in Man and Natural Selection," *Eugenics Review*, **57** (3):126-130 (London, England: The Eugenics Society, 1965).

*The fifth Darwin Lecture, arranged by the Institute of Biology, delivered in London on May 20, 1965.

cent if less serious defects are included. If spontaneous abortions are taken into consideration, the proportion of all zygotes affected would be found to be much higher still.

The spectrum of anomalies is extensive. It ranges from hidden defects of internal organs, heart, kidney or intestines, to visible malformations, like harelip, or dramatic deformities of limbs and even of the cerebrum. In the pioneer investigation by Malpas, gross defects of the brain had an incidence of 1 per cent.

Traditionally there has been a strong tendency to ascribe these disasters to the influence of an unfavourable environment and, particularly, to illness of the mother. Darwin took a guarded view. In *Variation of Animals and Plants under Domestication*, he stated that, although many congenital monstrosities were inherited (e.g. polydactyly, albinism, harelip and cleft palate), other malformations were probably due to non-inherited injuries in the womb or egg. There are several well-established examples in which external factors can be blamed.

As long ago as 1702 the observation was made by Düttel that, if a pregnant woman had smallpox, this disease could be transmitted to the foetus and the child would be born with severely pock-marked skin. The proof that another virus infection, rubella, in the mother could cause deafness and cardiac malformation in the child came comparatively recently through the work of Gregg in Australia. There are other, less well-defined, similar effects and some diseases, like tuberculosis, formerly credited with the power of producing malformed offspring, are now exonerated.

Physical trauma is a rare cause even if, in this group, we include damage by exposure to radiation. For radiation to produce a direct disturbance on foetal growth the dose must be large and such exposures are not likely to pass unnoticed in a civilized community. For example in Russell's experiments on pregnant mice, 200 Röntgen units were required. With poisons, the chain of cause and effect may be much more difficult to establish. The dangers of thalidomide were, for a surprisingly long period, unsuspected. Observations on pregnant animals have disclosed that a great number of chemical poisons can, in sufficient quantity, induce gross developmental anomalies in the offspring; but the effects in man need not be precisely parallel. Experimentally, vitamin deficiency and sometimes vitamin excess can also cause damage. The significant point is that, in classifying origins of malformation, allowance has to be made for possible environmental causes not yet identified. We may, however, with confidence, exclude the traditional superstition that maternal psychological impressions are significant. Microphthalmia cannot result, as Sedgwick reported in 1863, from the mother's being frightened by a ferret. Living close to a Chinese laundry does not cause mongolism. In ancient times, as far back as those of Hippocrates (460 B.C.), it was thought that, if any part of a parent was mutilated, it would be so likewise in the foetus. Darwin reviewed the evidence for this and found it inconclusive. We now reject it without hesitation.

Hereditary Deformities Caused by Gene Differences

In so far as a defect is environmentally determined it takes no part in the process of natural selection; that is to say, provided that genes do not predispose to a particular malformation, the case can be excluded from the present discussion. In a population of organisms which is entirely homogeneous genetically, there can be no selection in the Darwinian sense. In such circumstances, chances which favour surviv al or extinction would have no effect upon the genetical constitution of future generations. Genetical diversity, which Fisher called the "energy" of a species and whic] enables selection to act, arises in two ways. First by single-gene mutation and secondly by chromosomal aberration. The important point is that the self-reproducing properties of the nuclear material, which carries the hereditary instructions, ensure that any alteration tends to be permanent; the changed form is replicated.

Conditions which can be attributed to fresh single-gene mutations are rare. Typical instances are the detrimental dominant traits, like ectrodactyly and epiloia (Bourneville's syndrome). The milder condition, arising by mutation in an ancestor, though disfiguring, lasts for many generations. The more severe rarely continues for more than two. The force of selection is so strong in epiloia that the incidence of this disease in the population is controlled almost entirely by the rate of recurrent mutation, that is, about one in 60,000 per individual per generation. Epiloia is not strictly a congenital malformation because it is not visibly present at birth. More exact examples in this respect are achondroplasic dwarfism and acrocephalosydactyly (the syndrome of Apert).

Next to dominant malformations, selection pressure is greatest in sex-linked morphological defects. Again, only a few are known, e.g., one type of hydrocephaly and one kind of eye defect, microphthalmia. In such instances, as shown by Haldane in 1932, one-third of the cases arise by fresh mutation in each generation. The disease is perpetuated only when the results of mutation appear first in the child of a female carrier.

If a malformation is recessively inherited, the force of selection is much weaker. Several kinds of anencephaly and other central nervous system defects fall into this category. These persist in the population because the genes responsible for them are carried on from generation to generation by perfectly normal people. Defects caused in this way persist in populations all over the world. The local incidence can vary according to selective action for or against slight constitutional peculiarities in the heterozygous carriers. There might also be frequency differences caused by inbreeding. Anencephaly, which seems to be recessive in about half the cases, has a remakably uneven distribution throughout the world and sometimes the origin is non-genetic: all kinds of influences have been blamed, including too pure a water supply in which minerals are absent.

Even though selection against recessive malformations is weak they would eventually vanish, albeit exceedingly slowly, from the population unless there were mechanisms which neutralized the selective process. One of these which has already been mentioned, is recurrent mutation. The other is stable equilibrium of genes, first studied by Fisher, in which homozygotes are less fit than heterozygotes. A very slight advantage in the heterozygous carrier can compensate the loss of genes which occurs when the malformed recessive homozygote fails to survive. In the extreme theoretical case, although all homozygotes are deformed and lethal, the population can be in equilibrium because all heterozygotes are healthy. This phenomenon of completely balanced lethals is not known to exist in man but might be a possible explanation for a high prevalence of recessive malformations. If there were many alleles at a given locus, e.g., A, B, C, D, etc., and all homozygotes, AA, BB, CC, DD, etc., were lethal, the population could still survive on the large excess of normal heterozygotes, AB, BC, AC, AD, etc. For n alleles of equal frequency, $1/n$ would be the incidence of lethals; but, for a 1 per cent incidence of lethal deformities genetically balanced in this way, a set of 100 alleles would be required so this explanation seems rather improbable.

There are other ways in which this type of compensated selection may be effective and one of them is exemplified by weight at birth. Thus, both very light and very heavy infants experience risks much greater than the normal. Extreme types, moreover, are likely to represent homozygous states, less fit than the intermediate average types, which represent heterozygotes.

This kind of genetical equilibrium, in which the optimal population type is maintained by the sacrifice of unfavourable, that is to say, less fertile, extreme types, is called by Muller and by Crow and Kimura, "segregational" load as opposed to the "mutational" load of defects caused by spontaneous mutation. Expressed in another way, the high prevalence of malformations of the nervous system in man may be to some extent inevitable payment, at the lower end of the scale, for the maintenance of his high average level of cerebral development with its wide range of variation. The extreme upper limit of intelligence is also believed to be associated with some degree of infertility.

Malformations Associated with Chromosomal Aberration

In recent years, that is since 1958, it has become clear that a frequent source of malformation is chromosomal aberration. This fact could only be surmised before techniques of cell culture and chromosomal spreading were developed. Indeed, classical work on the mathematical theory of evolution, by Fisher, Haldane and Wright in the decade from 1922 to 1932, neglected almost entirely the kinds of mutation concerned with gross chromosomal changes. It was thought that these would rarely make significant contributions because they would lead to completely inviable states in all higher animals. Most authorities believed that embryos with aberrant chromosomal patterns could not develop beyond the very earliest stages, say, that of 8-cells. It was another blow to human self-esteem to find that, as with plants and insects, survival was possible with an imperfect set of chromosomes.

Some chromosomal mutations cause little or no disability. Females with three sex chromosomes, instead of the normal two, can be perfectly normal; so also perhaps are some males with two Y chromosomes instead of one. Otherwise any marked irregularity or deviation from the normal pattern is very liable to cause physical defect, often identifiable at birth. Mental defect is a very consistent accompaniment of these developmental disturbances.

Certain relatively common malformations of chromosomal origin are now well known. In most of these there is one extra chromosome, which has arisen in consequence of an error in meiosis at germ cell maturation as described by Ford; a homologous trio is produced instead of a pair in the cells of the zygote. In the plant, *Datura stramonium*, Blakeslee had found, many years ago, a different morphology for each of the possible trisomics for its twelve chromosomal pairs. In man, only four out of a possible twenty-three kinds have been clearly recognized so far, those associated with X, a chromosome in the 13 to 15 group, one in the 17, 18 group and one in the 21, 22 group. In the case of Down's syndrome, mongolism, believed to be caused by trisomy-21, the standard condition can actually be transmitted to the next generation in a manner rather like that of a single-gene dominant trait; but this is exceedingly unusual for such patients hardly ever become parents.

Individually, each of these standard trisomic types occurs in about one per 500 to one per 1,500 fertilized ova (zygotes) when allowance is made for the fact that they often cause foetal death and abortion. This rate is about 100 times as frequent as that of single-gene mutation. Selection against these trisomic types is, thus, very strong. Like single-gene dominant lethals, they are perpetuated only because the chromosomal non-disjunctional errors recur. The causes of Down's syndrome are connected with the ageing of the mother, that is to say, with the age of the maturing ovum which begins its cycle at the mother's own birth. Precisely the same effect is found in the other kinds of trisomy. During thousands of past generations there must have been selection against genes carried by mothers who, when they were young, gave birth to malformed offspring. Gradually the age of the parent at which non-disjunction would occur became further and further advanced. At late ages, where the generation time is nearly double the average, selection would necessarily be weaker and evolutionary improvements in the maturation mechanism would be made more slowly. Apart from this indirect selective action, standard trisomics have no effect on the genetical future of the population.

Besides alterations of number affecting one particular chromosomal pair, it is possible for all the pairs to be transformed into triplets. This results in cells with sixty-nine instead of forty-six chromosomes. If the peculiarity affects only a proportion of the cells of the body, say up to 50 per cent, it is compatible with survival though accompanied by mental and physical weakness. Pure cultures of triploid human cells have been obtained from numerous aborted embryos and from one live birth which survived only a few hours. Clearly there is strong natural selection tending to eliminate such individuals and, by implication, to a slight extent people like their parents also, if they should carry any genetical predisposition.

The incidence of tripoidy, again, is rather high judging by its occurrence in aborted foetuses, perhaps about one in 500 of all zygotes formed. The problem of its origin is somewhat similar to that explaining chimaeras, hermaphrodites and conjoined twins. Such freaks can arise through accidents at fertilization or shortly after. In so far as there are genetical factors which determine these events, selection will tend to make them rarer. However, it is usually believed that they occur in response to influences of a purely accidental nature perhaps demonstrating physical limitations of the reproductive mechanisms and, if this is true, it is understandable that they have not been eliminated from the species.

There is another, quite different, kind of chromosomal mutation or aberration found in association with varying degrees of malformation. This always involves breakage and reunion of fragments so as to produce morphological changes in the chromosomes. Individually their frequency of occurrence is rare, resembling that of mutation rates of single genes. Chromosomal deletion is one variety. Some developmental defects, so caused, are incompatible with survival, others, if the deletion is small, are only incompatible with reproduction. Microcephaly and heart defects are characteristic signs in such cases. Natural selection acts at once against these detrimental variations and they are eliminated from the population as they arise.

Two special types of chromosomal mutation, called *inversion* and *balanced translocation*, require special consideration. They do not directly harm the individual who carries them. However, they are likely to lead to malformation in the offspring of a carrier. Both types are caused by incorrect rejoining after two chromosomal fractures. Inversion can follow two breaks on the same chromosome and translocation, breaks on two different chromosomes.

It has been known for a long time that peculiar pedigrees shown by some kinds of familial malformations could be interpreted as the results of translocations or inversions rather than of irregularly dominant inheritance which they would otherwise suggest. Examples were shown by central nervous system malformations and by mongolism. Cytological researches now have made it clear that this kind of inheritance has an explicable and recognizable pattern.

Natural selection acts directly against the malformed offspring in such families and indirectly against the carriers of the inversions and translocations by diminishing effective fertility. The result is that such variations would, in general, be fairly rapidly eliminated from our species. However, there are two ways in which they might tend to persist after their entry into the population by mutation. First, there may be, as Dobzhansky has demonstrated in wild species of *Drosophila*, some special virility conferred on the heterozygous carrier of a chromosomal inversion which may be sufficient to neutralize deleterious effects on fertility. Secondly, in a small very closely inbred population, homozygotes for inversions or translocations may be produced and these could continue interbreeding without harm.

A considerable number of minor chromosomal differences are found among members of the human general population. In this way, as in the blood groups, man shows again an internal example of the polymorphism which is so characteristic of his external appearance. The study of chromosomal polymorphism, indeed, gives a hint as to how some evolutionary processes may have taken place. The translocation fusion of acrocentric chromosomes, illustrated in the carriers of Down's syndrome, is of a kind which has been observed in varieties within the same species of animals and plants. A normal human with forty-five instead of forty-six chromosomes is now almost a commonplace observation. In the fairly near future examples of normal people with only forty-four chromosomes will be found. If this same process happened twice over, that is to say, to two sets of chromosomes, a person could be born with only forty-two chromosomes and have a pattern not unlike that found in the *Rhesus* monkey. He would probably still look like a man and could be quite normal himself though a very high proportion of his grandchildren would be liable to suffer from developmental defects.

I think I have said enough to indicate the significance of studying human malformations in relation to natural selection. In many instances, particularly in those cases caused by chromosomal aberration, the selection is of a type which closely follows Darwin's ideas. To some extent the study of these phenomena helps us to understand how man has evolved and may be continuing to change very slowly. They also indicate the price which has to be paid, in terms of genetical defect, for the comparatively stable and successful state of development which the human race has now reached.

References

Blank, C. E. 1960. Apert's syndrome (a type of acrocephalosyndactyly): observations on a British series of 39 cases. *Ann. hum. Genet. Lond.* 24, 151.

Carr, D. H. 1963. Chromosome studies in abortuses and stillborn infants. *Lancet*, ii, 603.

Chu, E. H. Y., and Bender, S. 1961. Chromosome cytology and evolution in primates. *Science*, 133, 1399.

Crow, J. F., and Kimura, M. 1965. The theory of genetic loads. *Genetics Today*, 3, 495.

Darwin, C. 1859. *The Origin of Species*. London, G. Murray.

Darwin, C. 1868. *Variations of Animals under Domestication*. London, G. Murray.

Dobzhansky, Th. 1946. Genetic structure of natural populations. *Yrbk Carneg. Inst., Wash.* 43, 120.

Düttel, P. J. 1702. *De Morbis Foetum in Utero Materno*. Magdeburg, C. Henckelii.

Fisher, R. A. 1930. *The Genetical Theory of Natural Selection.* Oxford, Clarendon Press.

Ford, C. E. 1961. Human cytogenetics. *Brit. med. Bull.* 17, 179.

Gunther, M., and Penrose, L. S. 1935. The genetics of epiloia. *J. Genet.* 31, 413.

Haldane, J. B. S. 1932. *The Causes of Evolution.* London, Longmans Green.

Karn, M. N., and Penrose, L. S. 1951. Birth weight and gestation time in relation to maternal age, parity and infant survival. *Ann. Eugen. Lond.* 16, 147.

MacKenzie, H., and Penrose, L. S. 1951. Two pedigrees of ectrodactyly. *Ann Eugen., Lond.* 16, 88.

Malpas, P. 1937. The incidence of human malformations. *J. Obstet. Gynaec. Brit. Emp.* 44, 434.

Mørch, E. T. 1941. Chondrodystrophic dwarfs in Denmark. *Op. Dom. Biol. Hum. Hered.* 3.

Penrose, L. S. 1957. Genetics of anencephaly. *J. ment. Defic. Res.* 1, 4.

Penrose, L. S. 1961. Mongolism. *Brit. med. Bull.* 17, 184.

Russell, L. B., and Russell, W. L. 1954. An analysis of the changing radiation response of the developing mouse embryo. *J. Cell. Comp. Physiol.* 43, Supp. 1, 103.

Sedgwick, W. 1863. On the influence of sex in hereditary disease. *Brit. Foreign Med.-Chir. Rev.* 21, 448.

Wright, S. 1931. Evolution in Mendelian populations. *Genetics,* 16, 97.

9. Some Factors Influencing the Relative Proportions of Human Racial Stocks

FREDERICK S. HULSE

The study of the origin of human diversity is, and must remain, dependent upon data so inadequate as to permit of various quite legitimate conclusions. The study of human variation as it exists today is also based upon inadequate data, but this condition can be, and is being improved. Indeed, as we learn more of the nature of variation in living human populations, we can begin to judge better the relative plausibility of various theories concerning the origin of racial stocks, and the functional signficance of the differences which exist between them. In our concentration of interest upon the living, however, we must not neglect to pay attention to the data of the past, even though it be fragmentary. Only by viewing each type of evidence in relation to the other will we be able to construct a coherent and consistent account of how we humans came to be, as we indeed are, an exceedingly polytypic species.

It would seem to me that attention to the past, both historic and prehistoric, may be especially important to students of *human* biology and demography. Our species does really have a history, in a sense which other species do not, because of the influence which human activity has had in altering, as time goes on, the surroundings to which we must adapt. All forms of life, to be sure, alter their environment to some extent, but man is unique in the extent to which he can, and does, make drastic, and in recent millenia partially intentional alterations. For, as time has passed, we have gained in our ability to improve, as we fondly believe, the circumstances under which we live. This ability is the result in large part of the function, if not consciously the intention, of human culture, so well symbolized in the Book of Genesis as the Fruit of the Tree of the Knowledge of Good and Evil.

Early man was, as we may expect a species to be in the beginning, a rare beast. We have no idea, and no means of inferring, how long it took for our prehuman and semihuman ancestors to become numerous. I think that we may legitimately suspect, however, that even in their prehuman state our ancestors were not only more versatile but more clever than their contemporaries and competitors. Even as far

SOURCE. F. Hulse, "Some Factors Influencing the Relative Proportions of Human Racial Stocks," *Cold Spring Harbor Symposia on Quantitative Biology*, **22**: 3-45 (New York, 1957).

back as the early Miocene there existed such creates as Proconsul, with cranial capacities at least as large as those of living baboons.

By Pleistocene times humans were making tools of stone, of which the distribution is wide-spread enough so that there can be no doubt that men were living from the Atlantic to the Pacific in Eurasia and Africa. It is true that fossil remains of human beings from such early times are infrequently discovered. Even if they were never discovered, the presence of tools is proof that men existed, however; and the quantity of early Palaeolithic tools indicates that humans were becoming more numerous as time went on. Indeed it would appear that as tools were improved, population increased. Man was beginning, by tool use, to modify his environment to collect more food in less time.

However, for the first 99 per cent of his history, man remained a collector and hunter, exploiting the resources of his environment with great ingenuity to be sure, but producing few drastic alterations of a lasting nature, although Stewart (1956) believes that, by the use of fire, early man turned forests into grasslands, and this may well have been true in some areas, at least. Only towards the end of the Palaeolithic do we have evidence that some cultures had become more efficient at exploitation than had others. Birdsell (1953, p. 191) finds that among Australian aborigines, coastal tribes possessing dugout canoes have greater population densitites than have their coastal neighbors who lack such a means of exploiting the resources of the sea. It is not unreasonable to suppose that invention in the past produced similar results, although of course we cannot be sure. The exploitation of fire, and the use of clothing, certainly increased the geographical range of early man, although I cannot imagine that this was the result of intentional planning by stone-age engineers.

Nature and Culture

In general, a hunting-fishing-gathering economy, such as characterized Palaeolithic and Mesolithic times, does not afford much protection from such forces of natural selection as would operate upon any other mammal. The existence of culture adds a new factor to the environment, to which our ancestors had to adapt, but not very much was subtracted from the potency of existing hazards and stresses. Every advance in technology, no doubt, helped give added protection to members of the group concerned, but such advances were few. During these times, however, mankind did spread over all the continental areas of the world, and many of the off-shore islands as well. He was thus subjected to a great variety of climates, and had available to him the most diverse food-supplies. The fossil record is good enough to indicate that most of the racial types and stocks familiar to us today were already in existence by Mesolithic times, but as I pointed out in a previous paper (Hulse, 1955a) there is no reason to suppose that the numerical proportions of these varied populations, as they then existed, resemble the proportions which now exist.

Since the Neolithic revolution, of course, human population has increased enormously but this increase has been highly selective in nature. A few population groups have become extinct, some are no more numerous today than they were 10,000 years ago, a number are known to have increased and then decreased, while a relatively small number of groups have expanded enormously. Meanwhile, evolution has been proceeding, so that gene-frequencies characteristic of a given stock today may be quite different from those which characterized the genetic ancestors of the same stock at the end of the Pleistocene. This, of course, adds to the difficulty of reconstructing racial histories. Nevertheless, if we constantly bear in mind the fact that mankind is, and long has been, subject to cultural as well as natural selection, the relative probabilities of various hypotheses concerning racial history may be assessed with some degree of confidence.

According to Simpson (1950, p. 63) "among the basic adaptive processes involved in trends are the following:

1. Progressive improvement in efficiency or perfection of existing adaptive type.

2. Narrowing or specialization of adaptive type from more general to more specific, or ecologically from a wide to a narrow niche.

3. Converse broadening of adaptive type.

4. Change from one adaptive type to another."

Among humans the first of these processes has been, ever since culture began, the adaptation of the human organism to the necessities of culture. The evolution of human personality structure has been most important in this connection (Hallowell, 1950). The second, however, has been adaptation to more localized geographical conditions. The first trend has been universal, as shown by Dobzhansky and Ashley-Montagu in "Natural Selection and the Mental Capacities of Mankind" (1947), and has tended throughout the life of our species to promote continued unity both by parallel evolution and by continued, even if sporadic, hybridization. The second trend has promoted diversity within the species, and has become less and less important as man's control of his environment has broadened. Coon, Garn, and Birdsell in "Races" (1950) have dealt with some of the most interesting results of this divisive trend. Both of the works have stressed the survival value of given characteristics, rather than dealing with the mechanics of changes in genetic characteristics of which the mode of transmissions is reasonably well understood. Indeed, genetic data is, to a degree, avoided in the later study. This is quite understandable, since those characteristics which may most plausibly be imagined to be geographical adaptations have not, as yet, proved very amenable to genetic analysis.

Phenotype and Genotype

There are several reasons for this. It is clear that several and perhaps many genes affect stature and pigmentation, for instance. Changes in size are likely to be accompanied by allometric growth, and hence by changes in proportions. Furthermore, the human organism is remarkably resilient, and a considerable range of physical characteristics is possible with no change in genetic systems. Boas (1910), Shapiro (1939), and Lasker (1946, 1952) among others have demonstrated that the offspring of migrants to a new environment differ considerably from sedent members of the same populations. I, myself, have worked extensively in this field, with such different peoples as Japanese and Italian Swiss, and in consequence would like now to propose an additional statement concerning the word "Race" as it is used by geneticists. Dobzhansky (1950, p. 99) states that "Races may be defined as populations which differ in the frequencies of some genes. A population is a group of individuals cemented by intermarriage, and hence, sharing a common treasury of genes." A group need not be numerous to be considered a racial stock: It need only be genetically distinctive. I would add the statement that such a group may be expected to change, in phenotype, in a characteristic way when living under changed environmental circumstances. This is precisely what is revealed in the studies so far published on the offspring of migrants. In other words, we must explicitly recognize the well-known fact that a given gene which leads to phenotype A in one environment, leads to phenotype B, in another environment. Thus we may say that the plasticity of the human phenotype is directional rather than scattering or random. Definable changes in the environment will produce definable modifications in the phenotype. Changes in phenotype, then, do not necessarily reflect changes in genotype although they may do so. We have to examine each population with this in mind if we are to understand the significance of the diversity existing within the human species and of the relative numbers of racial stocks. We have to consider what we know about the history, as well as the present characteristics, both genetic and phenotypic, of each separate population being studied.

Let us take the population of Burma as an example. Here we find a variety of peoples, some numbering millions, others much less numerous. A recent study by Oschinsky (1957) gives us material for analysis. The highland Shan and the Chinese are lightest in skin color; the Chin, the Arakanese, the Tamils and inhabitants of the coastal Delta farmlands are darkest. The Chinese are the tallest, the Chin the shortest; the Chinese most brachycephalic, the Tamil the least. Almost all the Chinese have epicanthic eyefolds, whereas only 72 per cent of the Burmese, 60 per cent of the Mon and none of the Tamils have such folds. The Tamils have the lankiest build, the Chins and Chinese the burliest. The Burmese themselves, intermediate in all these characteristics, are most numerous, occupying the lowlands. Chinese and Tamils are largely concentrated in and near cities, the former largely in commerce, the latter in manual labor. The other groups mentioned are mostly hill folk. How is this to be interpreted? Not at all, without some knowledge of the history of each group; knowledge of local ecology is not enough. Fortunately we know that both Chinese and Tamils are newcomers, whose homelands are far apart; there are clear selective advantages to being lean and dark in South India, where the Tamils have lived for thousands of years. Well documented historic and protohistoric movements brought the Chin, Shan, and Burmese from northern highlands at various times. The Mon have linguistic affinities with aboriginal Indian and Cambodian tribes. There can be no doubt that the Burmese are intermediate in physical characteristics because they have been longer than the other peoples from the north in the hot lowlands, but we have as yet inadequate data to tell us whether their adaptation is mostly the result of phenotypic plasticity, of adaptive selection of genes, or of mixture with local people already well adapted. The percentage of epicanthic eyefolds suggests strongly that much mixture with the Mon has taken place.

The Burmese owe their numerical superiority, however, not to their physical fitness, but to the fact that they occupy the lands best suited to rice culture, have been willing to engage in rice culture, and have been able to maintain a higher degree of social cohesion than the other peoples of Burma. Human culture as well as natural ecology have been important in promoting the growth of their population. In recent times the even better organized and less physically adapted Chinese have been moving into Burma as well as other Southeast Asia countries. It will be interesting to see what the relative proportions of racial stocks in Burma will become in the future. If the natural environment alone were the controlling factor, we might expect that genetic systems similar to the older strata of inhabitants would eventually prevail because of natural selection. Since we humans employ technology and social organization to combat the forces of nature, I doubt whether this will happen. But we cannot foresee the future extent of these protective devices.

Time and Technology

It seems to be generally accepted among physical anthropologists that racial stocks originate not only by adaptation to local circumstances, preferably accompanied by some degree of isolation, but also by hybridization. So far as I know, however, Coon, Garn and Birdsell, in "Races" (1950) were the first to list, among such stocks, populations whose origin by hybridization could be historically documented. In any study of the relative proportions of human racial stocks, now or in the past, the circumstances surrounding hypothetical hybridizations in the past must be examined as carefully as possible. Time and technology, as well as location, need to be taken into account. The resources necessary to support a numerous in contrast to a scanty population, among humans, must be considered in relation to the tools, and the customs, prejudices and social organizations of the people concerned. This is true no matter how, when or where any racial stock may have come into existence. Neo-Hawaiians have simply not yet had time to become numerous, for instance: Eskimos, despite a degree of both technical and bodily adaptation so high as to evoke the admiration of all who know them, are even less numerous, despite a long history, because of their location. Resources are so meager, and the food supply so scanty, that technological improvements are superfluous if not dangerous.

The mere fact of unfavorable location has surely kept some other populations from becoming numerous, whereas favorable locations have at least encouraged population increase. It has long been noted by anthropologists that more centrally located groups are those most likely, other things being equal, to give and take ideas from without, while the reverse is true of those groups living in remote places. And with ideas may come genes, a process of which I gave an example in "Blood Types and Mating Patterns among Northwest Coast Indians" (1955b).

During the overwhelmingly greater part of man's span of existence, the area which we now call the Near and Middle East was centrally located with reference to the total area inhabited by humans. It is still the crossroads between Europe, Africa and Asia. In the days when the only way to get somewhere else was to walk, this was even more true than today. It may possibly be due to this fact that the idea of food production, as distinct from simple collection, appears to have originated in this area. One consequence of food production was, of course, an increase in population. Estimates vary as to how much of an increase in population took place, and as to how rapidly it took place. But there can be no doubt that such an increase occurred, and that, as the idea of farming spread, the more densely occupied area expanded also.

As we all know, the spread of an idea or a technique does not require the migration of a people. Agricultural practices undoubtedly spread faster and further than did communities of agriculturalists. Tribes as well as customs do move from place to place, however, and we have ample evidence that migration occurred during protohistoric times, both out of and into the areas where food production began. An assured food supply breeds wealth, whether among Northwest Coast Indian fishermen, or Near Eastern farmers. New arts and crafts develop, including the art of war, and fertile lands are worth stealing from one's backward neighbors, who, one comes to believe, don't know how to use them anyway. So we find that time and time again tribes and nations with more efficient techniques seize and even come to inhabit the areas previously held by so-called savages. We have only to think of how we replaced the American Indian to realize what has happened in other places at other times.

Central Locations and Refuge Areas

As the better areas come to be thickly settled by more technologically advanced peoples, who become more numerous as a result of their techniques, less desirable regions only are left for the offspring of the aborigines. Such refuge areas are of various sorts; deserts, mountain valleys, remote uplands, dense jungles, distant peninsulas and islands come to mind as obvious examples. The distribution of linguistic stocks throughout the world can give us an idea of the location of such spots. A multitude of diverse and apparently unrelated stocks are to be found, jumbled up together very often—but not always—in out of the way places such as the Kalahari Desert, Lower California, highland New Guinea, northern and northeastern Siberia, the Caucasus; whereas vast and, typically, contiguous areas are held by peoples of Niger-Congo, Indo-European, Ural-Altaic or Sinitic stocks. It would be absurd to argue that one can infer racial stock from linguistic family. The analogy

MAP 1. Distribution of peoples within Burma.

between the spread of language and the spread of genes can none the less be made. People are much more likely to move in some social group than singly—if not by tribes, than by families. People are much more likely to speak the language of their parents than any other. And when people move in large and cohesive groups they are likely to bring their language with them. There are documented cases of linguistic barriers to gene-flow (Hulse, 1957). Mixtures of peoples of course takes place, but so do mass migrations, and expulsions and massacres of natives.

One immediate, and up to present times continuing result of technological advance has, then, been an increasing disproportion in the numbers of the various racial stocks in the world. As I pointed out two years ago (Hulse, 1955a) there is no reason to suppose that blonds were more numerous than Bushman 10,000 years ago. The fact that there are now 100,000 blonds for every living Bushman is no sign of superior biological equipment among the blonds. Yet the disproportion exists as a biological fact.

There are further consequences. The more numerous peoples of the world are better able to intermingle extensively, and do so. New racial stocks have originated, and continue to originate among them by hybridity. In the refuge areas inhabited by the remnants of other tribes there is less opportunity for this process to occur. The habitat is likely to be extreme, and the technology retarded, so that selection continues to favor those best adapted to natural circumstances rather than to culturally created ones. Thus we find the so-called "archaic" physical features of Australian aborigines, for instance, to be quite expectable. At the same time, there is little incentive, as well as little opportunity, for an outsider to marry into such a group. Plenty of young Hollanders 200 years ago in South Africa married Hottentot girls, whose families owned large herds of cattle, but matings with Bushmen, who owned nothing, appear to have been rather casual.

The breeding populations of peoples inhabiting refuge areas are likely to be quite small. Natural barriers between settlements or bands are often reinforced by tabus or at least prejudices, so that the index of consanguinity is likely to be very high. In Swiss valleys where I have worked, I have found that a majority of marriages are still contracted between fellow-villages. Most of these villages have less than 400 inhabitants, so that, despite the injunctions of their religion against cousin marriage, everyone is related to everyone else. Many villages are reputed, in the local folklore, to breed true with respect to certain body characteristics. Everyone knows, for instance, that the people of Moghegno have big noses—and, as a matter of fact, they do! Homozygosity increases under such conditions. Genetic drift is effective. Any new mutations are kept at home, and, in time, serve to differentiate gene-frequencies between one village and the next. If this is true of the Canton of Ticino, in Switzerland, in the heart of Europe, how much more true must it be in really remote refuge areas.

Small populations are also particularly subject to hazards which may lead to their utter extinction, and this is especially true if their technology be retarded and the environment a harsh one. The entire population of Southampton Island, Eskimos, starved to death during the winter of 1902 (Kroeber, 1948, p. 377). This small band had earlier lost knowledge of how to build kayaks, so that they had no means of leaving the island after having hunted down all available land game. Typhoons have depopulated some South Sea Islands, although of course contagious diseases brought by outsiders have done a much more effective job in reducing Polynesian population as a whole. Sometime, in a purely fortuitous manner, the genetic system of a group, and consequently its racial characteristics, may be altered by such a disaster. Early in the 18th Century a typhoon washed away all but seventeen of the inhabitants of Puka Puka (Beaglehole, 1938, p. 386). These few survivors were, it is said, all of the lower class. Thus, the island lost not only the traditions associated with upper-class status, but also, I strongly suspect, most of the genes for tall stature, for the modern population is notably shorter than in other Polynesian islands (Shapiro, 1942, pp. 141-69).

Technology and Resources

The basic advantage, in the struggle for survival, possessed by peoples having technologies more advanced than those of their neighbors, has already been mentioned. The relationship between technology and natural circumstances needs to be considered also. Unless one's manufactured equipment is suitable for exploiting local resources and combating local hazards, it will not be advantageous. We have yet to better the Eskimos' equipment for survival in the Arctic, for instance. Medical knowledge and supplies are just now beginning to be adequate for the job of keeping a European alive in many parts of west Africa. In the same way, certain inventions had to be made before large areas of the earth's surface could be inhabited at all, by anybody. Some racial stocks could not even have come into existence without the aid of appropriate technology.

The Americas appear to have been free of human habitation until sometime during the last glaciation, probably closer to the end than the beginning of this cold period. From archaeological data, we have no suggestion of the use of clothing until this same period, nor of any interest in food supplies from the water, such as might have led to the invention of water-craft. Lack of data is no disproof, but I find it highly suggestive that humans did appear in the Americas only after the devising of techniques suitable for arctic survival, such as clothing, and quite possibly ice-fishing, pit-dwellings and even primitive types of boats. This fits well with the working hypothesis accepted by most anthropologists that the Americas were populated by way of Siberia and Alaska, and would account for the arrival of the earliest migrants during a cold rather than a warm period.

Once arrived, these ancestors of American Indians found themselves in the fortunate position of having no competition and 16,000,000 square miles of territory to exploit. Even if only a few bands wandered across, there was little to prevent unlimited population increase and territorial expansion for many generations. It seems to me that we may have here an example of people having been forced into a refuge area, evolving their special characteristics there, and then suddenly having almost unlimited space and resources made available to them. Glass (1954) and Brues (1954) have both suggested that genetic systems established, by drift, in a small population isolate, will not be lost if that population later becomes numerous. The lack of such genes as cde (r) and B, among American Indians, can be explained on this basis. The explanation of any racial characteristic on the basis of genetic drift is, to be sure, rather dangerous. But human populations, because of technology and social organization, can recover from near extinction to an extent which would be impossible among other living creatures: If drift is effective in any creature, it should be in man.

In any case, the comparative—though far from absolute—uniformity among

American Indians can best be explained by the hypothesis of immigration via the Arctic. Many generations must have been spent in areas where the pressure of selection would have been for adaptation to cold, and we note that even in tropical parts of America some of these adaptions have been retained. As Haldane (1956, p. 6) writes: "From the absence of a very dark indigenous race in tropical America we can suggest that the evolution of such a race requires over 10,000 (sic) years." Bodily proportions of even tropical jungle Indians remain Mongoloid as well, which would add confirmation to this statement. Appropriate mutations do not occur just because it would be convenient for a racial stock to become better adapted. Even more interesting to the student of the relative numbers of human racial stocks is the fact that American jungles appear never to have supported very dense populations, even though the tribes were for the most part agricultural. The lack of metal tools, and over vast expanses of territory, of even adequate supplies of stone, could well have added to the difficulties of the Indians in such places as the Amazonian jungles. Since the importation of Negro slaves to America, their off-spring have in many cases succeeded where the aborigines failed, and have become very numerous.

Interestingly enough, the plants brought into cultivation by American Indians are, with the exception of potatoes and quinoa, almost exclusively tropical in origin. This is not true of a great many Old World crops. We find that agriculture, although practiced for thousands of years in Middle America, was unable to spread very far north, and then only in attenuated form. Really dense populations certainly existed in Mexico (Cook and Simpson, 1948) and the Andean regions as much as a thousand years ago, but the areas which we find most useful could not be so well exploited by the technology known to the Indians. Consequently the early English settlers in America rather easily disposed of their aboriginal opponents, and population replacement occurred; whereas the Spanish, for the most part, moved in at the top level of society instead. As a result, there is now a Ladino racial stock which is quite numerous, but a relatively small group of Northern European-American Indian hybrid origin. The historical events which have led to this situation are basically due to the relationship between technology and resources.

The discovery and exploitation of the Americas by Europeans affected not only the relative numbers of racial stocks in the New World, but in Europe as well. The looted wealth of the Indies stimulated trade, and this had certain effects which will be discussed later. Just as important, however, in promoting population growth in some areas rather than others was the importation of the potato. This is one of the few plants cultivated by American Indians which does well in chilly climates, and within a few centuries potatoes had become a staple crop in many parts of northern and northeastern Europe. During the same period, for other social causes, agricultural practices were so improved in many ways in these same areas that a much larger food supply was brought into existence. Ample winter fodder, for instance, was provided so that greater numbers of cattle could be kept alive throughout the year, and flocks and herds increased. So did the human population. Countries as far apart as Ireland (Lack, 1954, p. 176) and Russia have been affected by this technological advance, but whereas the increase in population in Ireland came to an abrupt stop as a result of the potato blight a century ago, and has since been halved, the northeast European countries have continued to increase rapidly in population to this day. In so far as differences in the proportions of genes distinguish these populations from others, the relative proportions between racial stocks have shifted as a result. In fact, there are such differences, as for instance, in the frequency of the allele for blood-group B. Manuila presents a map, in "Distribution of *ABO* genes in Eastern Europe" (1956, p. 585) showing that the incidence of the gene *B* reaches 15 per cent or more throughout the Soviet Union, except for Lithuania and western Latvia, and also throughout prewar Poland. Table 1 shows how much more rapidly these areas have grown, during the past 350 years, than has the rest of Europe. It must be emphasized that the increased incidence of blood-type B in Europe, resulting from this population shift, may not be taken as evidence of adaptive value for blood-type B.

Table 1. Population Growth in Europe Since 1600 A.D., by Areas of Lower and Higher Frequency of the Gene for Blood-Type B

	B 15% or More	B Less than 15%	Total
1600	15,000,000	85,000,000	100,000,000
1950	180,000,000	375,000,000	555,000,000
Increase %	1,100%	341%	445%

Just as the Americas would appear to have become habitable only after invention of Arctic survival gear, so the Polynesian islands, among others, could not have been reached until adequate shipping and navigational skills had been developed. Despite Mr. Heyerdahl (1952) neither I nor most anthropologists regard balsa rafts as adequate shipping. Outrigger canoes, we may suspect, are not among the earliest forms of shipping, and Polynesia appears to be the last great area of the world to have been inhabited by man. Furthermore, in almost all cases, skill at cultivating the appropriate food plants was required if the islands were to be permanently occupied, because of the paucity of indigenous resources. Only during the last few thousand years did the culture complex necessary for the discovery and settlement of these far flung and often tiny spots of land come into existence.

    That the technical skills of this population were quite adequate for the circumstances is well demonstrated, however, by the astonishingly dense populations attested by all early European visitors for most of the island groups. How numerous the actual migrants into Polynesia may have been we have no way of knowing, but the increase in numbers of such well-documented groups as the hybrid descendents of the Mutineers of the Bounty (Shapiro, 1929) in the course of a century and a half indicates that a few dozen large canoe loads might have brought enough ancestors.

As yet we know all too little of the genetic characteristics of this racial stock, although body build, facial type and pigmentation have been amply studied. It is clear that their resemblance to the stocks now common in the Indonesian area from which they must have come is not as great as might have been expected. The Polynesians have been in Polynesia for too brief a time, and increased in numbers too rapidly, for selection to have altered their genetic system too radically since their arrival. It may be that in some of their characteristics, such as size, they differ from Indonesians only phenotypically (Coon, 1954, p. 286). It may be, too, that a rather drastic selection of types took place early in their migration, or even before. They have, of course, differentiated to some degree, from one archipelago to another (Shapiro, 1944), yet still form an easily recognizable racial stock.

Their technology and social organization, quite adequate for developing their territories while undisturbed, were not able to protect them against the ravages of diseases inadvertently introduced by Europeans and Asians during the past few generations, and population declined during the 19th century, while other racial stocks were rapidly increasing in numbers. The depopulation of the islands of the Pacific has been described and discussed in numerous works and I do not propose to enlarge upon it at this time. Certainly the causes have been various, differing from one archipelago to another. Hunt and others, in "The Depopulation of Yap" (1954) provide a recent and in my opinion highly plausible explanation for the sharp decline in population in that Micronesian island. In some other areas, as for instance the Marquesas, an intentional refusal to breed by the local population is alleged (Linton, p. 137, in Kardiner, 1939). Linton also points out, in the same work, the extraordinary sex ratio of five men for every two women, accompanied by a situation which may be described either as group marriage or a most unusual degree of male permissiveness in sexual matters. From his description, it is clear that the sexual role of women was stressed almost to the exclusion of the maternal role. A situation of this sort certainly encourages population decline, even without the alleged intentional refusal to breed.

Mating Patterns and Preferences

Even if we remain skeptical with respect to this example from the Marquesas, the fact remains that mating patterns can and do affect the incidence of genes, and the relative proportions of different populations. Human mating patterns are always assortative, never random, because humans behave in a cultural as well as a biological manner. Although customs regulating sexual behavior vary to a degree which members of any particular society may find extraordinary, there are always customs of one sort or another. Even sexual misbehavior appears to be controlled by custom. It may well be and often is true that the particular genetic item being examined does not appear to be affected by assortative mating. Most of the work in human genetics, it must be remembered, has concerned characters, such as bloodtypes, of which their possessors are unaware. One cannot develop a prejudice or a preference concerning such a character. Nonetheless, there are some general and many particular regularities in human mating systems which certainly affect geneflow.

In the first place, all societies have rules which encourage outbreeding to some degree. Parent-child matings and brother-sister mating are so exceedingly rare that no student of human genetics need consider them. Marriage to cousins is far more common and especially to certain sorts of cousins. The particular type of cousin regarded as most desirable varies from one society to another. Pride, power, prestige or property are commonly associated with assortative mating of this variety, and it may serve as a barrier to gene-flow. As a rule, however, some degree of exogamy is encouraged by the rules which govern mating among humans. The genetic consequence of this fact is the promotion of gene-flow. I suspect that this is one of the many reasons that we have remained one species.

Tendencies, and sometimes stated rules, for endogamy also exist. As I pointed out above, village endogamy is common in Switzerland. It has also been common in many other areas, both in Europe and elsewhere. In Japan, it is regarded as best to marry into a neighboring village, or at least another hamlet (Embree, 1938, p. 92), but cousin marriage is much more frequent than among ourselves (Neel *et al.*, 1949). The conservation of family property is involved in this example, and, since adoption is often used for the same purpose one cannot be too sure how much genetic consanguinity accompanies the cultural endogamy.

Since fellow-villagers and even cousins are more accessible, because of physical propinquity, than are outsiders, the two types of preferential mating just mentioned do not serve to distinguish us from creatures lacking culture. Caste, however, is a creation of culture, an invention of the human imagination, and yet it has also served as a barrier to gene-flow. Members of a caste, whether in Bombay or Natchez or Buckingham Palace, do their best to prevent their females from mating with males of a so-called "lower" caste. Much more often than not, their attempts are successful, and therefore have genetic effects upon the future population. Males have been less easily restrained, so that this genetic barrier is in no sense absolute, but the violence of sanctions employed against misbehavior has in fact kept racial stocks, living side by side, quite different in gene-frequencies for prolonged periods of time.

Human culture also affects the survival rate of members of different castes. Economic as well as ritual differences separate them. If, as often happens, one caste is economically depressed, or even enslaved, its death rate is likely to be exceptionally high. During the first half of the 19th century, for instance, more slaves were imported into Cuba, during some decades, than were alive at the end of the decade

(Hulse, ms.). Almost as many slaves were imported, up to 1853, when importation ceased, as there were Negroes alive in Cuba by that year. At the same time, the population classified as white increased more than seven fold in seventy years. Table 2 gives a more precise breakdown of these figures. Although freed or escaped Negro slaves have competed successfully with Indians in tropical America, any natural advantages of climatic adaptation which they may possess have not been

Table 2. Population Growth and the Importation of Slaves into Cuba
During the Colonial Period

	Population		Slave	Slaves Imported Since Previous Count
	White	Free Colored		
1792	107,000	75,000	90,000	85,000
1817	257,000	116,000	199,000	200,000
1827	311,000	107,000	287,000	80,000
1841	418,000	153,000	432,000	112,000
1861	793,000	226,000	377,000	94,000

Total importation of slaves: 571,000.
Total colored population by 1861: 603,000.

enough to counteract their unfavorable economic position vis-a-vis populations of European ancestry, where such populations existed in any numbers. Improvements in public sanitation and nutrition during the present century appear to be changing the situation, but, for at least as long as slavery existed, this was not the case. A study of the United States Census figures for the period before large-scale immigration from Europe began indicates that here, too, Negroes were not increasing as rapidly as whites. The proportion of Negoes fell, between 1790 and 1840, from about one-fifth to little more than one-sixth of the total as is shown in Table 3. Slavery kills.

At the same time, there is often, at least, a tendency for the lower caste to approximate the higher one in some observable characteristics. It is well known that great numbers of American Negroes are partly European in ancestry, because the caste barrier to gene-flow has not been absolute. Since aesthetic ideals are set by the members of the dominant caste, persons showing one or more Caucasian characteristics are regarded as desirable mates. Henriquez, in "Family and Color in Jamaica" (1953) has described the situation in that Caribbean Island very well. The proportion of Caucasians is, and has been since the beginning of the nineteenth century, less than 10 per cent of the total population. Of the present population, only 3 per cent are Caucasian, but "Colored" as distinct from "Negroes" number more than one quarter of the whole. This situation has been brought about by the fact that the Colored are favored both socially and economically. Their children have a

Table 3. Population Growth in the United States During the Period
Preceding Heavy European Immigration, by Thousands

	1790	1800	1810	1820	1830	1840
White	3,172	4,304	5,862	7,867	10,532	14,190
%	80.7	81.1	81.2	81.5	81.9	83.2
Free colored	59	108	186	234	320	386
%	1.5	2.0	2.4	2.5	2.5	2.2
Slave	698	893	1,191	1,538	2,009	2,487
%	17.8	16.9	16.4	16.0	15.6	14.6
Total	3,930	5,306	7,240	9,638	12,866	17,069
White increase since previous count		1,132	1,558	2,005	2,665	3,658
%		36	36	34	34	35
Slave increase since previous count		195	298	347	471	478
%		28	33	25	31	24

better chance of survival than Negro children, and those among them who approximate more closely to Caucasian features or pigmentation have a better chance of good jobs and mates similar to themselves. This process of sexual selection has apparently been going on for several generations, and is producing a biological result: an increase in the percentage of genes for Caucasian features. It would be most interesting to discover whether there has been any similar effect upon the frequencies of blood types such as R_o which differentiate European from African populations so markedly. Since, up to the present time, the very existence of blood-type genes has been practically unknown to the people of Jamaica, it would seem impossible that cultural preferences could have altered their frequency. We do know that the incidence of the gene for sickle-cell is rather low among Jamaicans (Mourant, 1954) but this could be due to the tribes of origin, in Africa, of the Jamaican slaves.

Members of dominant castes enjoy other advantages, too, both genetically and demographically. Young males among them are in a position to contribute a far greater number of genes to the lower caste than their numbers would warrant; and their hybrid offspring are rather more likely than usual to survive, not only because of social favor or economic advantages as exemplified in the case of Jamaica, but also because of greater immunity to such new diseases as may have been introduced by the conquering caste. Of course, if the conquering caste does not bring in new diseases, but is itself lacking in immunity to local afflictions, this will not be the case. We Europeans, for instance, appear to have introduced more virulent contagions to American and Pacific Island aborigines than we received in return. In Central Africa and Asia the reverse would appear, from the results, to have been true.

Nevertheless, where polygyny is established and favored, certain males may be expected to contribute to the gene-pool much more than do others. If the genetic systems of a conquering and a conquered group differ, there should be a closer approximation to the genetic system of the conquerors than would be expected by simply counting the numbers of each group and assuming that the genetic contribuion of each group is proportional to its number. Just as a very few Caucasians in Jamaica have inadvertently created a large "Colored" group, so, in areas where the proportion of the dominant group has been larger, an even bigger population of hybrids will be created. Meier (1949, p. 235) for instance, found that among students at a Negro college in Mississippi, only one-sixth were unaware of either white or American Indian ancestry. A good majority knew of white ancestry. In both of these areas polygyny has been, as it were, kept under discreet wraps. In areas where it has been open and favored, the same results can be expected more rapidly.

In considering the effect of mating systems upon populations, one must consider also the span of active reproductive life, and this, too, is affected by cultural factors. Among many peoples, marriage and mating begin at an early age; among others, mating may begin early, but marriage is postponed. Among some, both, at least insofar as reproduction is concerned, are put off. In Ireland, for instance, a man does not marry until his father is willing to retire (Arensberg, 1937). Up to this time he is considered a boy. In some Swiss and other Teutonic peasant groups, the same rule exists (Lowie, 1954, p. 148). I have found that in the Canton of Ticino, out of 1,063 men who were 40 to 45 years old, only 778 had ever married. Yet in all these areas, the rate of illegitimate births is very low. Unless a man marries a woman much younger than himself—as the Swiss apparently do, but as the Irish apparently do not—this postponement of marriage will inevitably result in such a reduced birthrate as we find in Ireland, and a few other countries. Among some of the European peasant communities, only one son is expected to marry at all. Of course, the other sons may migrate. In this case, although the population of the homeland may remain stable, the numbers of the racial stock to which that population belongs may increase very rapidly indeed.

Culture and Resources

It has often been said that social practices which restrict the birth rate become customary in response to a situation in which the land will not support a greater population. The extent to which land will support population is in itself a creation of culture, both in its technological and ideological aspects, however. How much does one's culture lead one to expect? And what variety of material goods, as well as what quantity, are regarded as necessary? Different cultures give different answers to these questions (Spoehr, 1956) and the cultural aspirations of any people are subject to change without notice. Twenty years ago most American demographers were convinced, with good reason, that population growth in the United

States was about to cease. But fashions have changed. Having learned contraceptive techniques, our young people decided that it would be nice to have big families anyway, and they continue to procreate at a rate which has made all previous calculations of population growth seem ridiculous. There has clearly been a shift in ideology, or at least in attitudes, within the past generation, which no one had reason to anticipate. As our knowledge of cultural dynamics improves, we may some day learn to predict such shifts, but we have not been able to do so up to now.

It is obvious, of course, that certain traditions, as for instance traditional food-preferences, will have an effect upon population size. Rice culture is an example. As practiced for thousands of years in East and Southeast Asia, it required much labor, but gave excellent returns if properly done. Relatively dense populations could be supported on the proper sort of land if everyone worked hard (Wittfogel, 1956, p. 159). Thus the racial stocks inhabiting the regions where rice culture was practiced soon became numerous, and spread into neaby regions, both displacing and hybridizing with the earlier inhabitants. Tribes or families unable or unwilling to accept the way of life required by rice culture were displaced; those who were both willing and able increased in numbers, or at times were hybridized or even absorbed. Li Chi, in "The Formation of the Chinese People" (1928) has traced the expansion and movement of the Chinese during the past two thousand years in China, in a manner which illustrates this process very well. It must be remembered, of course, that the Chinese at the beginning of their history were cultivators, not of rice, but of other grains. They did, however, practice intensive cultivation with large scale irrigation, in a manner which produced very similar demographic results. And it was in the rice growing areas that their population expanded to the greatest extent, and from which, too, most of the overseas migration of Chinese into still more southerly parts of Asia has occurred.

At the same time, expansion to the north has been hampered, among East Asiatic farming peoples, by their specializations. In Japan, for instance, until modern times, Hokkaido was neglected because of the difficulty of growing rice there. Where rice would not do well, Japanese farmers did not care to go. The sharp cultural cleavage between pastoralist and gardener in North China hampered expansion to the north by Chinese populations, too. Resources which to us appear worth exploiting seemed valueless to men whose cultural perceptions were different. Thus the current of population growth, in East Asia, has been set in a southerly direction, whereas in Europe it moved to the northwest. In both cases, it is true, racial stocks at one time insignificant in numbers have become very numerous.

In some cases, however, the cultural preconceptions of a people have prevented numerical expansion, and in others their technological practices have led to population decrease. Except in New Guinea, for instance, pygmies are always hunters and collectors of jungle produce, having, as a rule, trade relations with neighboring groups of agriculturalists. So long as they continue this way of life, they can never become numerous. Many agricultural practices have led to such a degree of soil exhaustion or erosion that regions once populous later declined in population. Cook, in "Soil Erosion and Population in Central Mexico" (1949) documents this state of affairs for that area very thoroughly. Certain parts of the Mediterranean basin, especially in the Near East, have undergone the same fate. The recent world-wide expansion of technological improvements, including not only industrialization but easy transportation, has of course, served to protect some population groups which could not possibly maintain their numbers by means of local resources alone.

Migration

The improvement in means of transportation is perhaps the greatest single technological factor determining the present distribution and the relative numbers of racial stocks, while the course of political history is the greatest non-technological factor. As I pointed out in a previous paper (Hulse, 1955a), the population of British origin has increased eight times as fast, during the last four centuries, as has the population of the world, and the population originating from the area of blondness in North and West Europe at least four times as fast as the average. This exceptionally large increase has been largely due to overseas migration, although the introduction of new foods, which has been mentioned above, has been important too. Migration overseas from Europe began only after improved shipping and navigation methods were invented, and only after political organization in European nations had developed to a degree adequate for dealing effectively with aboriginal populations in overseas areas.

Whether inadvertently or by design, the British found themselves in possession of thinly populated areas. During the seventeenth century a few score thousand English, Irish and Scotch came to North America, and during the next century a few hundred thousand more, altogether a very small fraction of the total population at home. With no effective opposition from the local Indians, with ample resources of food, and probably reduced incidence of contagious diseases because of the lack of urban crowding, the immigrant population more than doubled with each generation. Although the migration of non-British Europeans into the United States has numbered many millions, the offspring of the earlier British migration were already so numerous, when the later migration began, that they have maintained their initial advantage in numbers up to this day. More than a generation ago, to insure their continued advantage, they imposed, by political action, a quota system of immigration. One does not have to approve of political action of this sort in order to realize that it does have an effect upon population proportions. One does not have to prefer blonds to realize that they have, in fact, become a larger proportion of the world's population during the past few centuries. It has been from the area where the alleles for blondness are most frequent that overseas migration has been most easy, and it has been in these areas that industrialism and urbanization, with consequent extra population increase, have proceeded furthest.

Political restrictions upon migration have often had as their intention the control of the "Racial" characteristics of the population, both in this country and in a number of others; no matter how clumsily, nor with how little knowledge of genetics, they may have been devised. But in some cases the results have been quite unintentional, and the selection inadvertent, for controls in the past have been more frequently religious or ideological than "Racial" in purpose. A very great many of the migrants to the British colonies were members of minority religious groups who felt themselves oppressed at home. By what appears to me a purely fortuitous correlation, such groups were especially common in those areas, such as East Anglia and Ulster, where blonds are more numerous than elsewhere. The migrants were not, therefore, a random sample of the home population. Similar circumstances have existed in other countries and at other times, to cause migrants to deviate to some degree from the norm. To discover the type and the extent of such a deviation, one must examine the circumstances of each migration separately.

Soon after migration started from Spain to the Indies, for instance, it was decreed that persons of Moslem or Jewish ancestry were forbidden to go to these colonies. Southern Spain, from which ships sailed to America, was that part of the Iberian peninsula most recently recovered from the Moors, and a great proportion of the native inhabitants had either Moslem or Jewish ancestry. The upper classes, however, were more largely descended from soldiers from the northern part of Spain. Historical records (Catalogo de pasajeros a Indias, Vol I., 1930) show the the migrants were drawn in the main from the upper classes, and a study which I made some years ago (Hulse, ms.) shows also that Cubans of Spanish rather than of mixed ancestry appoximate, in physical characteristics, the upper class rather than the lower class of Andalusians. Both white Cubans and upper class Andalusians are, on the average, less deeply pigmented than are lower class Andalusians, for instance. There was no prejudice against brunettes on the part of the Spanish lawgivers. Fear and distrust of religous dissidence, however, resulted in the selection of the less brunette as migrants. It is interesting to note that even under tropical conditions, such as those of Cuba as contrasted to Spain, natural selection has not, apparently, led to a decrease in the alleles for blondness. Technological progress would appear to give very good protection even to light skinned people, against those aspects of the tropical environment which probably have favored deeper pigmentation in earlier times.

A physical type is not a racial stock, but selection and isolation may transform what was simply a type into a stock, just as hybridization may eliminate a stock, but leave types (Hulse, 1943). Whenever, either inadvertently or by design, migrants are a select group rather than a random sample of the population from which they were derived, the possibility for the emergence of a new racial stock is created. Should the migrant group remain genetically isolated, such a stock will come into existence. If circumstances lead to a considerable increase in the numbers of such a group it will even be recognized as a racial stock. In the course of human history, many such groups have had only a brief existence, but others have flourished. I suspect that both American Indians and Polynesians originated in this manner. In modern times travel is so easy that the necessary isolation no longer exists, unless it be reinforced by social or political sanctions, either internally or externally imposed. Such groups as the Sephardic Jews after their expulsion from Spain and the French Canadians have remained separate by their own choice. I should very much like to know how genetically distinctive the French Canadians are, if at all, from Frenchmen in France.

Some Problems

Even more, I would like to look into the future, and be able to foresee the genetic consequences of the creation of the state of Israel. Here is a laboratory for the study of human genetics which has no equal. If, as I maintain, the cultural history of mankind has had a strong effect upon the relative numbers as well as the special characteristics of the various racial stocks, this should be clearly seen in the case of the Israelis. Here is a group derived from what has been historically a religious community, but a community which has during much of its history functioned as a caste, and a caste may become as genetically isolated as an island. The barriers erected by ideology, as has been demonstrated, can be very effective in preventing gene-flow, or in channeling it in one direction. The dispersion of the Jews into many countries has clearly forestalled the emergence of a single Jewish racial stock, which might otherwise have come into existence as a result of caste endogamy, and the probably strong selective forces for adaptation to urban life.

The mutiplicity of castes in India adds to the difficulty of analyzing the genetic structure of that vast area, and, consequently, of studying the relative numbers of any racial stocks to be found there. It is clear that we cannot consider Sikhs and Veddas, for instance, as representatives of the same racial stock. But should we consider each caste as a separate stock? That also appears to absurd. Whereas in Europe we can prepare maps showing gene-frequencies for blood types and at least the phenotypic distribution of pigmentation, a third dimension, that of caste, has to be considered in attempting to visualize the situation in India. Again, since some castes have better economic opportunities than others, and perhaps even other advantages in the struggle for survival and increase, can we expect them to fluctuate in numbers independently of one another? Or, since their economics are mutually dependent, are their relative numbers likely to be more or less stable? All that I feel confident of is that the impact of culture upon genetics and upon the relative numbers of populations genetically distinct, has been at least as profound in India as elsewhere in the world, and probably more than in most areas.

Summary and Conclusion

I have attempted, in this paper, to suggest some of the more obvious factors which need to be taken into account if one tries to ascertain the reasons for the numerical proportions of human racial stocks. Further examples to illustrate these factors come readily to mind. The discussion of other factors has not been attempted, in part because other scholars have already presented them so effectively already, and in part because of my desire to stress and to bring together in one paper, a number of factors which have been, perhaps, rather neglected. Coon (1950), Angel (1950), and Kluckhohn and Griffith (1950) in their valuable contributions to the Cold Spring Harbor Symposium seven years ago all pointed up the importance of culture in the study of human genetics and human evolution. Carr-Saunders (1936, pp. 44-45) has mentioned the effects upon relative proportions of racial stocks of differential rates of population increase during the past three centuries. Lorimer and others, in "Culture and Human Fertility" (1954) have contributed much to our knowledge of some of the relationships between demography and culture. Much deeper exploration of the many ways in which human culture and genetics interact in the creation and expansion—or contraction—of human racial stocks remains to be done, however. It will be noted that most of the factors which I have discussed have been cultural, and that culture has guided, as it were, the relationship between raw or untamed nature and man. Resources for sustenance are available to man only through technology. The refuge area of one millenium becomes the center of human life in the next because of technological progress. Mating systems are determined by social systems, so that genetic isolates can and do exist interspersed throughout densely populated areas. Arbitrary standards, as often as not completely irrational, determine man's food habits, political behavior, esthetic standards and social structure. Men of all tribes today appear equally capable of accommodating themselves to the oddest demands of culture, and thereby taking advantage of the opportunities opened up to mankind by culture. But the cultural imperatives of one society may, and usually do, differ from those of another in many respects. Population expansion may then be restricted and will certainly be guided by the specific culture of the population concerned.

Since culture changes, humanity has a history. Culture alters the environment, and thus, even without migration, people find themselves living under new circumstances which may permit population expansion or enforce population decrease. Far more often than not, culture protects human life, and promotes population growth, but the extent to which this is accomplished differs from tribe to tribe. Between the dawn of the Neolithic and the present century, the difference in the protective efficiency of different cultures has been sharply accentuated, so that some population groups, especially those of Northern and Western European origin, have increased strikingly at the expense of others during the past few centuries. Earlier, such populations as those of the Near East and of East Asia have expanded because of technological advantages. During the present century, with the spread of modern technology throughout the world, still other groups have been increasing rapidly in numbers.

What the future will bring is unpredictable. We may be sure of but one thing. Whatever changes do take place will be brought about by the cultural rather than the natural aspects of man's environment.

References

Angel, J. L., 1950, Population size and microevolution in Greece. *Cold Spr. Harb. Symp. Quant. Biol.*, 15: 343-351.

Ashman, R., 1950, Origins of blood groups ABO and the European Mongoloid problem. *Amer. J. Phys. Anthrop.*, 8 n.s. 427-452.

Arensberg, C. M. 1937, *The Irish Countryman.* Great Britain: P. Smith.

Beaglehold, E. and P., 1938, Ethnology of Puka Puka. *B. P. Bishop Museum Bulletin* 150. Honolulu. The Museum.

Birdsell, J. B., 1953, Some environmental and cultural factors influencing the structuring of Australian aboriginal populations. *Amer. Nat.*, 87: 171-207.

Brues, A. M., 1954, Selection and polymorphism in the A-B-O blood groups. *Amer. J. Phys. Anthrop*, 12 n.s.: 559-597.

Carr-Saunders, A.M., 1936, *World Population.* Oxford, Clarendon Press.

Catálogo de Pasajeros a Indias, 1930, Volumen I (1509-1533) Ministerio de Trabajo y Prevision. Madrid, Espasa-Calpe.

Cook, S. F., 1949, Soil erosion and population in central Mexico. *Ibero-Americana:* 34. Berkeley, Univ. Calif. Press.

Cook, S. F. and Simpson, L. B. 1948, The population of Mexico in the sixteenth century. *Ibero-Americana:* 31. Berkeley, Univ. Calif. Press.

Coon, C. S., 1950, Human races in relation to environment and culture with special reference to the influence of culture upon genetic changes in human population. *Cold Spr. Harb. Symp. Quant. Biol.*, 15: 247-258.

    1954, Climate and Race. Smithsonian Report for 1953: 277-298. Washington, Smithsonian Institution.

Coon, C. S., Garn, S. M., and Birdsell, J. B., 1950, *Races.* Springfield, Charles C Thomas.

Dobzhansky, Th., 1950, The genetic nature of differences among men. In: *Evolutionary Thought in America,* ed. S. Parsons: 86-155. New Haven, Yale Univ. Press.

Dobzhansky, Th., and Montagu, M. F. Ashley, 1947, Natural selection and the mental capacities of mankind. *Science,* 106: 587-590.

Embree, J. F., 1939, *Suye Mura, a Japanese village.* Chicago, Univ. of Chicago Press.

Glass, B., 1954, Genetic changes in human populations, especially those due to gene-flow and genetic drift. *Ad. Genetics* 6: 95-139.

Haldane, J. B. S., 1956, The argument from animals to man: an examination of its validity for anthropology. *J. Roy. Anthrop. Inst.*, **86**: 1-14.

Hallowell, A. I., 1950, Personality structure and the evolution of man. *Amer. Anthrop*, **52**: 159-173.

Henriquez, F. M., 1953, *Family and Color in Jamaica*. London, Eyre and Spottiswoode.

Heyerdall, T., 1952, *American Indians in the Pacific*. London, G. Allen and Unwin.

Hulse, F. S., The comparative physical anthropology of Cubans and Spaniards. Unpublished dissertation.
  1943 Physical types among the Japancese. *Peabody Museum Papers*, **20**: 122-133.
  1955a Technological advance and major racial stocks. *Hu. Biol.*, **27**: 184-192.
  1955b Blood-types and mating patterns among northwest coast Indians. *Southwest J. Anthrop.*, **11**: 93-104.
  1957 Linguistic barriers to gene flow: blood types of the Swinomish, Okanagon and Yakima Indians. *Amer. J. Phys. Anthrop.*, 15 n.s.

Hunt, E. E. Jr., Kidder, N. R., and Schneider, D. M. 1954, The depopulation of Yap. *Hu. Biol.*, **26**: 21-51.

Kluckhohn, C., and Griffith C., 1950, Population genetics and social anthropology. *Cold Spr. Harb. Symp. Quant. Biol.*, **15**: 401-408.

Lasker. G., 1946, Migration and physical differentiation: a comparison of immigrants with American-born Chinese. *Amer. J. Phys. Anthrop.*, 4 n.s.: 273-300.
  1952, Environmental growth factors and selective migration. *Hu. Biol.*, **24**: 262-289.

Li Chi, 1928, *The Formation of the Chinese People: An Anthropological Inquiry*. Cambridge, Harvard Univ. Press.

Linton, R., 1939, Marquesan culture. In *The Individual and His Society*, ed. by A. Kardiner: 137-196. New York, Columbia Univ. Press.

Lack, D., 1954, *The National Regulation of Animal Numbers*. Oxford, Clarendon Press.

Lorimer, F. and others, 1954, *Culture and Human Fertility*. UNESCO.

Manuila, A., 1956, Distribution of ABO genes in eastern Europe. *Amer. J. Phys. Anthrop.*, 14 n.s.: 577-588.

Meier, A., 1949, A study of the racial ancestry of the Mississippi college Negro. *Amer. J. Phys. Anthrop.*, 7 n.s. : 277-240.

Mourant, A. E., 1954, *The Distribution of the Human Blood-groups*. Oxford, Blackwell Scientific Publications.

Neel, J. V., Kodani, M., Brewer, R., and Anderson, R. C., 1949, The incidence of consanguineous matings in Japan. *Amer. J. Hu. Genet.*, 1: 156-178.

Oschinsky, L., 1957, Personal communication.

Shapiro, H. L. 1929, Descendents of the Mutineers of the Bounty, Memoirs, *B. P. Bishop Museum*, **11**: 3-106.
  1939. *Migration and Environment*, London, Oxford Univ. Press.
  1942, The anthropometry of Puka Puka. *Anthrop. Papers Amer. Mus. Nat. Hist.*, **38**: 141-169.
  1943, Physical differentiation in Polynesia. *Peabody Mus. Papers*, **20**: 3-8.

Simpson G. G., 1950, Some principles of historical biology bearing on human origins. *Cold Spr. Harb Symp. Quant. Biol.*, **15**: 55-66.

Spoehr, 1956, Cultural differences in the interpretation of natural resources. In *Man's Role in Changing the Face of the Earth:* 93-102. Chicago, Univ. of Chicago Press.

Stewart, O. C., 1956, Fire as the first great force employed by man. In *Man's Role in Changing the Face of the Earth:* 115-133. Chicago, Univ. Of Chicago Press.
Wittfogel, K. A., 1956, The hydraulic civilizations. In *Man's Role in Changing the Face of the Earth:* 162-164. Chicago, Univ. of Chicago Press.
United States Census Bureau, 1853, Report of the Seventh Census. Washington.

Discussion

L. D. SANGHVI, Human Variation Unit, Indian Cancer Research Center, Bombay, India: Dr. Hulse made a reference to the genetic diversity of the castes in India. We have some additional data on this question and we are finding a greater diversity than was reported by us earlier. At that time our material consisted of four endogamous groups belonging to the Marathi-speaking people in Western India. To this we have added six more endogamous groups speaking the same language and including some backward castes. We have also started studies on the tribal population in that area and have completed investigations on one. We hope to continue this work to cover some more tribal groups.

In these investigations, we have studied the A_1A_2BO, MN and Rh blood groups with subtypes, secretion of ABH substances in saliva, taste reaction to phenylthiocarbamide (PTC) and color blindness. Sickling tests were carried out on some of the castes and the tribal group. Information regarding consanguineous marriages was also recorded. The sample size for each group was about 200 individuals as in the earlier work.

The diversity in terms of phenotypic frequencies was something as follows. The blood groups A and B varied from 20 per cent to 35 per cent, subgroup A_2 making up anywhere from 5 per cent to almost 50 per cent of all A's. The blood group N showed a range from 11 per cent to about 30 per cent. The frequency of R_1R_1 phenotype varied from 31 per cent to 69 per cent. The secretion of ABH substances in saliva showed a range from 71 per cent to 88 per cent of secretors. Individuals who could taste PTC varied from 50 per cent to 65 per cent. Sickling test was positive in 17 per cent of the individuals of the tribal group, in one per cent of the backward caste and absent in the other three investigated.

F. S. HULSE: The further information concerning genetic characteristics of Marathi-speaking peoples, provided by Dr. Sanghvi, illustrate all too well the taxonomic difficulties faced by students of race in man. One can indeed speak more meaningfully of racial characteristics than of races as entities. Although the multiplicity of endogamous castes in India is extreme, it demonstrates beautifully the extent to which culture has affected human biology, which has been the main point of my paper. It creatures other than man, the percentage frequency of a certain allele may be, perhaps, taken as an index of its adaptive value, but as is demonstrated by Dr. Sanghvi's examples, we dare not make this assumption for man.

Selected Bibliography Including References Cited

PART II

Bilsborough, A. 1969. Rates of Evolutionary Change in the Hominid Dentition. *Nature*, 223: 146-149.

Blum, H. 1961. Does the Melanin Pigment of Skin Have Adaptive Value? *Quarterly Review of Biology*, 36: 50-63.

Brues, A. 1968. Mutation and Selection—A Quantitive Evaluation.*American Journal of Physical Anthropology*, 29: 437-439.

Brues, A. 1966. Probable Mutation Effect and the Evolution of Hominid Teeth and Jaws. *American Journal of Physical Anthropology*, 25: 169-170.

Campbell, B. 1963. Quantitative Taxonomy and Human Evolution. Pp. 50-74 in Washburn, S. (ed.). 1963. *Classification and Human Evolution*. Aldine, Chicago.

Clark, P., and Spuhler, J. 1959. Differential Fertility in Relation to Body Dimensions. *Human Biology*, 31: 121-137.

Dahlberg, A. 1963. Dental Evolution and Culture. *Human Biology*, 35: 237-249.

Damon, A., and Thomas, R. 1967. Fertility and Physique—Height, Weight and Ponderal Index. *Human Biology*, 37: 5-13.

Dobzhansky, T. 1963. Possibility that *Homo sapiens* Evolved Independently 5 Times is Vanishingly Small. *Current Anthropology*, 4: 360, 364-367.

Furusho, T. 1964. Relationship Between the Stature of Parents and the Mortality of Their Children. *Japanese Journal of Human Genetics*, 9 (1): 18-34.

Furusho, T. 1964. Relationship of Stature of Infertility Miscarriages and Fetal Deaths: A Preliminary Report. *Japanese Journal of Human Genetics*, 9 (2): 100-109.

Greene, D. 1970. Environmental Influences on Pleistocene Hominid Dental Evolution. *BioScience*, 20: 276-279.

Haldane, J. B. S. 1949. Suggestions as to the Quantitative Measurement of Rates of Evolution. *Evolution*, 3: 51-56.

Harrison, G., and Peel, J. (eds.). 1969. *Biosocial Aspects of Race*. Blackwell Scientific Publications, Ltd.

Hemmer, H. 1969. A New View of the Evolution of Man. *Current Anthropology*, 10: 179-180.

Hulse, Frederick. 1955. Technological Advance and Major Racial Stocks. *Human Biology*, 55: 184-192.

Hulse, F. 1957. Exogamie et Hétérosis. *Archives Suisses d'Anthropologie Generale*, 22: 103-125.

Hulse, F. 1967. Selection for Skin Color Among the Japanese. *American Journal of Physical Anthropology*, 27: 143-156.

Hunt, E., Jr., 1959. The Continuing Evolution of Modern Man. Pp. 245-254 in Woolridge, C. (ed.) 1959. *Genetics and Twentieth Century Darwinism. Cold Spring Harbor Symposia on Quantitative Biology Vol. 24.* The Biological Laboratory, Cold Spring Harbor, N.Y. 321 pp.

Keith, A. 1969. Theories Concerning the Evolution of Man's Posture. *Clin. Orthop.* 62: 5-14.

Kurten, B. 1959. Rates of Evolution in Fossil Mammals. *Cold Spring Harbor Symposia on Quantitiative Biology*, 24: 205-215.

Livingstone, F. 1969. The Founder Effect and Deleterious Genes. *American Journal of Physical Anthropology*, 30: 55-59.

Livingstone, F. 1969. Polygenic Models for the Evolution of Human Skin Color Differences. *Human Biology*, 41: 480-493.

Loomis, W. F. 1967. Skin-Pigment Regulation of Vitamin-D Biosynthesis in Man. *Science*, 157: 501-506. Letters with rejoinder, *Science*, 159: 652-653 (1968).

Montagu, A. 1963. What is Remarkable About Varieties of Man is Likenesses, Not Differences. *Current Anthropology*, 4: 361-364.

Napier, J. 1967. The Antiquity of Human Walking. *Sci. American*, 216: 56-66.

Post, R. 1969. Deformed Nasal Septa and Relaxed Selection: II. *Social Biology*, 16: 179-196.

Reed, T. E. 1969. Caucasian Genes in American Negroes. *Science*, 165: 762-768.

Schreider, E. 1964. Ecological Rules, Body-Heat Regulation, and Human Evolution. *Evolution*, 18: 1-9.

Schreider, E. 1967. Body-height and Inbreeding in France. *American Journal of Physical Anthropology*, 26: 1-4.

Shapiro, H. 1943. Physical Differentiation in Polynesia. *Harvard University Peabody Museum Papers*, 20: 3-8.

Stewart, T. D. 1960. A Physical Anthropologist's View of the Peopling of the New World. *Southwestern Journal of Anthropology*, 16: 259-273.

Turner, C., II. 1969. Microevolutionary Interpretations from the Dentition. *American Journal of Physical Anthropology*, 30: 421-426.

VanValen, L. 1963. Selection in Natural Populations: Human Fingerprints. *Nature*, 200: 1237-1238.

Weidenreich, F. 1945. The Brachycephalization of Recent Mankind. *Southwestern Journal of Anthropology*, 1: 1-54.

PART III

Natural Selection in Relation to Disease

The ecological shift from a hunting-fishing-food gathering society to a sedentary agricultural society that mankind began making several thousand years ago provided him with a much more stable food supply. There was a price associated with this ecological shift, however. Man's success in increasing and stabilizing his food supply by adopting agriculture made relatively dense human populations possible. As a result infectious diseases, facilitated by overcrowding, became one of the major agents of natural selection with respect to man.

Haldane has argued that from an evolutionary standpoint this was a misfortune. Haldane (1956-1957) states that:

"Infectious diseases have doubtless spread a great many previously rare genes through human populations. The fact that they were rare means that they mostly lowered the fitness of our ancestors. They probably lower our own, except in environments where the organism against which they confer resistance is fairly common. This is certainly true of the gene for sickle cell anemia."

In his paper, "Do Advances in Medicine Lead to Genetic Deterioration," which is included here in the section on the future genetic composition, Medawar argues that this is indeed the case. If the genes that confer resistance to infectious diseases predispose their carriers to constitutional or degenerative diseases such as anemia, ulcers, cancer, and arthritis, then the relaxation of selection via infectious diseases will result in evolutionary changes considered beneficial by man.

Man appears to be well on his way to gaining appreciable control over virtually all of the infectious diseases via a more adequate and varied human diet, better sanitation, and specific treatments such as vaccination. However the resulting evolutionary changes will not be considered to be beneficial if man loses his ability to control infectious diseases.

145

It is important to remember that infectious diseases have played a significant role in altering the course of history (cultural change) in addition to bringing about genetic change in man. Epidemics have decimated native populations in many parts of the world throughout history, thereby aiding invading colonists and armies and defeating armies on the verge of conquest (Zinsser, 1935; Stearn and Stearn, 1945; Dubos, 1965).

The papers on smallpox as a selective agent affecting the ABO blood groups and on malaria as a selective agent favoring sickle-cell hemoglobin suggest how infectious diseases have affected and are affecting the evolution of human populations.

A question that has continued to puzzle biologists is whether senescence—the degenerative phenomena (such as cancer and cardiovascular disorders) that accompany aging—is a direct product of natural selection. Several biologists have proposed theories to explain how natural selection could favor senescence (Williams, 1957; Hamilton, 1966; Guthrie, 1969). The paper by Nye discusses how natural selection might favor degenerative diseases affecting individuals who are largely past their reproductive peaks as safety mechanisms which "could ensure that a community did not become overburdened with relatively less productive individuals."

10. ABO Blood Groups and Smallpox in a Rural Population of West Bengal and Bihar (India)*

F. VOGEL and M. R. CHAKRAVARTTI

The problem was reexamined whether there is a relationship between the ABO blood groups and incidence, severity and outcome of smallpox. In persons suffering from smallpox, and living in rural areas of West Bengal and Bihar, India, severity and outcome of the disease were registered, and the ABO blood groups were determined. Healthy siblings were used as controls for incidence. Four series were available:

1. 200 suffering patients and 200 controls examined during summer 1965 in the Burdwan district, West Bengal.

2. 237 suffering patients and 228 controls examined during spring 1966 in West Bengal and Bihar. 415 of the cases contained in series 1 and 2 had never been vaccinated; the mortality was about 50%.

3. 402 survivors of a 1964 epidemic, who were examined during the 1965 survey in the Burdwan district, and 350 control siblings.

4. 147 survivors of the 1966 epidemic, who were examined during the 1966 survey in the same area.

Relative incidence of smallpox was shown to be much higher in suffering patients of groups A and AB, when compared with the healthy siblings (x = 6.09; χ^2 (DF =1) = 128.92). Within the patient group, severe forms and fatal outcome were significantly more frequent in groups A and AB as compared with groups B and O.

The surviving cases showed a much higher frequency of groups B and O than the fresh cases of the same area, indicating a much higher mortality of A and AB persons also in these series.

Within the group of survivors, severe smallpox scars were significantly more frequent in groups A and AB.

SOURCE: F. Vogel and M. Chakravartti, "ABO Blood Groups and Smallpox in a Rural Population of West Bengal and Bihar (India),"*Humangenetik,*3: 166-180 (Germany, 1966).

*This work was supported by WHO Grant Nr. G 3/181/15, by the Indian Statistical Institute, Calcutta, and the Deutsche Forschungsgemeinschaft.

The combined results of an examination of 986 smallpox patients and 778 controls gave conclusive evidence in favour of a strong disadvantage of groups A and AB towards the smallpox infection. However, as comparison with other series shows, the influence of the blood group seems to be confined to severe manifestations in unvaccinated persons who are living under natural, primitive conditions without medical care.

Some immunological implications of the result are discussed.

1. The Problem

Vogel, Pettenkofer and Helmbold (1960) discussed the hypothesis that smallpox could show a more severe course and a higher mortality in patients of blood groups A and AB as compared to groups B and O. This prediction was based on rabbit experiments of Pettenkofer and Bickerich (1960). Immunizing anti-A rabbits with egg-grown vaccina virus, these authors observed a definite increase of anti = A titre. As appropriate controls without virus did not render this antibody increase, it was concluded that vaccinia virus,—and by inference also variola virus,—contained antigen similar to the human A. In this case, a milder course of smallpox would be expected in patients of blood groups B and O, as their anti-A antibodies would react with the virus during the viraemia state.

This argument was challenged by Springer and Wiener (1962) who cited reports from the literature indicating that some egg material was found to contain A-like material. Reports on A-like (Forssmann) antigen in chicken material have been cited frequently by Springer. Experiments carried out by Harris, Harrison, and Rondle (1963) seemed to lend support to the opinion of Springer and Wiener that Pettenkofer's result might be due to the A antigen content of the egg material. Pettenkofer, on the other hand, showed in repeated experiments that the egg material used in this laboratory was free of A-antigen. This problem seems not to be settled so far, further experiments are urgently needed. The present report, however, is concerned with the statistical aspects of the problem only.

In the meantime, statistical results had begun to accumulate. Pettenkofer et al., 1962; (see also Vogel, 1961) presented data showing a much higher incidence of pathological vaccination reactions, primarily encephalitis, in persons with blood groups A and AB as compared to groups B and O, as expected on the basic hypothesis. In a series of Indians with smallpox scars, Helmbold found a much higher incidence of severe, confluent scars in groups A and AB, whereas mild, discrete scars were more frequent in group B and O. Series of fresh smallpox cases were examined by Helmbold (see Vogel, 1963) with relatively weak positive results, and by Downie et al. (1966) and Sukumaran et al. (1966) with negative results. These three series which had been examined in big city hospitals in India and Pakistan contained several hundred patients each, and among them many patients had been vaccinated some time in their life. Unfortunately, the exact number of vaccinated persons is only given in Downie's paper; here, it is very high. Mortality was relatively low in all three series. The results will be discussed later on in greater detail. A small series from Africa, examined by Garlick, which was claimed also to contain negative evidence, has never been published and hence, has to be omitted from any serious discussion.

Comparisons of groups of persons with smallpox scars with the population showed a certain increase of group B and O as compared with A and AB (Livingstone, see Vogel, 1963; Azevedo et al., 1964). This could point to a higher mortality in the last-mentioned groups. Bernhard (1966) looking for indications of possible selective effects of differential mortality due to smallpox in the population of India and Pakistan found a significantly negative rank correlation between frequency of of gene A and smallpox mortality. This result is in concordance with our basic hypothesis.

2. Material and Methods of our own Examinations

The experience with the more or less inadequate series published so far, as well as with their contradictory results has led us to the following consequences for the planning of further work:

a) As we were primarily interested in natural selection, the disease had to be studied under conditions, in which natural selection was still strongly at work. This means that the patients to be studied must have as little modern medical care as possible. By this condition, the big infectious diseases hospitals with their modern facilities were excluded. The examinations had to be confined to patients living at home or staying in small rural hospitals.

b) In order to avoid uncontrollable biases in the ascertainment of patients a great effort had to be made towards full ascertainment of all known patients suffering at a given time in a special region. This condition excluded the big cities, in which many patients are deliberately hidden, and confined our efforts to villages and little towns.

c) Previous vaccinations of the patients had carefully to be registered. Effects in unvaccinated patients were most interesting. It turned out that—with few exceptions—almost all patients examined in the villages and towns had never been vaccinated.

d) As the Indian population is highly stratified according to castes, religion and ethnic group, selection of suitable controls had to be considered very carefully. It was decided to match, if possible, every fresh patient with a sibling or—in rare cases—another near relative who lived together with him, but was not affected with smallpox. This matching tends to maximize the statistical consequence of any blood group effect on incidence.

e) As mortality is the most important factor for natural selection it had to be registered directly in the homes of the patients, and no secondary information could be relied upon. These considerations led to the following research plan:

First, the areas in which smallpox was going on, were ascertained in cooperation with the state Deputy Directors of Health, responsible for the smallpox registration in Calcutta (West Bengal) and Patna (Bihar). Then, the chief health officers of the affected districts were visited and a program for seeing the places with smallpox patients was designed. In India, a district is the largest subunit of a state. The health officers and health assistants in charge of the smallpox patients established the contacts with the families concerned and helped to overcome socio-psychological opposition against blood sampling. In India, the person suffering from smallpox is considered to be specially devoted to the mother goddess (Sitala), and must not be touched.

Whereas at the beginning, many families were reluctant to permit blood sampling from the patients, blood could finally be taken from all of them without any exception. This, however, is due to the fact that we only took some drops of blood from the finger tips. Venepuncture would have been refused in most cases. Hence, besides of the A_1A_2BO system, other polymorphisms could not be included in the program. In many of the severe cases, blood sampling was made difficult by the extreme exsiccation of the patients due to the heat (38–40°C), and the dry winds in combination with the high body temperature. It might be asked whether the high mortality in these areas could be lowered by sufficient water and elektrolyte intake.

The blood was taken up in test tubes with 3.8% sodium citrate and transported to the night quarters of the group, where the determinations were carried out according to standard technics. In 1965, commercial Anti-A, Anti-B, and Anti-AB sera were used; in 1966, Anti-A_1 and Anti-H sera were also available.[1]

At the time of the blood sampling, the patients were classified according to a 4 scale degree with regard to the virulence of the attack (T_1 = discrete; relatively mild disease; T_2 = confluent; relatively severe; T_3 = confluent; very severe; T_4 = haemorrhagic). Type T_4 was not observed; however, occasional local haemorrhagies were counted as T_3. Some weeks later the families were visited again in order to find out, whether the patient had survived his attack or not.

The 1964 and 1965 series were examined during spring and summer 1965 by M. R. Chakravartti, the 1966 series were examined during spring 1966 by F. Vogel together with M. R. Chakravartti.

The 1964 and 1965 series were examined in the Burdwan district, West Bengal. In 1966 the following districts of the State of Bihar were visited: Purnea; Darbhanga; Gaya; Patna; and Muzaffarpur which are situated in the Eastern part of

[1]We wish to thank the Dr. Molter GmbH., Heidelberg, for providing sera at a very much reduced charge.

India. In West Bengal, the districts of Bankura, Purulia, and Birbhur were examined (South Eastern part of West Bengal).

The population covered in this survey was mainly from three groups: a) High caste Hindus, b) low caste Hindus, and c) Muslims.

a) The high castes are mainly the Brahmins (both Radhi and Varendra). These are two endogamous groups with a number of exogamous clans or gotras. They held the highest rank in the Indian society. Persons of these castes were examined in the multi-caste villages. Most of the Brahmins were agriculturists or minor businessmen in the local market. Besides the Brahmins, some Vaidyas (the physician caste group) and some Kayasthas (the occupational and the traditional merchant caste groups) were also examined. People from these higher castes were mainly agriculturists and a little bit literate. According to V. Eickstedt and Guha they are classified as Mediterranean and Eastern Brachycephals with a lot of racial admixture. All these caste groups are basically endogamous and consist of exogamous gotras (clans) which are derived from some common ancestry.

b) More of the patients belonged to the low castes. Within our survey, they comprised the Sudras (Artisan class). There are two divisions among them: Some of them are touchables and others are untouchables, (Tál-chal and Tál-achal). Groups of them include the Bagdis, Kaibartas, Chamars, Tele, the fishermen-community, Barui, the betel-leaf community, Tanti, the weavers a.s.o. They are basically agriculturists besides their traditional and occupational practices. Most of them are illiterate and are ignorant of the modern life, hygienic conditions, and sanitations. They are distributed in multi-caste villages and sometimes live side by side with the high castes of those areas. The low-caste group show some anthropological characters which can be derived from the aboriginal population. They frequently have Australoid features. They, too, have some ancestric gotras.

c) The Muslim populations covered in the two states contributed an especially high number of cases to our series. Generally, the Muslims live along with the low-caste Hindus in the same villages, but their settlements will be situated together in a corner of the village. Basically, they are a trader-group, but many of them are working in agriculture. They are divided into two sects—the Siahs and the Sunnis. Long back in their history, they had been Hindus, but during the Muslim regime, they got converted into Muslims. They are most unhygienic, ignorant of the basic rules of sanitation, and prevention of epidemic diseases. It would be opposite to their customs to allow the vaccinators to touch their women. If one person in the village is affected, they generally do not take care of preventive measures like segregation. Frequently we found affected and healthy children not only in the same room, but also on the same bedstead, if there was any bedstead at all, and in other cases together on the clay floor. The Muslims of these regions have somewhat Australoid features, and most of them prior to their conversion were low Caste Hindus.

3. Results

The material available consists of 4 series:

Series No. 1: 200 fresh unvaccinated cases examined in 1965. For each of these cases, one healthy sibling was examined as control.[2]

Series No. 2: 237 fresh cases examined during spring 1966. Of these patients, 22 had been vaccinated before, the others were unvaccinated. 228 unaffected siblings or occasionally other near relatives were used as controls. For the remaining 9 cases, no suitable controls were available.

Series No. 3 contains 402 patients which had suffered from smallpox during 1964, but survived, and were examined during summer 1965, together with 350 controls.

Series No. 4 contains 147 surviving cases of the 1966 epidemic which were examined at the occasion of the fresh cases survey (Series No. 2).

A) Incidence in Fresh Cases

Table 1 shows the results of all 986 patients examined and of the 778 controls. In Table 2, relative incidence is analyzed for the fresh cases series (No. 1, No. 2) in comparison with their controls. x (A + AB = B + O) is very high, indeed, and fully significant according to any standards. x (A:O) is somewhat higher than x (A:B), in both series, but x (B:O) is not significant. In addition to the proven big disadvantage of the gene A bearers, there is the possibility of a much smaller advantage of O as compared with B persons, which, however, cannot be proven with the material available. As Table 1 shows, there is no important sex difference in the disadvantage of A and AB persons. The figures for blood groups A_2 and A_2B are too small to prove or disprove any difference against groups A_1 and A_1B. In series No. 2 vaccinated and unvaccinated patients can be compared. The over-all number of vaccinated persons however is very low, and no conclusions can be drawn.

B) Severity of the Disease in Fresh Cases

Table 3 shows the data as to severity of the disease. There is a strong and significant tendency towards more severe forms in group A and AB as compared with B and O. In spite of some apparent differences in classification between 1965 and 1966—series 1 shows more discrete (T_1)-cases—the blood group correlation is in the same order of magnitude in both series.

Additional comparisons were not carried out, as the material is too small.

---

[2] A preliminary report on this series has been published by Chakravartti et al. (1966).

Table 1. Blood Group Distribution of Smallpox Patients and Controls

| | | | | Blood Group | | | | | |
| | | | | A | | | | AB | |
Series No.:	Pat./ Contr.	♂/♀	N	A$_1$	A$_2$	B	O	A$_1$B	A$_2$B
1	Pat.	♂	84	43		23	16	2	
(1965, fresh)		♀	116	63		25	21	7	
	Contr.	♂	84	17		26	38	3	
		♀	116	21		40	46	9	
2a	Pat.	♂	135	57	7	32	11	25	3
(1966, fresh patients		♀	80	26	11	14	12	15	2
unvaccinated)	Contr.	♂	129	7	2	69	45	6	0
		♀	78	7	3	32	31	4	1
2b	Pat.	♂	10	4	0	3	3	0	0
(1966, fresh patients		♀	12	5	0	4	3	0	0
vaccinated)	Contr.	♂	12	3	0	4	4	1	0
		♀	9	2	1	3	3	0	0
3	Pat.	♂	216	54		90	62	10	
(1964, survived)		♀	186	59		79	40	8	
	Contr.	♂	212	79		72	46	15	
		♀	138	52		39	42	5	
4	Pat.	♂	80	22	1	31	18	7	1
(1966, survived)		♀	67	10	8	25	16	8	0

C) Mortality in Series 1 and 2

Table 4 shows the mortality data of series 1 and 2. Again, there is a strong and fully significant deviation in the same direction: Mortality in groups A and AB is much higher than in groups B and O. This deviation is also borne out, when A is compared with O or B. On the other hand, there is no indication of any difference between groups B and O. The data are too small for a decision whether a somewhat lower mortality of group A$_2$ as compared with A$_1$ is a chance result only, or whether it reflects a real difference. Remarkable, though only based on a very small number, is that 8 of 9 vaccinated patients of group A$_1$ died, whereas all 13 vaccinated patients of groups B and O survived.

The most important side-result, however, is the very high mortality figure itself. Of the 415 unvaccinated patients of both series no less than 217 (52.3%) died. This could be due to an especially severe epidemic. More probably, however, the usual

Table 2. Relative Incidences of Smallpox

Comparison Series No. 1	x	χ^2 (DF = 1)	P Series No. 2	x	χ^2 (DF = 1)	P
A + AB : B + O	4.059	41.642	$\approx 10^{-10}$	9.758	94.844	$< 10^{-10}$
A : O	6.333	45.617	$< 10^{-10}$	12.59	66.744	$< 10^{-10}$
A : B	3.836	25.191	$\approx 10^{-6}$	8.966	62.316	$< 10^{-10}$
B : O	1.297	0.900	0.32	1.405	1.549	0.25

Comparison Series No. 1 + 2	x	χ^2 (DF = 1)	P	χ^2 of Heterogeneity	P of Heterogeneity
A + AB : B + O	6.09	128.92	$< 10^{-10}$	6.09	0.015
A : O	8.534	109.618	$< 10^{-10}$	2.743	0.1
A : B	5.774	82.671	$< 10^{-10}$	4.836	0.03
B : O	1.35	2.406	0.14	0.043	0.83

Table 3a. Severity of the Disease

Series No.:	Category	N	A — A_1	A — A_2	B	O	AB — A_1B	AB — A_2B
1 (1965, fresh)	T_1	95	40		30	23	2	
	T_2	94	58		17	14	5	
	T_3	11	8		1	0	2	
2a (1966, fresh, unvaccinated)	T_1	37	10	4	17	2	4	0
	T_2	121	47	13	23	17	19	2
	T_3	57	26	1	6	4	17	3
2b (1966, fresh, vaccinated)	T_1	5	2	0	3	0	0	0
	T_2	10	3	0	4	3	0	0
	T_3	7	4	0	0	3	0	0

statistics which are based on material from big hospitals and contain many vaccinated cases (for references see Herrlich, 1960) do not show the full "natural" mortality of the disease, either due to beneficial effects of medical care provided in these hospitals, or due to some biases in their sample of patients. Our results seem to confirm a conclusion which Herrlich (1960) derived from the age and sex distribution of the mortality that the mortality, of unvaccinated cases without more elaborate medical care is about 50% in India.

Table 3b. Relative Incidence of Severe $(T_2 + T_3)$ vs. Mild (T_1) Cases, Series 1 and 2 (Fresh Cases)

Comparison	Series 1 (1965)			Series 2 (1966)		
	x	χ^2 (DF = 1)	P	x	χ^2 (DF = 1)	P
A + AB : B + O	2.879	13.478	0.0003	2.475	6.872	0.01

Comparison	Series 1 + 2			χ^2 of Heterogeneity	P of Heterogeneity
	x	χ^2 (DF = 1)	P		
A + AB : B + O	2.70	19.52	10^{-5}	0.83	0.33

Table 4a. Smallpox Mortality

Series No.:	Mortality	N (% died)	Blood Group					
			A		B	O	AB	
			A_1	A_2			A_1B	A_2B
1 (1965, fresh)	survived	97 (48.5%)		43	30	22	2	
	died	103 (51.5%)		63	18	15	7	
2a (1966, fresh; unvaccinated)	survived	116 (54.0%)	29	13	36	19	17	2
	died	99 (46.0%)	54	5	10	4	23	3
2b (1966, fresh; vaccinated)	survived	14 (63.6%	1	0	7	6	0	0
	died	8 (36.4%)	8	0	0	0	0	0

D) Series 3 and 4 (Surviving Cases)

Whereas the most direct information comes from the series of fresh cases, some additional conclusions can also be drawn from series of surviving cases. Their smallpox scars can be used for classification of mild and severe attacks.

Besides, their over-all blood group distribution can be compared with the distribution of fresh cases in the same area: If the surviving cases are the remainder of

Table 4b. Relative Incidence

Comparison	Series 1			Series 2		
	X	χ^2 (DF = 1)	P	X	χ^2 (DF = 1)	P
A + AB : B + O	2.451	9.343	0.002	7.286	34.899	$\approx 5 \cdot 10^{-9}$
A : O	2.149	3.869	0.05	9.738	15.771	≈ 0.0001
A : B	2.442	6.799	0.01	6.700	22.412	$\approx 2 \cdot 10^{-6}$
B : O	0.880	0.077	0.76	1.453	0.338	≈ 0.7

Comparison	Series 1 + 2				
	X	χ^2 (DF = 1)	P	χ^2 of Heterogeneity	P
A + AB : B + O	3.975	38.655	$\approx 10^{-9}$	5.587	0.015
A : O	3.46	18.473	$\approx 2 \cdot 10^{-5}$	1.167	0.25
A : B	3.902	25.965	$\approx 6 \cdot 10^{-7}$	3.246	0.08
B : O	1.043	0.0122	≈ 0.9	0.403	0.55

higher mortality in groups A and AB, they are expected to show a "negative picture": Groups B and O will be more frequent than among the fresh cases.

Two series are available (Table 1): Series No. 3 contains patients which had suffered from smallpox during 1964, but survived, and were examined during summer 1965. Classification as to severity was done from their smallpox scars. These patients came from the same area as series No. 1. During the 1964 epidemic, 749 attacks had been registrated, 347 (46.3%) of which ended fatal. The 402 surviving patients could all be examined. 350 siblings were used as controls.

Series No. 4 contains 147 surviving cases of the 1966 epidemic which were examined at the occasion of the fresh cases survey (series No. 2). No effort was made as to completeness and no controls were examined.

Table 5 shows relative incidences of fresh as compared with surviving cases from the same areas. Series 1 is compared with series 3, and series 2 is compared with series 4. A strong prevalence of A and AB in fresh cases emerges, indicating again a higher maotality in A and AB.

In Table 6, the surviving patients are classified according to the severity of the disease as infered from the extent of their smallpox scars. Again, there is a striking and significant deviation in the direction expected. The relatively high heterogeneity between the two series might very well be due to the somewhat different classification in 1965 and 1966, which has been mentioned above.

Now, the surviving cases of series 3 will be compared with their controls (Table 7). The comparisons are carried out in the opposite direction, the main comparison being (B + O); (A + AB). There is a moderate, but significant increase of x, indicating a somewhat higher incidence of group B and O in the surviving patients as

Table 5. Blood Group Distribution in Fresh Cases as Compared
with Surviving Cases of the Same Areas.
Relative Incidences

Comparison	Series 1 vs. 3			Series 2 vs. 4		
	X	χ^2 (DF = 1)	P	X	χ^2 (DF = 1)	P
A + AB : B + O	2.799	33.334	$\approx 10^{-8}$	2.985	25.278	$\approx 8 \cdot 10^{-7}$
A : O	2.586	16.378	$\approx 8 \cdot 10^{-5}$	3.146	13.488	$\approx 5 \cdot 10^{-4}$
A : B	3.303	31.698	$\approx 2 \cdot 10^{-8}$	2.835	15.464	$\approx 2 \cdot 10^{-4}$
B : O	0.783	0.940	≈ 0.32	1.110	0.108	≈ 0.75

Comparison	Combined Comparisons			χ^2 of Heterogeneity	P of Heterogeneity
	X	χ^2 (DF = 1)	P		
A + AB : B + O	2.872	58.588	$< 10^{-10}$	0.054	≈ 0.82
A : O	2.778	29.613	$\approx 6 \cdot 10^{-8}$	0.253	≈ 0.7
A : B	3.111	46.955	$< 10^{-10}$	0.207	≈ 0.75
B : O	0.897	0.058	≈ 0.82	0.990	≈ 0.35

compared with the controls. This means, that in this series the effect of differential
mortality seems to have been stronger than the effect of differential incidence.

4. Dicussion

Statistical analysis of the ABO distribution in the four series examined has confirmed our prediction that persons with blood group A and AB have a disadvantage
when exposed to smallpox. This disadvantage has two main components:

1. As relative incidence figures show, these individuals run a much higher risk to
acquire the disease when exposed to it. According to current opinion (Herrlich,
1960) virtually all humans can be infected, provided that they are sufficiently exposed. Evidence in favour of this opinion comes from populations infected for the
first time. It has to be taken in mind, however, that the population examined by us
has been living for uncounted centuries in an endemic smallpox centre for which
annual outbreaks, especially during spring, are characteristic. Usually, these outbreaks will start from different places in different years. It is one of the most striking experiences for everybody who works in affected places, that always only a
minority of the population catches smallpox. Even if some hundred outbreaks are
observed, the majority of the population remains unaffected in spite of the fact
that many persons have immediate and repeated contact with suffering patients,

Table 6. Surviving Cases (Series 3 and 4); Classification According
to Severity of the Disease

Series	Category	N	A A_1	A_2	B	O	AB A_1B	A_2B
3 (1964, survived)	T_1	215	34		101	74	6	
	T_2	187	79		68	28	12	
	T_3	0	0		0	0	0	
4* (1966, survived)	T_1	65	13	2	26	18	5	1
	T_2	69	19	7	22	14	7	0
	T_3	12	0	0	7	2	3	0

*This table only contains 146 cases. One case was erroneously not classified.

Relative Incidences $T_2 + T_3$ vs. T_1

	Series 3			Series 4		
Comparison	X	χ^2 (DF = 1)	P	X	χ^2 (DF = 1)	P
A + AB : B + O	4.147	38.811	$\approx 10^{-9}$	1.676	2.216	0.35

	Combination of Series 3 and 4				
Comparison	X	χ^2 (DF = 1)	P	χ^2 of Heterogeneity	P of Heterogeneity
A + AB : B + O	3.154	36.267	$\approx 5 \cdot 10^{-9}$	4.760	0.03

Table 7. Blood Group Distribution in Surviving Patients
vs. Controls (Series 3)

Comparison	X	χ^2 (DF = 1)	P
B + O : A + AB	1.570	8.803	0.0027
O : A	1.344	2.405	0.12
B : A	1.765	10.28	0.0015
O : B	0.761	2.067	0.16

and that they have never been vaccinated. There can be no doubt any more, that among the factors which diminish the smallpox risk for these persons and under these conditions, blood groups B and O are of major importance.

2. Even when they do suffer from smallpox, however, persons of groups B and O have a better chance to get a relatively mild attack, and above all, to survive. This conclusion is supported by evidence from two different sources, a direct and an indrect one:

Fresh cases were shown to survive much more frequently when belonging to groups B and O, and series of surviving cases show more persons with groups B and O, than series of the fresh cases from the same area.

The result of our examinations seems to differ from the results recently published by three groups:

1. Helmbold et al. (see Vogel, 1963) examined 599 smallpox patients from the Infectious Diseases Hospital in Karachi (Pakistan). When severe and mild cases were compared, there was a certain deviation in the direction expected, which, however, was relatively small and not significant. x (A + AB : B + O) = $1.248; \chi^2$ = 1.527). In this material, blood group B seemed to behave like A, not like O, and x (A + B + AB : O) was $1.400; \chi^2$ = 4,086. Mortality, however, was low (54 = 9.02%), and Dr. Helmbold has recently informed us that the material contained many vaccinated persons. Unfortunately, the exact number of prevaccinations could not be traced any more.

2. Downie et al. (1966) examined a series of 330 smallpox patients, successive hospital admissions of the Infectious Diseases Hospital in Madras, India, during a part of the year 1963. First they compared the blood group distribution of these patients with a local control series and did not find any difference. Then, they classified their patients according to an elaborate system into six severity scales. They failed to find any difference in the blood group distribution. Mortality (29 deaths = 8.79%) was low and did not depend on the blood group. The great advantage of this series is, however, that the number of prevaccinated patients was recorded. 61 patients only did not show vaccination scars, 54 of which belonged to groups O and B!

3. The third comprehensive series of fresh cases available to date is that of Sukumaran et al. (1966). 461 cases of the Kasturba Hospital, Bombay, were classified according to severity using a 4-scale system. No relationship with the blood group could be established. In 401 cases, the outcome was recorded, and it proved to be fatal in 67 (16.71%).

Again, there was no relationship with the blood group. Neither was there any correlation in a series of 296 severe cases who required blood transfusions and of whom 45 only survived. The overall blood group distribution among the patients seemed to be similar to series from the general population in different parts of India, including Bombay.

The authors mention that part of their patients had been vaccinated some time, but we were informed (personal communication of Dr. Sanghvi), that exact figures as to the percentage of vaccinations are not available.

The question arises why the series mentioned rendered a more or less negative result, whereas in our own material, the deviation found was so strikingly positive. At present, this question cannot be answered definitely, but four main differences between these three series on the one hand and our material on the other hand should be kept in mind:

1. All other series come from big hospitals. However, the patients admitted to these hospitals comprise a sample only of the whole population of smallpox patients. As to the sampling process, we can only be sure of one thing: that it is not at random. Importance and direction of the biases involved can hardly be estimated. Our patients, on the other hand, were collected during an attempt towards full ascertainment of all suffering cases at a given time in a special area.

A second disadvantage of the hospital series is that the outcome might be influenced by therapy in spite of the fact that specific chemotherapy is not available.

2. As to relative incidence, blood group distribution among the patients was compared by the other authors with usual control series of the same areas. This method works fairly well in European or US American populations and in problems like relationship of blood group and cancer.

Very early, however, some critics pointed to possible biases due to stratification in the population concerned.

In order to avoid these biases, it was recommended to use healthy siblings as controls (Clarke and Sheppard, 1957). In India, which has a highly stratified population, this source of error has to be considered very carefully. Only one possible bias might be mentioned here: Smallpox patients are mainly coming from endemic areas. If there is a selective effect of smallpox on the frequency of gene A, these areas are expected to have low A frequencies. Besides, it has recently been demonstrated by Bernhard (1966) that this is actually true: there is a significant negative rank correlation between frequency of gene A and prevalence of smallpox in India and Pakistan.

Now, if incidence is higher in group A, but the patients are mainly coming from areas with generally very low A frequency, whereas the controls represent the population average, these two effects might cancel out, giving a spuriously negative net result.

The method to take the severely exposed, but healthy siblings as controls, as we did, provides optimal conditions for detection of a blood-group specific difference, while biases due to stratification are avoided efficiently.

3. There is a third important difference between the series mentioned and ours: These series contain a good proportion of persons who had been vaccinated sometimes before in their lives. With exception of 22 persons, all our fresh cases, and

also most of our surviving cases, had never before been vaccinated. In Downie's series, on the other hand, most of the patients were recorded to be prevaccinated. Unfortunately, in Helmbold's and Sukumaran's series, the exact number of prevaccinated patients can not be traced any more, but both authors mention an admixture of vaccinated persons.

4. The fourth difference is possibly connected with the third one: mortality is much lower in these series than in ours. In two series, the mortality figure is lower than 10%, in the third one it is about 16%. In our material, on the other hand, about 50% ended fatally. It must be asked, whether this difference is caused by the prevaccinations in the other series—possibly in connection with hospital care—or whether the virulence of the epidemics was different. Generally, differences in the virulence of smallpox epidemics are well known (Herrlich, 1960).

In order to examine this problem, we had decided also to include vaccinated patients in our 1966 survey. Unfortunately, the number of cases available (22) is too small for any definite conclusion. However, of the nine vaccinated cases with blood group A_1, no less than 8 died, whereas all 13 vaccinated persons of group B and O survived (Table 4a). It can at least be said that this result does not point to vaccination as the main factor which removes the blood group effect on mortality. However, vaccination of our patients might have been carried out long back in the past Whichever the exact physiological reason might be, the strong relationship between blood groups A and AB and incidence, severity and outcome of smallpox, as demonstrated in the patients by our group, seems to be confined to severe epidemics with high mortality in unvaccinated populations, living under natural conditions.

Differences pointing into the same directions were also seen, when series of surviving cases were examined according to severity of scaring. One series of 378 former patients was examined by Helmbold in India 1961.[3] There was a definite and significant deviation in the direction expected: Persons with blood groups A and AB had significantly more severe, confluent scars, whereas discrete, mild scaring was more frequent in groups B and O.

The persons investigated were mainly industrial workers from Nagpur. Material which has recently become available to Helmbold (personal communication), and which has been collected in Poona among soldiers, did not show the deviation. In this connection, examinations have to be mentioned shortly, in which the over-all blood group distribution of a series of pox-scared patients is compared with the normal population distribution. Three comparisons are available.

1. Our comparison of patients and controls in series No. 3 (Table 7), which give a significantly higher incidence of groups B and O among the patients.

2. A small series of Livingstone from Africa (see Vogel, 1963) which shows the same deviation, and

[3]The result in 358 of these persons has already been mentioned in Pettenkofer et al. (1962), Vogel (1961, 1963).

3. A widely quoted series of Azevedo et. al. (1964) which showed a (small) deviation in the same direction. Apparently, Azevedo et al. had carried out their examination under the wrong assumption that we had done the same and reported in our 1962 paper (Pettenkofer et al.). Apparently, they did not realize that we had only compared severe and mild scars within a group of poxscared persons. They arrived at the widely quoted conclusion that their data could serve to disprove our predictions. On the other hand, we concluded (Vogel, 1963) that a higher incidence of B and O among scared persons had to be expected, as they are the remainder of selection. From our present investigation, however, it turns out, that no prediction whatsoever is possible as to blood groups frequencies among surviving patients. These frequencies are influenced by two factors working in opposite directions: The higher risk of A and AB to be manifestly infected tends to increase the number of A and AB cases among the survivers, whereas the high mortality of A and AB tends to decrease this frequency. Hence, comparisons between persons with scars and their basic population are generally of little value; they can only serve to answer the question whether the effect of the blood group on incidence or its effect on mortality is stronger. This, however, might depend on accidental factors.

Returning to the conclusion that the blood group influence seems to be confined to severe epidemics in unvaccinated people, there is an interesting result on blood group and vaccination which points into the same direction. In severe, pathological vaccination reactions a strong influence of the blood group was demonstrated: Groups A and AB were found to be more than twice as frequent, as one would expect if the blood group would have no influence (Pettenkofer et al., 1962). The 'normal' local vaccination reaction, on the other hand, seems not to be influenced by the blood group. This was shown in painstaking examinations by Bourke et al. (1965). Besides of the success of the vaccinations, many other parameters were examined, among them scar size, sickness, headache, body temperature, pains, sore throat, a.s.o. For all of them the result was negative—with one interesting exception: There was a significantly higher incidence of pain under the unvaccinated arm in groups A and AB, possibly indicating a stronger lymphohaematogenic dispersion of the virus.

If the 'normal' and pathological vaccination reactions are compared, it turns out again that the extreme, severely pathological situation (encephalitis) is strongly influenced by the blood group, whereas the usual, local, mild reaction is not.

The present paper is concerned with statistical evidence only, not with the immunological implications of the correlations demonstrated. However, some tentative conclusions as to the possible immunological mechanisms might be drawn. According to current views (vgl. Herrlich, 1960), infection occurs via the respiratory system, and probably the lungs. A short time later, during a first viraemia, the virus is scattered over the reticuloendothelial system. There, during the incubation period, virus multiplication takes place. A second viraemia marks the end of the incubation period and the beginning of the "initial state". The second viraemia is much more massive than the first one; skin, mucous membranes and organs are affected. With the beginning of the pustulous eruption, viraemia is terminated in mild and moderate cases. In severe cases, however, and especially in fatal ones, a prolonged viraemia was observed (Downie, quoted from Herrlich).

The ABO blood groups seem to influence probability of manifest infection as well as severity of the general and local manifestation and the outcome in severe epidemics. This could most easily be explained, if a humoral factor would be involved. The immunological response of the organism has two main components: One humoral and one localized in the cells and tissues. The normal local vaccination reaction, for example, depends (almost) exclusively on the tissue component and is independent from the humoral component. This is demonstrated convincingly by the fact that even persons with agammaglobulinaemia show a normal vaccination reaction (Barandun et al., 1959).

On the other hand, there can be no reasonable doubt that intensity and persistence of the viraemia depends on the humoral component. If for the fatal outcome of smallpox as well as for severe complications after vaccination, the extent and persistence of viraemia are of some importance, the influence of the blood group especially on these extreme situations could most easily be explained by the assumption that the humoral factor of immunity is influenced by the blood group whereas the local tissue factor is not.

This hypothesis would also explain the higher incidence in group A and AB in a severely exposed population: Infections with relatively few virus particles could easier be overcome completely during the first viraemia, if there would be a humoral factor attacking the virus.

Now, in our initial hypothesis, the serum anti-A, which is present in groups B and O, was postulated to influence the smallpox infection. Hence, this hypothesis would fit excellently into the picture emerging now from statistical evidence. It would also explain that possibly the blood group influence is not seen in prevaccinated persons: Small amounts of neutralizing antibodies were found 10–20 years after the first vaccination (Downie, quoted from Herrlich, 1960). It would not be surprising, if the blood group antibodies would only have an influence in absence of any other specific antibodies.

Any attempts to correlate the blood group effect with the tissue component of immunological response, on the other side, run into unsurmountable difficulties due to the proven independence of the local vaccination response from the blood group.

In the light of these considerations, new experiments seem to be urgently needed to test the problem whether Pettenkofer's conclusion is correct, that the virus contains an A-like antigen. An obvious approach would be to grow the virus on human placenta tissue cultures of defined blood groups, as there would be no discussion as to possible A-antigens in the culture medium. During 1965, these experiments were started in Heidelberg. Unfortunately, the virologist in charge (K. Bingel) died suddenly before finishing the experiments.

Some remarks might be appropriate from the point of view of population genetics: A selective disadvantage of gene A which is as strong as the one demonstrated, must have influenced gene frequencies in a population in which smallpox has been endemic for a long time. It might even be asked why there is still an appreciable fraction of A genes in this population. The obvious answer is that other selective forces might have counteracted smallpox. As we are now aware of blood group correlations of many other diseases, among them also some infections, and as there are other selective factors in blood groups, it has become increasingly difficult to separate their effects and to explain actual gene frequencies in special populations. However, as was already mentioned, Bernhard (1966) has recently collected literature data about ABO gene frequencies in various parts of India and Pakistan, which were based on almost 100.000 blood samples, and compared them with regional frequencies of smallpox mortality reported in the "Welt-Seuchen-Atlas, Bd. 3". There was a strong and significant (ϱ = —0.643, p < 0.01) negative rank correlation between frequency of gene A and overall smallpox mortality, as expected from our data.

The authors have to thank Prof. B. P. Adhikari for his kind advice. Dr. P. Bose, Deputy Director of Health Services, West Bengal, Calcutta, and Dr. S. C. Sinha, Deputy Director of Health Services, Bihar, Patna, as well as the District Health officers have given invaluable support to our investigations. The help of Mr. Gourdas as cardriver. Mr. Devrat, as cook, and Mr. Manik Roy, as boy, was very useful during the field work.

References

Azevedo, E., H. Krieger, and N. E. Morton: Smallpox and the ABO blood groups in Brazil. *Amer. J. hum. Genet.* 16, 451-454 (1964).

Barandun, S., H. Cottier, A. Hassig and G. Riva: Das Antikörpermangelsyndrom. *Helv. med. Acta* 26, Suppl. (1959).

Bernhard, W.: Über die Beziehung zwischen ABO-Blutgruppen und Pockensterblichkeit in Indien und Pakistan. *Homo* 17, 111-118 (1966).

Bourke, G. J., and N. Clarke: Smallpox vaccination and Serum Anti-A levels. *Acta genet.* (Basel) 11, 123-456 (1964).

———— ———— and B. H. Thornton: Smallpox vaccination: ABO and rhesus blood groups. *J. med. Genet.* 9, 93-156 (1965).

Clarke, C. A., and P. M. Sheppard: The ABO blood groups in duodenal ulcer. A study of sibships. *Acta genet.* (Basel) 6, 570-574 (1957).

Chakravartti, M. R., B. K. Verma, T. V. Hanurav, and F. Vogel: Relation between smallpox and the ABO blood groups in a rural population of West Bengal. *Humangenetik* 2, 78-80 (1966).

Downie, A. W., G. Meiklejohn, L. St. Vincent, A. R. Rao, B. V. Sundaha Babu, and C. H. Kempe: Smallpox frequency and severity in relation to A, B, and O blood groups. *Bull. Wld Hlth Org.* **33**, 623 (1965).

Eickstedt, E. V.: *Rassenkunde und Rassengeschichte der Menschheit.* Stuttgart: Enke 1934.

Harris, R., G. A. Harrison, and C. J. M. Rondle: Vaccinia and human blood group A substance. *Acta genet.* (Basel) **13**, 44-57 (1963).

Helmbold, W.: personal communication.

Herrlich, A., and A. Mayr: Die Pocken. Erreger, Epidemiologie und klinisches Bild. Stuttgart: Thieme 1960.

Pettenkofer, H. J., and R. Bickerrich: Über Antigen-Gemeinschaften zwischen menschlichen Blutgruppen ABO und den Erregern gemeingefährlicher Krankheiten. *Zbl. Bakt., I. Abt. Orig.* **179**, 433 (1960).

_____ B. Stöss, W. Helmbold, and F. Vogel: Alleged causes of the present-day world distribution of the human ABO blood groups. *Nature* (Lond.) **193**, 444-446 (1962).

Springer, G. F., and A. S. Wiener: Alleged causes of the present-day world distribution of the human ABO blood groups. *Nature* (Lond.) **193**, 444-446 (1962).

Sukumaran, P. K., H. R. Master, J. V. Undevia, V. Balakrishnan, and L. D. Sanghvi: ABO blood groups in active cases of smallpox. *Ind. J. med. Sci.* **20**, 119-122 (1966).

Vogel, F.: Zur Theorie der natürlichen Auslese im ABO-Blutgruppen-System. *Proc. Sec. Int. Congr. Human Genetics, Roma 1961.*

_____ Neuere Untersuchungen zur Populationsgentik der ABO-Blutgruppen, Homo-Bericht über die 8. Tagung der Dtsch. Ges. f. Anthropologie 1963, S. 143-161.

_____ Blood groups and natural selection. *Proc. 10th Congr. Int. Soc. Blood. Transf., Stockholm 1964,* p. 268-279.

_____ H. J. Pettenkofer and W. Helmbold: Über die Populationsgenetik der ABO-Blutgruppen. 2. Mittlg.: Genhäufigkeit und epidemische Erkrankungen. *Acta genet.* (Basel) **10**, 267-294 (1960).

11. Polymorphism and
Natural Selection in
Human Populations

A. C. ALLISON

I should like to begin by considering the abnormal hemoglobins, which provide much the best evidence of polymorphism and natural selection in man. At the Cold Spring Harbor Symposium in 1955 I put forward the following propositions:

(1) The homozygous sickle-cell condition is virtually lethal in Africa. In view of the high frequencies of the sickle-cell gene in many regions, the rate of elimination of the gene could not be compensated by recurrent mutation.

(2) Balanced polymorphism has resulted because the sickle-cell heterozgote is at an advantage, mainly as a consequence of protection against falciparum malaria.

(3) Malaria exerts its selective effect mainly through differential viability of subjects with and without the sickle-cell gene between birth and reproductive age, and to a much lesser extent through differential fertility.

(4) High frequencies of the sickle-cell gene are found only in regions where falciparum malaria is, or was until recently, endemic.

(5) In most New World Negro populations, frequencies of the sickle-cell gene are lower than would be expected from dilution of the African gene pool by racial admixture. This is probably the result of elimination of sickle-cell genes without counterbalancing heterozygous advantage.

(6) In regions where two genes for abnormal hemoglobins co-exist, and interact in such a way that individuals possessing both genes are at a disadvantage (such as the Hb_β^S and Hb_β^C or Hb_β^T), then as a result of selection these genes will tend to be mutually exclusive in populations.

In the nine years that have elapsed most of these statements have been challenged, but enough evidence has now accumulated to establish beyond reasonable doubt that, insofar as these statements are testable, they are substantially correct. The evidence will be summarized briefly and somewhat dogmatically; a more

SOURCE: A. Allison, "Polymorphism and Natural Selection in Human Populations," *Cold Spring Harbor Symposia on Quantitative Biology*, Vol. XXIX (New York, 1964), pp. 137-149.

extended discussion with full documentation is to be found in a book on *Polymorphism in Man* (Allison and Blumberg, 1965).

Much recent information on abnormal hemoglobins has been summarized at a meeting held earlier this year at Ibadan, Nigeria, under the auspices of the Council for International Organizations of Medical Sciences, the proceedings of which will shortly be published.

Lethality of the Homozygous Sickle-Cell Condition in Africa

It is now generally agreed by workers in Africa that sickle-cell disease is common in childhood, occuring at about the expected frequency of homozygotes in the populations concerned (see Vandepitte and Stijns, 1964; Watson-Williams and Weatherall, 1964). The data do not support the suggestion of Nance (1963) that the incidence of sickle-cell disease might be much less than expected from the heterozygote frequency because of $Hb_\beta{}^{AS}$, duplications. Although such duplications may occur, they must be uncommon, as the absence of Hb-A in many studied cases of genotype $Hb_\beta{}^S/Hb_\beta{}^C$ is sufficient to illustrate.

There is also agreement that most African sickle-cell homozygotes die in childhood (see Vandepitte and Stijns, 1964). A few subjects with S + F hemoglobin patterns (mainly genetic variants of sickle-cell disease) survive to adulthood and reproduce with difficulty (see Fullerton, Hendrickse and Watson-Williams, 1964). Nevertheless, it can be said that the fitness of the sickle-cell homozygote under African conditions is close to zero. Even the upper limits of mutation at the Hb_β locus estimated by Vandepitte et al. (1955) and Frota-Pessoa and Wajintal (1963)—which are certainly in excess of the true figure—are far too low to replace the loss of sickle-cell genes from populations.

Malaria in Sickle-Cell Heterozygotes

The considerable amount of work that has been carried out on this subject has been widely misinterpreted, as the quotations in some textbooks (e. g., Harris, 1959; Wagner and Mitchell, 1964) will show. At the beginning of the investigation (Allison, 1954a) it was pointed out that only results on young children have any validity because of the powerful effects of acquired immunity in children of school age and adults. Before this point was appreciated, attempts to confirm the observations on schoolchildren and adult populations failed, and these results have been passed from review to review and so fossilized in the literature. Eventually the point sank in, and series of observations on the right age groups were made, which tell the story plainly enough for all to see.

In Table 1 all the published observations that have been made on malaria in susceptible African children with and without the sickle-cell trait are collected together. The only data that have been omitted are those in older children where the authors themselves give reasons for concluding that acquired immunity is having a powerful effect (i. e., in Nigerian children and children from Northern Ghana over five years of age). Where individual groups in the same investigation were small (groups A and B of Foy, Brass, Moore, Timms, Kondi, and Olouch, 1955; three groups of Walters and Chwatt, 1956; and two groups of preschool children of Garlick, 1960), the figures have been pooled, which simplifies the analysis but does not affect the overall result. The observations have been analyzed by the method of Woolf (1955) which allows comparisons to be made between populations with different gene frequencies and attack rates.

In column 7 of Table 1 the relative incidence of *P. falciparum* infections in children with and without the sickling trait is expressed relative to an incidence of unity in the sickling children. It will be clear that the relative incidence is always above unity, the weighted mean being 1.46. The next column but one has the χ^2 values of the differences, which are significant at the 5% level in 8 out of the 10 groups. There is a probability of much less than 1 in 1,000 that the total difference from unity would occur by chance; and the heterogeneity between groups, despite the differences in conditions under which the observations were made, is no more than would be expected by chance.

In Table 2 the incidence of heavy *P. falciparum* infections is compared in sickling and nonsickling subjects. The differences are even more striking: the weighted mean relative incidence is 2.17 and the difference from unity is very highly significant ($\chi^2 = 51.379$ for 1 d.f.), with only very slight heterogeneity between groups. Other data that cannot be analyzed in the same way also show considerable protection by the sickle-cell trait (e. g., the difference in parasite levels above and below $1000/\mu l$ in Congolese children studied by Vandepitte, 1959, gives $\chi^2 = 18.9$ for 1 d.f.; p < 0.01). Since there is evidence both from Asia and Africa (Figs. 1 and 2) that mortality from falciparum malaria is related to parasite counts, the results in Table 2 provide powerful evidence that sickle-cell trait carriers are more likely to survive in malarious environments than nonsickling children. Direct evidence that this is so comes from available data on malarial mortality summarized in Table 3. The probabilities of obtaining the observed results by chance are calculated from the binomial distribution and combined by the method of Fisher (1959), giving $\chi^2 = 46.4$ (p < 0.001). It is difficult to see what further proof any reasonable person could expect.

Table 1. *P. falciparum* Parasite Rates in African Children

Authors	Subjects and Age in Years	Sickle-Cell Trait		Non-Sickle-Cell Trait		Relative Incidence (1)	Weight	Woolf χ²	Probability
		Falciparum	Total	Falciparum	Total				
(1) Allison (1954a)	Uganda, <6	12	43	113	247	2.18	7.58	4.60	0.05 > p > 0.02
(2) Foy et al. (1955)	Kenya, <6	131	241	154	241	1.49	28.81	4.53	0.05 > p > 0.02
(3) Raper (1955)	Uganda, <10	73	191	494	1,009	1.55	38.26	7.36	0.01 > p > 0.001
(4) Colbourne and Edington (1956)	S. Ghana	42	173	270	842	1.47	27.10	4.05	0.05 > p > 0.02
(5) Colbourne and Edington (1956)	N. Ghana, <5	11	15	165	177	5.00	2.32	6.01	0.02 > p > 0.01
(6) Walters and Chwatt (1956)	Nigeria, <5	162	213	680	890	1.02	31.24	0.01	p > 0.99
(7) Edington and Laing (1957)	N. Ghana, <6	13	19	109	127	2.79	3.24	3.42	0.10 > p > 0.50
(8) Garlick (1960)	Nigeria, <6	51	91	245	342	1.98	16.95	7.93	0.01 > p > 0.001
(9) Allison and Clyde (1961)	Tanganyika, <5	77	136	272	407	1.54	24.38	4.60	0.05 > p > 0.02
(10) Thompson (1962, 1963)	S. Ghana	34	123	176	593	1.10	20.52	0.20	p > 0.50

(1) Incidence of *P. falciparum* infections in non-sickle-cell trait groups relative to unity in corresponding sickle-cell trait groups.

Weighted mean relative incidence = 1.46.

Difference from unity, $\chi^2 = 29.2$ for 1 d.f., $p < 0.001$.

Heterogeneity between groups $\chi^2 = 13.5$ for 9 d.f., $0.20 > p > 0.10$.

Table 2. Incidence of Heavy P. falciparum Infections in African Children

Authors	Classification of Infection	Sickle-Cell		Non-Sickle-Cell		Relative Incidence (1)	Weight	Woolf χ^2	Probability
		Heavy Infections	Total	Heavy Infections	Total				
(1) Allison (1954a)	Group 2 or 3	4	43	70	247	3.86	3.38	6.16	$0.02 > p > 0.01$
(2) Foy et al. (1955)	Heavy	21	241	38	241	1.96	11.99	5.43	$0.02 > p > 0.01$
(3) Raper (1955)	$>1000/\mu l$	35	191	374	1,009	2.63	25.49	23.74	$p < 0.001$
(4) Colbourne and Edington (1956)	$>1000/\mu l$	3	173	57	842	4.11	2.79	5.59	$0.02 > p > 0.01$
(5) Colbourne and Edington (1956)	$>1000/\mu l$	5	15	75	177	1.47	3.07	0.46	$p > 0.50$
(8) Garlick (1960)	$>1000/\mu l$	25	91	147	342	1.99	14.91	7.06	$0.01 > p > 0.001$
(9) Allison and Clyde (1961)	$>1000/\mu l$	36	136	152	407	1.66	20.71	5.27	$0.05 > p > 0.02$
(10) Thompson (1962, 1963)	$>5630/\mu l$	3	123	42	593	3.05	2.72	3.38	$0.10 > p > 0.05$

(1) Incidence of heavy P. falciparum infections in non-sickle-cell trait groups relative to unity in corresponding sickle-cell trait groups. Weighted mean relative incidence = 2.17. Difference from unity $\chi^2 = 51.379$ for 1 d.f. $p < 0.001$. Heterogeneity between groups $\chi^2 = 5.719$ for 7 d.f., $0.7 > p > 0.5$.

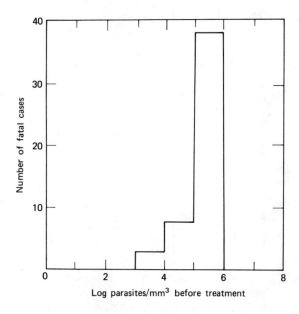

FIGURE 1. Number of fatal cases of falciparum malaria in
relation to the parasite count in peripheral blood before
treatment (data of Field, 1949).

Table 3. Deaths from Malaria in Relation to the
Sickle-Cell Trait in African Children

Author	Subjects	No. of Deaths	No. with Sickle-Cell Trait	Incidence of Sickle-Cell Trait in Population	Probability
Raper (1956)	Uganda (Kampala)	16	0	0.20	0.028148
J. and C. Lambotte-Legrand (1958)	Congo (Leopoldville)	23	0	0.235	0.0021095
Vandepitte (1959)	Congo (Luluaborg)	23	1	0.25	0.115938
Edington and Watson-Williams (1964)	Ghana (Accra)	13	0	0.08	0.33826
Edington and Watson-Williams (1963)	Nigeria (Ibadan)	29	0	0.24	0.00034953

$\chi^2 = 46.4$ (10 d.f.), $p < 0.001$.

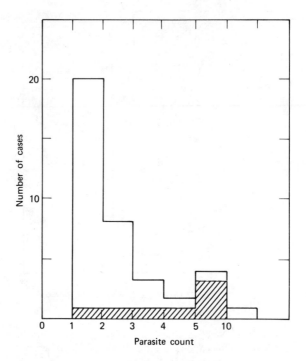

FIGURE 2. Total number of Congolese children with parasite counts x 10^5 per cu. mm. observed by Vandepitte and
Delaisse (1957) and number of fatal cases (hatched). There
is a progressive increase in the proportion of fatal cases as
the parasite count increases.

Differential Viability vs.
Differential Mortality

Resistance against malaria might favor the sickle-cell heterozygote either by increasing the viability of this genotype between birth and reproductive age or by
increasing the relative fertility of sickle-cell trait carriers. Because it is difficult to
demonstrate any effect of the sickle-cell trait on malaria in adults. Allison (1954a)
concluded that the main selective effect was exerted by differential mortality in
young children before appreciable immunity against malaria developed.

Allison (1956) obtained direct evidence in support of this interpretation. In the
Musoma district of Tanganyika the frequency of sickle-cell heterozygotes in the

adult population (38.4%) was significantly higher than that in the corresponding population of young infants. A number of other workers have found higher frequencies of sickle-cell trait carriers in older than younger age groups. Available data were summarized by Allison (1964): the unweighted mean ratio of trait incidences in schoolchildren and adults to young children is 1.17 to 1, the difference from unity being highly significant. Although these results have to be treated with some reservation because of various possible sources of bias, they provide strongly suggestive evidence in support of the differential viability hypothesis. In an extensive study of the progeny of 4,700 Congolese families, Burke, de Bock, and de Wolf (1958) also found highly significant evidence of a lower mortality among children carrying the sickle-cell trait than other children.

The case for differential fertility among woman as a result of placental malarial infection has been argued by Livingstone (1957). As Rucknagel and Neel (1961) have pointed out, in a population with 35% sickle-cell trait carriers equilibrium can be attained if the fertility of sickle-cell trait carriers (both sexes) is 1.97 times that of normal subjects and fertilities are additive in the sexes. If the excess fertility applies only to females, the difference has to be increased accordingly. Quite considerable data from Africa, summarized by Allison (1964) shows that nowhere is there any finding of enough increased female fertility of sickle-cell heterozygotes to account for the persistence of the gene. The largest series investigated (Burke et al., 1958) showed that the fecundity of sickling mothers is not superior to the mean in the Congo.

Among the Black Caribs of British Honduras, Firschein (1961) found a fertility ratio of sickle-cell trait to normal mothers of about 1.45 to 1, which is sufficient to maintain the sickle-cell gene even in the absence of differential mortality due to malaria; very few adult males were included in this study, and no attempt was made to estimate differential viability.

The most reasonable conclusion is that differential female fecundity is at most a minor contributory factor to the persistence of the sickle-cell gene in Africa, but may be more important where the malaria transmission rate is lower, as in British Honduras.

Distribution of the Sickle-Cell Gene

One of the original arguments in support of the malaria hypothesis was the distribution of the sickle-cell gene in Africa and elsewhere (Allison, 1954b). The extensive surveys that have since been carried out have all supported this view, with the exception of one preliminary claim (Foy, Brass, and Kondi, 1956) which seems to have been based on a technical error. Thus, in the Congo (Hiernaux, 1962), Tripolitania (Modica, Levadiotti, and Sorrenti, 1960), Greece (Barnicot, Allison, Blumberg, Deliyannis, Krimbas, and Ballas, 1963), Arabia (Lehmann, Maranjian, and Maurant, 1963) and India (see Shukla and Solanki, 1958; Mital, Parok, Sukumaran, Sharman, and Dave, 1962), high frequencies of sickling are confined to regions where malaria is or was endemic.

Frequencies of the Sickle-Cell Gene in the New World

Interpretation of the results of selection in New World populations is complicated by their uncertain origin. Slaves were obtained between 1520 and 1800 from areas widely scattered along the West Coast of Africa, from Cape Verde to Angola. Historical evidence indicates that the greatest number came from the neighborhood of Ghana, and this is supported by the relatively high frequencies of Hb-C in New World Negro populations. Over much of West Africa the incidence of the sickle-cell trait is of the order of 20%, and it is unlikely that the frequencies in the populations transported to the New World were less than this. Lower rates are found in Liberia and the neighborhood, but it is known that few slaves were taken from these regions; and this is confirmed by the exceedingly low frequencies of Hb-N in the Americans.

The second source of uncertainty is the degree of non-Negro admixture in the New World populations. This problem is more easily resolved by the use of suitable genetic markers, as in the recent study of Workman, Blumberg, and Cooper (1963). A comparison was made of 15 polymorphic traits in American White and Negro populations living in a rural Southern United States community with frequencies observed in West African Negroes and other American Negro and White populations. Estimating the total amount of gene migration, m, from the Whites to the American Negroes (assuming no selection), the polymorphic traits fell into two groups. In the larger group, including several red-cell blood-group antigens, the estimates of m (0.1 to 0.2) are consistent with the hypothesis that migration alone can account for the differences in gene frequencies between the West African and American Negro populations. In the second group, significantly higher estimates of gene migration were obtained (0.46 to 0.69 in the case of the sickle-cell gene). It was concluded that these resulted from both gene migration and different adaptive values of the traits in West African and American environments. Watson-Williams and Weatherall (1964) have also concluded that the frequencies of the sickle-cell gene in American Negroes are too low to be accounted for by non-Negro admixture. These results are of considerable interest because they suggest that a selective change in the genetic structure of a human population, that is to say, evolution, can take place within the short historical span of some three hundred years.

The frequencies of the sickle-cell gene in different West Indian populations are summarized in Table 4. Jonxis (1959) drew attention to the higher frequencies of

Table 4. Incidence of Sickle-Cell and Hemoglobin C Genes in
West Indian Populations

Population	No. Tested	% Hb-S Trait	% Hb-C Trait	Reference
Honduras, Black Caribs	705	23.3	2.6	Firschein (1961)
Curacao	1502	7.2	8.0	Jonxis (1959)
Surinam, Kabel	519	16.8	5.2	Jonxis (1959)
Surinam, Moengo	172	20.3	4.7	Jonxis (1959)
Surinam, Stoelman's Island	275	11.3	3.3	Jonxis (1959)
Surinam, Djukas	343	15	3	Liachowitz et al. (1958)
Jamaica	1018	10.9	3.1	Went (1957)
Jamaica		11.5	1.8	J. Parker-Williams (unpublished)
St. Vincent,	748	8.7	2.7	same
Dominiq	664	9.5	1.5	same
Barbados	912	7.0	4.6	same
St. Lucia	825	14.0	3.8	same

the sickle-cell trait in Surinam (which is malarious) than in Curacao (which is not). The sickle-cell trait is also very common in the Djukas of Surinam and the Black Caribs of British Honduras, both living in highly malarious environments. J. Parker-Williams (unpublished) has carried out comparisons of populations living in different West Indian islands. Significant higher frequencies of the sickle-cell gene were found in St. Lucia and Jamaica (malarious) than Barbados (nonmalarious). In St. Lucia significantly higher frequencies of sickle-cells were found in Gros Islet, which was highly malarious, than in the Soufriere district, with a low incidence of malaria. The frequencies of Hb-C were not significantly different in all the regions studied. Although these results must be interpreted with caution because of the uncertain origin of the populations concerned, they are consistent with the view that malaria has played a substantial part in maintaining high frequencies of the sickle-cell gene in some parts of the New World.

Interaction of Allele Genes
For Hemoglobin Variants

Allison (1955) drew attention to the special case of two alleles, such as Hb_β^S and Hb_β^C, which occur in the same region and interact to produce a heterozygote Hb_β^S/Hb_β^C of lowered fitness. There has been some doubt about the fitness of this genotype, and one approach to this problem is examination of large numbers of adults to see whether the corresponding phenotype occurs at frequencies different from Hardy-Weinberg expectation. Watson-Williams and Weatherall (1964) reported the results of hemoglobin electrophoresis of specimens from 12,387 Yoruba blood donors from Ibadan, Nigeria. Only a slight deficiency of the S + C type was observed; but, as Allison (1964) pointed out, these results must be interpreted with caution, since the blood donors were nearly all male and were, to a considerable extent, selected through relationship to anemic subjects. When an unselected panel of donors was analyzed, a considerable deficiency of S + C was found, and since it is quite clear that Hb^S/Hb^C females are unduly prone to complications in pregnancy (Fullerton et al., 1964), there is no doubt that this genotype is less fit than the mean.

From theoretical considerations Allison (1955) predicted that the Hb_β^S and Hb_β^C genes ought to be mutually exclusive in West African populations. Observations made shortly afterwards by Edington and Lehmann (1956) and Allison (1956) showed that this was so: high frequencies of Hb_β^C were found in Northern Ghana, but low frequencies of Hb_β^S; to the South and East frequencies of Hb_β^S increased while those of Hb_β^C fell. Allison (1956) concluded that most populations could be in or near a state of equilibrium resulting from selection. However, Neel, Hiernaux, Linhard, Robinson, Zuelzer, and Livingstone (1956) found that some populations living in Liberia and Guinea have low incidences of both Hb-S and Hb-C, and could not be in equilibrium, and the interpretation was further complicated by the fact that in the latter regions β-thalassemia is also relatively common (Oleson, Oleson, Livingstone, Cohen, Zuelzer, Robinson, and Neel, 1959; Neel, Robinson, Zuelzer, Livingstone, and Sutton, 1961).

In general, comparison between Fig. 4 and Figs. 3 and 6 shows a striking difference. Were there no unfit interaction heterozygotes, positive correlations with frequencies of different abnormal hemoglobins and glucoes-6-phosphate dehydrogenase (G-6-PD) deficiency might be expected. In contrast, the data for Hb-S and Hb-C remain strongly suggestive of mutual exclusion. The populations with low frequencies of both S and C are anthropologically distinct from other West African

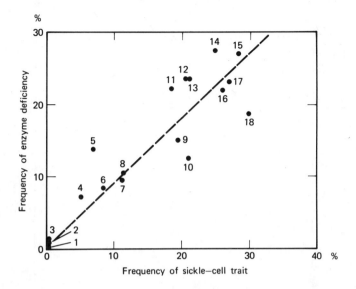

FIGURE 3. Frequencies of the sickle-cell and glucose-6-phosphate
dehydrogenase deficiency traits in African populations. Combined
data of Motulsky (1960) and Allison (1960).

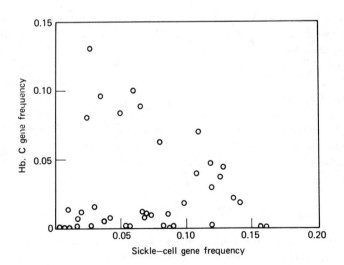

FIGURE 4. Frequencies of sickle-cell and hemoglobin C
genes in West African populations.

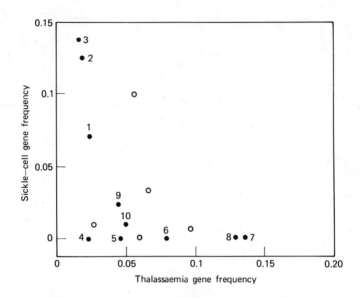

FIGURE 5. Frequencies of the sickle-cell gene plotted against frequencies of the β-thalassemia gene in different Greek populations
(Barnicot et al., 1963, ●; Stamatoyannopoulos
and Fessas, 1964, o).

populations (Livingstone, 1958; Cabannes, 1964), and it is possible that selection
has not yet raised the frequencies of abnormal hemoglobin genes to near equilibrium values in these cases.

Similar problems are posed by the presence of both sickling and β-thalassemia in
Greece. Here again the interaction heterozygote (Hb_β^S/Hb_β^T) is at a disadvantage,
and it is found that where sickling frequencies are high, thalassemia frequencies are
low, and vice versa (Fig. 5). In Southeast Asian populations Hb-E and β-thalassemia
are relatively common and interact to produce a somewhat disadvantageous heterozygote, and it would again be of interest to have accurate population and fitness
data; the data of Flatz and Sundaragiati (1964) again suggest mutual exclusion.

Blood Groups and Natural Selection

At the 1955 meeting I pointed out that we still have no explanation for the persistence of the Rh blood-group polymorphism in the face of selection against the
heterozygote, and that the inverse relationship of ABO incompatibility and Rh

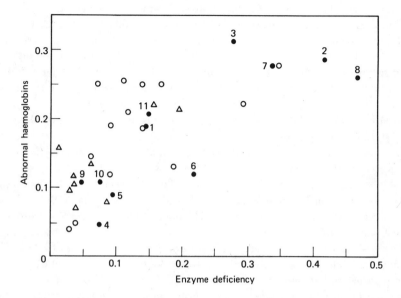

FIGURE 6. Frequency of glucose-6-phosphate dehydrogenase deficiency plotted
against combined phenotype frequency of sickling and β-thalassemia in various
Sardinian and Greek populations [data of Bernini et al., 1960 (o); Allison et al.,
1963 (●); Stamatoyannopoulos and Fessas, 1964 (△)].

immunization could provide a mechanism which would help to stabilize one of
these loci if the other were independently stabilized. These points remain valid, and
nothing done during the past nine years has thrown any light on the matter.

G orman (1964) has recently suggested that selection against the Rh-negative
gene may be operating indirectly through malaria. His argument is that there are
genetically-controlled differences in the capacity to form antibodies and that peoples living in malarious environments may be, in general, superior antibody formers.
Hence a higher proportion of Rh- incompatible matings would result in hemolytic
disease of the newborn, and there would be more intense selection against the Rh-
negative gene. Unfortunately, the facts are in exactly the opposite direction. Hemolytic disease due to Rh incompatibility is exceedingly rare in tropical countries such
as Nigeria, but somewhat more common in African populations living in nonmalarious regions, such as South Africa. The argument from distribution is also weak.
Very low frequencies of Rh-negatives are found in Chinese, Japanese, Eskimos, etc.,
and G orman's argument that these may never have had the Rh-negative gene is beg-
ging the question.

Further data have accumulated establishing beyond reasonable doubt that possession of particular ABO blood groups predisposes individuals to certain diseases (see Fraser Roberts, 1959; Clarke, 1961; Allison and Blumberg, 1965). Unfortunately, the biochemical factors underlying these susceptibility differences are unknown, and the associations tell us very little about the action of selection on the blood-group genes. This is so because in most cases the diseases take their toll after reproductive age (see Reed, 1960), and because it is not known whether the associations are different for homozygous and heterozygous A or B subjects.

One of the problems that has attracted attention is the relationship of blood-group status to susceptibility to infection by micro-organisms. The higher susceptibility of nonsecretors of ABO, and subjects of blood group A, to rheumatic fever and its complications, may be due to their higher susceptibility to virulent streptococcal infection (Van Ryjsewijk and Goslings, 1963). A remarkably clear-cut result was obtained with influenze A2 virus by McDonald and Zuckerman (1962): those infected with the virus, which was new to the population of Air Force recruits studied, showed a highly significant relative excess of group O and corresponding deficiency of group A, as compared with controls in all three regions (Table 5). This was not seen with other strains of influenza, but the results were complicated by prior exposure and immunity to those infections. The marked difference of susceptibility to influenza A2 according to blood group status is of interest because influenza, complicated by bacterial infection, could certainly have brought about selection during reproductive age in the past.

Table 5. Relative Incidence of Blood Groups O and A in Patients with
Influenza A2 as Compared with Controls by Geographical Region
(McDonald and Zuckerman, 1962)

Region	No. in Disease Series	Relative Incidence O : A	χ^2
1	313	1.699	17.929
2	316	1.315	4.926
3	72	1.463	2.130
Total	701		$\overline{24.985}$ $p (< 0.00001)$

χ^2 : Difference from unity, 22.869; D.F. = 1. Heterogeneity, 2.196; D.F. = 2.

It has been known for many years that certain bacteria, protozoa, and helminths have antigens that are related to blood-group substances. These facts have suggested to a number of workers that particular organisms might be resisted unequally by subjects possessing different blood groups, either because of "naturally occurring" antibodies or because of inability to produce antibodies to antigens resembling those possessed by the host as a result of immunological tolerance. This approach has been taken to extremes by Pettenkofer and his colleagues (Pettenkofer and Bickerich, 1960; Vogel, Pettenkofer, and Helmbold, 1960) who have suggested that the present day distribution of ABO blood groups in human populations can be explained largely through selection exerted by epidemics of plague and smallpox. The arguments are too involved to discuss here; they are considered in detail by Allison and Blumberg (1965). Suffice it to say that the claims are altogether unjustified by the scanty and unsatisfactory evidence provided.

This does not, of course, invalidate the whole approach, which has much to commend it, as well as some support from animal experiments. Thus Rowley and Jenkin (1962) found that "opsonins" which increase the uptake by mouse phagocytic cells, and probably intracellular destruction, of *Salmonella typhimurium* CS are present only at low concentration in mouse serum but at much higher concentrations in rat and pig sera. This was thought to be due to the fact that the bacterium and mouse tissues share antigenic components, so that the mice are unable, because of immunological tolerance, to produce the appropriate opsonins. Some experimental evidence in support of this view was obtained. Other examples of heterophile immunological reactions, occurring apparently nonspecifically between antigens and antibodies, are reviewed by Jenkin (1963) and Boyden (1963), who emphasize that sharing of antigens between hosts and parasites may affect susceptibility to disease.

In general, we need have no doubt that selection *is* operating on human bloodgroup genes, and observations in animals indicate that these effects may be quite strong. Thus, several small inbred lines of chickens have remained polymorphic for one or more blood-group systems. All of the highly inbred White Leghorn lines studied by Gilmour (1962) still show segregation of at least one to four bloodgroup loci. The levels of inbreeding reached are the result of from 21 to 26 generations of full-sib matings, corresponding to computed coefficients of 98.9 to 99.6%. Briles (1960) sampled 73 different closed populations (inbred up to coefficients of 86%) and found segregation at the B locus in all but two of them: and most were polymorphic for other blood-group systems as well. This type of evidence argues powerfully for heterozygous advantage, because in the absence of selection many of the populations should have become homozygous for the blood-group alleles. Briles and Gilmour have produced detailed evidence that in a number of instances heterozygotes are at an advantage in terms of hatchability and viability.

In this connection, Morton et al. (1965) have obtained evidence that several transferrin genotypes in chickens have different probabilities of surviving. The reasons for the differences are unknown, but the findings point up the weakness of the claim by Martin and Jandl (1959) that selection may operate through inhibition by transferrin of the multiplication of micro-organisms. This inhibition is an example of the iron-chelating effect previously described by Schade (see 1961) and no convincing evidence of a difference between effects of normal and variant forms of transferrin in this respect has been forthcoming.

Regarding the possible effects of selection on blood-group antigens at the gametic level, Edwards, Ferguson, and Coombs (1964) were unable to demonstrate A or B antigens on human spermatozoa of nonsecretors by the mixed agglutination and mixed antiglobulin reactions, and Holborow, Braun, Glynn, Hawes, Gresham, O'Brien, and Coombs (1960), and Allison and Bishop (unpublished) found none by the fluorescent antiglobulin technique. Hence the claims of Güllbring (1957) and Shahani and Southam (1962) that the A and B antigens segregate on separate spermatozoa must be treated with reservation. Matsunaga and Hiraizumi (1962) presented data indicating that heterozygous AO and BO fathers transmitted more than the expected 50% of O-bearing sperm to their children. Since neither sperm incompatibility nor reproductive compensation appeared to account for the observations, it was concluded that they demonstrate prezygotic selection operating in the ABO blood groups of man, probably due to meiotic drive or sperm competition. This study has been criticized by Novitski (1962) and Reed (1963), and the criticisms only partially answered by Hiraizumi (1963). The recent evidence of Hiraizumi (1964) presented in this volume, shows no indication of prezygotic selection.

Selection and the Sex Ratio

As everybody knows, in man equal numbers of homogametic and heterogametic zygotes would be expected, but vital statistics from many countries show a preponderance of male births, usually about 105 males to 100 females. Over millions of births this excess cannot be due to chance, and its significance must be considered. The proportion has not been constant, as Parkes (1963) has reminded us. The sex ratios of births in England and Wales from 1841 to the present time are summarized in Fig. 7. The decrease in proportion of males during the first 30 years of this period is not understood; perhaps registration was incomplete. For the next 40 years the ratio hovered between 104 males per 100 females and then at the time of the First World War rose to about 105. A subsequent decrease was followed by another sharp peak coincident with the Second World War.

The sex ratio changes in another way. In the middle of the last century, the excess of males at birth disappeared by the age of 5 years because of the differential mortality of males over this period, and thereafter females were in excess. By the turn of the century mortality had decreased and sex ratios were about equal up to the age group 15-19 years; thereafter, owing to differential mortality and emigration falling sharply to below 90 males per 100 females in the 25-29 age group (Fig. 8). At the present time, with the further decrease in child mortality, the excess of males found at birth is being preserved much longer, up to the age group 25-29. Even now, females exceed males from the age-group 35-39 onwards, and the

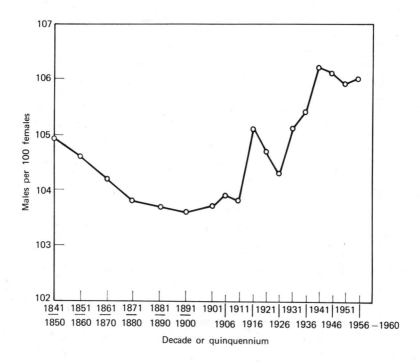

FIGURE 7. Sex ratio for live births in England and Wales,
1841-1960 (from Parkes, 1963).

decrease in relative numbers of males becomes very sharp after the age 55. The sex
ratio among births is also inversely correlated with parental age, and the low age of
parents during the two World Wars probably explains the coincident peaks of male
birth frequencies.

Thus there is abundant evidence that mortality after birth is higher among males
than females. Until recently, there was an even more striking difference among stillbirths, and it has been widely accepted that this is true also of abortuses (Crew,
1952). However, Tietze (1948) and McKeown and Lowe (1951) have pointed out
that it is difficult to sex abortuses reliably by morphology, so that the evidence for
a very high primary sex ratio in man is inadequate. There is scope here for cytological study, since human sex chromatin is detectable at 10-12 days' gestation (Park,
1957).

Evidence from animals indicates the existence of a high primary sex ratio and
proportionately greater loss of male fetuses during pregnancy. Crew (1952) estimated that the sex ratio of pigs at conception was about 150. Nuclear sexing has

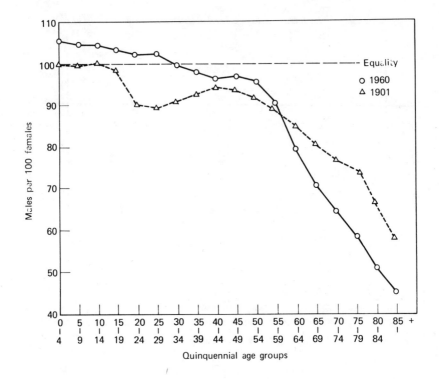

FIGURE 8. Sex ratio according to age (England and Wales, from Parkes, 1963).

shown a high sex ratio before implantation in the golden hamster (Lindahl and Sundell, 1958), which implies that there is a considerable differential mortality *in utero* since the sex ratio at birth is about unity.

The sex ratio represents, in fact, a most interesting polymorphism determined by a simple genetic switch mechanism and subject to selection. The substantial excess of males at conception found in some species (and possibly present in man) might be due to formation of a higher proportion of functional Y-bearing than X-bearing spermatozoa, to some advantage of the former in the female genital tract, or to some other more obscure mechanism. The higher mortality of male than female fetuses is presumably the expression of sex-linked or sex-limited factors reducing viability. Some social consequences of the current trend in sex ratios are discussed by Parkes (1963).

There are other reports of deviation of the sex ratio. The most important is the indication that the sex ratio of offspring is altered by parental exposure to X-irradiation (Lejeune and Turpin, 1957; Schull and Neel, 1958; Tanaka and Okhura,

1961). The ratio was increased after paternal, and decreased after maternal exposure, which suggests that some daughters of exposed fathers and some sons of exposed mothers failed to develop because of damage to genes on the parental X-chromosomes. This is the main indication of a genetic—as opposed to a somatic—effect of X-irradiation in man.

The second series of reports concerns departures from randomness of the sexes of successive children. Renkonen, Mäkelä, and Lehtovaara (1961) reported that the birth of a boy renders a woman less likely to give birth to further boys in the future. Reviewing Finnish and French data, Edwards (1962) states that there is good evidence for a correlation between the sexes of successive children. One interpretation of the findings of Renkonen et al., if substantiated, would be that bearing a male fetus could immunize a woman against a male (Y) antigen, which might harm subsequent male fetuses or inactivate a proportion of Y-bearing sperm before fertilization. Such immunization can be achieved in homografting experiments. However, McLaren (1962) found no change in the sex ratio among offspring of inbred mice deliberately immunized against the male antigen.

General Comment

The evidence for selection acting on the sickle-cell gene is an order of magnitude better than that for selection operating on other abnormal hemoglobins and G-6-PD deficiency, which is, in turn, a good deal better than that acting on most other human polymorphic systems. We can, however, be quite confident that selection *is* acting on the blood groups, the sex ratio, and probably other polymorphisms, although of the underlying mechanisms we are sadly ignorant. The last ten years have seen posed the problems and provided some of the solutions, but progress in this field is not furthered by the uncontrolled speculations that have become common. Observations on animals provide powerful evidence for selection acting on blood-group and transferrin polymorphisms, and by analogy we may suppose that similar effects are operating in man. But well-controlled, precise evidence on human populations is needed before this can be accepted as true. The existence of human polymorphisms controlled by selection is established, but much more work is needed to establish how widespread these effects are and the magnitude of the selective forces involved.

References

Allison, A. C. 1954a. Protection by the sickle-cell trait against subtertian malarial infection. *Brit. Med. J.*, i: 290.

_____,1954b. The distribution of the sickle-cell trait in East Africa and elsewhere and its apparent relationship to the incidence of subtertian malaria. *Trans. Roy. Soc. Trop. Med. Hyg.* 48: 312.

_____. 1955. Aspects of polymorphism in Man. *Cold Spring Harbor Symp. Quant. Biol.*, 20: 230.

_____. 1956. The sickle-cell and haemoglobin C genes in some African populations. *Ann. Hum. Genet.*, 21: 67.

_____. 1960. Glucose-6-phosphate dehydrogenase deficiency in red blood cells of East Africans. *Nature*, 186: 431-432.

_____. 1964. Population genetics of abnormal haemoglobins. In *Council for Intern. Organ. for Med. Sci. Symposium on Abnormal Haemoglobins*. Blackwell, Oxford. In press.

Allison, A. C., B. A. Askonas, N. A. Barnicot, B. S. Blumberg, and C. Krimbas. 1963. Deficiency of glucose-6-phosphate dehydrogenase in Greek populations. *Ann. Hum. Genet.*, 26: 237.

Allison, A. C., and B. S. Blumberg. 1965. *Polymorphism in man*. Little, Brown and Co., Boston.

Allison, A. C., and D. F. Clyde. 1961. Malaria in African children with deficient glucose-6-phosphate dehydrogenase. *Brit. Med. J.*, i: 1346-1348.

Barnicot, N. A., A. C. Allison, B. S. Blumberg, G. Deliyannis, C. Krimbas, and A. Ballas. 1963. Haemoglobin types in Greek populations. *Ann. Hum. Genet.*, 26: 229-236.

Bernini, R. N., V. Carcassi, B. Latte, A. G. Motulsky, L. Romei, and M. Siniscalco. 1960. Indagini genetiche sulla predispositione al favismo. III. Distribuzione della frequenze geniche per il locus gd oin Sardegna. Interazione con la malaria e la talassemia al livello populazionistico. *Recond. Acad. Natl. dei Lincei*, 199: 1-18.

Boyden, S. V. 1963. Cellular recognition of foreign matter. *International Review of Experimental Pathology*, 2: 311. Academic Press, New York.

Brilles, W. E. 1960. Blood groups in chickens, their nature and utilization. *World's Poultry Sci.*, 16: 223.

Burke, J., C. de Bock, and C. de Wolf. 1958. La drépanocytémie simple et l'anémie drépanocytaire au Kwango (Congo Belge). Memoires in 8° de l'Acad. Roy. Sciences Colon. Nouv. Ser. 7: 1.

Cabannes, R. 1964. Distribution of haemoglobin variants. In *Council for Intern. Organ. for Med. Sci. Symposium on Abnormal Haemoglobins*. Blackwell, Oxford. In press.

Clarke, C. A. 1961. Blood groups and disease. *Progr. Med. Genet.*, 1: 81.

Colbourne, M. J., and G. M. Edington, 1956. Sickling and malaria in the Gold Coast. *Brit. Med. J.*, i: 784.

Crew, F. A. E. 1952. The factors which determine sex, p. 741-792. In Parkes (ed.), *Marshall's Physiology of Reproduction*. Longmans, London.

Edington, G. M., and W. N. Laing. 1957. Relationship between haemoglobins S and C and malaria in Ghana. *Brit. Med. J.*, ii: 143.

Edington, G. M., and H. Lehmann. 1956. The distribution of haemoglobin C in West Africa. *Man,* 56: 34.

Edington, G. M., and G. J. Watson-Williams. 1964. Sickling, haemoglobin C, glucose-6-phosphate dehydrogenase deficiency and malaria in Western Nigeria. *Council for Intern. Organ. for Med. Sci. Symposium on Abnormal Haemoglobins*. Blackwell, Oxford, In press.

Edwards, A. W. F. 1962. Genetics and the human sex ratio. *Adv. Genet.* 11: 239.

Edwards, R. G., L. C. Ferguson, and R. R. A. Coombs. 1964. Blood group antigens in human spermatozoa. *J. Reprod. Fertil.* 17: 153-161.

Field, J. W. 1949. Blood examination and prognosis in acute falciparum malaria. *Trans. Roy. Soc. Trop. Med. Hyg.*, 43: 33.

Firschein, I. L. 1961. Population dynamics of the sickle-cell trait in the Black Caribs of British Honduras. *Am. J. Hum. Genet.*, 13: 233.

Fisher, R. A. 1959. p. 99. In R. A. Fisher, *Statistical methods for research workers.* Oliver and Boyd, Edinburgh.

Flatz, G., and B. Sundaragiati. 1964. Malaria and haemoglobin E in Thailand. *Lancet*, ii: 385.

Foy, H., W. Brass, R. A. Moore, G. Timms, A. Kondi, and T. Olouch. 1955. Two surveys to investigate the relation of the sickle-cell trait and malaria. *Brit. Med. J.*, ii: 1116.

Foy, H., W. Brass, and A. Kondi. 1956. Sickling and malaria. *Brit. Med. J.*, i: 289.

Frota-Pessoa, O., and A. Wajntal. 1963. Mutation rates of the abnormal hemoglobin genes. *Amer. J. Hum. Genet.*, 15: 123.

Fullerton, W. T., J. P. de V. Hendrickse, and E. J. Watson-Williams. 1964. Haemoglobin SC disease in pregnancy. *Council for Intern. Organ. for Med. Sci. Symposium on Abnormal Haemoglobins.* Blackwell, Oxford. In press.

Garlick, J. P. 1960. Sickling and malaria in South West Nigeria. *Trans. Roy. Soc. Med. Hyg.*, 54: 146.

Gilmour, D. G. 1962. Blood groups in chickens. *Ann. N. Y. Acad. Sci.*, 97: 166.

Gorman, J. G. 1964. Selection against the Rh-negative gene by malaria. *Nature*, 202: 676.

Güllbring, B. 1957. Investigation on the occurrence of blood group antigens in spermatozoa from man, and serological demonstration of the segregation of characters. *Acta Med. Scand.*, 159: 169.

Harris, H. 1959. *Introduction to human biochemical genetics.* 2nd. ed. University Press, Cambridge, 310 p.

Hiernaux, J. 1962. Donnes génétiques sur six populations de la Republique du Congo. *Ann. Soc. Belge Med. Trop.*, 40: 339.

Hiraizumi, Y. 1963. Assumption in tests for meiotic drive. *Science*, 139: 406.

Holborow, E. J., P. C. Braun, L. E. Glynn, M. D. Hawes, G. A. Gresham, T. F. O'Brien, and R. R. A. Coombs. 1960. The distribution of the blood group A antigen in human tissues. *Brit. J. Exp. Path.*, 41: 430.

Jenkin, C. R. 1963. Heterophile antigens and their significance in the host-parasite relationship. *Adv. Immunol.*, 3: 351.

Jonxis, J. H. P. 1959. The frequency of haemoglobin S and C carriers in Curacao and Surinam. *Council for Intern. Organ. for Med. Sci. Symposium on Abnormal Haemoglobins.* Blackwell, Oxford. In press.

Lambotte-Legrand, J., and C. Lambotte-Legrand. 1958. Notes complémentaires sur la drépanocytose, 11. Sicklémie et malaria. *Ann. Soc. Belge. Med. Trop.*, 38: 45.

Lehmann, H., G. Maranjian, and A. E. Mourant. 1963. Distribution of sickle-cell haemoglobin in Saudi Arabia. *Nature*, 198: 492.

Lejeune, J., and R. Turpin. 1957. Mutations radioinduites chez l'homme et dose de doublement. Sur la validité d'une estimation directe. *Compt Rend.*, 244: 2425.

Liachowitz, C., J. Elderlein, I. Gilchrist, H. W. Brown, and H. M. Ranney. 1958. Abnormal hemoglobins in the Negroes of Surinam. *Amer. J. Med.*, 24: 19.

Lindahl, P. E., and G. Sundell. 1958. Sex ratio of the golden hamster before uterine implantation. *Nature*, 182: 139.

Livingstone, F. B. 1957. Sickling and Malaria. *Brit. Med. J.*, i: 762.

———. 1958. Anthropological implications of sickle-cell gene distribution in West Africa. *Amer. Anthrop.*, 60: 533.

McDonald, J. C., and A. J. Zuckerman. 1962. ABO blood groups and acute respiratory virus disease. *Brit. Med. J.*, ii: 89.

McKeown, T., and C. R. Lowe. 1951. The sex ratio of stillbirths related to cause and duration of stillbirths. An investigation of 7,066 stillbirths. *Human Biol.*, 23: 41.

McLaren, A. 1962. Does maternal immunity to male antigen affect the sex ratio of the young? *Nature*, 195: 1323.

Martin, C. M., and J. H. Jandl. 1959. Inhibition of virus multiplication by transferrin. *J. Clin. Invest.*, 38: 1024.

Matsunaga, E., and Y. Hiraizumi. 1962. Prezygotic selection in the ABO blood groups. *Science*, 135: 432.

Mital, M. S., J. G. Parok, P. K. Sukumaran, R. S. Sharma, and P. J. Dave. 1962. A focus of sickle-cell gene near Bombay. *Acta Haemat.*, 27: 257.

Modica, H., M. Levadiotti, and A. M. Sorrenti. 1960. Incidenza della sicklemia, progressa malaria e distribuzione razziale dell'oasi costiera di Tauroga. *Arch. Itai. Sci. Med. Trop. Parasitol*, 41: 595-603.

Morton, J. R., D. G. Gilmour, E. M. McDermid and A. L. Ogden. 1965. Association of blood group and protein polymorphisms with embryonic mortality in the chicken. *Genetics*, 51: (1), in press.

Motulsky, A. G. 1960. Metabolic polymorphisms and the role of infectious diseases in human evolution. *Human Biol.*, 32: 28.

Nance, W. E. 1963. Genetic control of hemoglobin synthesis. *Science*, 141: 123.

Neel, J. V., J. Hiernaux, J. Linhard, D. D. Robinson, W. W. Zuelzer, and F. B. Livingstone. 1956. Data on the occurrence of hemoglobin C and other abnormal hemoglobins in some African populations. *Amer. J. Hum. Genet.*, 8: 138.

Neel, J. V., A. R. Robinson, W. W. Zuelzer, F. B. Livingstone, and H. E. Sutton. 1961. The frequency of elevations in the A_2 and fetal hemoglobin fractions in the natives of Liberia and adjacent regions with data on haptoglobin and transferrin types. *Amer. J. Hum. Genet.*, 13: 263.

Novitski, E. 1962. Meiotic drive. *Science*, 137: 861.

Olesen, E. B., K. Olesen, F. B. Livingstone, F. Cohen, W. W. Zuelzer, A. R. Robinson, and J. V. Neel. 1959. Thalassemia in Liberia. *Brit. Med. J.*, i: 1385.

Park. W. W. 1957. The occurrence of sex chromatin in early human and macaque embryos. *J. Anat.*, 91: 369.

Parkes, A. S. 1963. The sex ratio of human populations. P. 91. In G. Wolstenholm (ed.). *CIBA Foundation Symposium on Man and His Future*. Churchill, London.

Pettenkofer, H. J., and R. Bickerich. 1960. Über Antigen-gemeinschaften zwischen den menschlichen Blutgruppen ABO und den erregern gemeingefährlicher Krankheiten. *J. Bakt. Parasitol.*, 179: 433.

Raper, A. B. 1955. Malaria and the sickling trait. *Brit. Med. J.*, i: 1186-1189.

———. 1956. Sickling in relation to morbidity from malaria and other diseases. *Brit. Med. J.*, i: 965.

_____. 1959. Further observations on sickling and malaria. *Trans. Roy. Soc. Trop. Med. Hyg.* 53: 110

Reed, T. E. 1960. Polymorphism and natural selection in blood groups. In *Genetic polymorphisms and geographic variations in disease.* U. S. Public Health Service, Washington.

_____. 1963. Assumptions in tests for meiotic drive. *Science,* 139: 408.

Renkonen, K. O., O. Mäkelä, and R. Lehtovaara. 1961. Factors affecting the human sex ratio. *Ann. Med. Exp. Biol. Fenn.,* 39: 173.

Roberts, J. A. Fraser. 1959. Some associations between blood groups and disease. *Brit. Med. Bull.,* 15: 129.

Rowley, D., and C. R. Jenkin. 1962. Antigenic cross-reaction between host and parasite. *Nature,* 193: 151.

Rucknagel, D. L., and J. V. Neel. 1961. The hemoglobinopathies. *Progr. Med. Genet.,* i: 158.

Schade, A. L. 1961. The microbiological activity of siderophilin, p. 261. In *Protides of Biological Fluids,* Eisevier, Amsterdam.

Schull, J., and J. V. Neel, 1958. Radiation and the sex ratio in man. *Science,* 128: 343.

Shahani, S., and A. L. Southam. 1962. Immunofluorescent study of the ABO blood group antigens in human spermatozoa. *Amer. J. Obstet. Gynec.,* 84: 660.

Shukla, R. N., and B. R. Solanki. 1958. Sickle-cell trait in Central India. *Lancet,* i: 297.

Stamatoyannopoulos, G., and P. H. Fessas. 1964. Thalassemia and glucose-6-phosphate dehydrogenase, sickling and malarial endemicity in Greece. A study of five areas. *Brit. Med. J.,* i: 875-879.

Tanaka, K., and K. Okhura. 1961. Genetic effects of radiation in man: a study on the sex ratio in the offspring of radiological technicians. *Proc. 2nd Int. Conf. Human Genetics, Rome.*

Thompson, G. R. 1962. Significance of haemoglobins S and C in Ghana. *Brit. Med. J.,* i: 682.

_____. 1963. Malaria and stress in relation to haemoglobins S and C. *Brit. Med. J.,* ii: 976.

Tietze, C. 1948. A note on the sex ratio of abortions. *Human Biol.,* 20: 156.

Vandepitte, J. 1959. The incidence of haemoglobinoses in the Belgian Congo. In *Council for Intern. Organ. for Med. Sci. Symposium on Abnormal Haemoglobins.* Blackwell, Oxford. In press.

Vandepitte, J., and J. Delaisse. 1957. Sicklémia et paludisme. *Ann. Soc. Belge Med. Trop.,* 37: 703.

Vandepitte, J., and J. Stijns. 1964. Haemoglobinopathies in the Congo (Leopoldville) and the Rwanda Burundi. In *Council for Intern. Organ. for Med. Sci. Symposium on Abnormal Haemoglobins.* Blackwell, Oxford. In press.

Vandepitte, J., W. W. Zuelzer, J. V. Neel, and J. Colaert. 1955. Evidence concerning the inadequacy of mutation as an explanation of the frequency of the sickle-cell gene in the Belgian Congo. *Blood,* 10: 341.

Van Rijsewijk, M. G. H., and N. E. O. Goslings. 1963. Secretor status of streptococcus pyogenes group A carriers and patients with rheumatic heart disease or acute glomerulonephritis. *Brit. Med. J.,* ii: 542.

Vogel, F., H. J. Pettenkofer, and W. Helmbold. 1960. Über die Populationsgenetik der ABO Blutgruppen. *Acta Genet.,* 10: 267.

Wagner, R. P., and H. K. Mitchell. 1964. *Genetics and Metabolism,* 2nd ed. John Wiley, New York.

Walters, J. H., and L. J. Bruce-Chwatt. 1956. Sickle-cell anemia and falciparum malaria. *Trans. Roy. Soc. Trop. Med. Hyg.*, 50: 51.

Watson-Williams, G. J., and D. J. Weatherall. 1964. The laboratory characterization of haemoglobin variants. *Council for Intern. Organ. for Med. Sci. Symposium on Abnormal Haemoglobins.* Blackwell, Oxford. In press.

Went, L. N. 1957. Incidence of abnormal haemoglobins in Jamaica. *Nature*, 180: 1131.

Wilson, T. 1961. Malaria and glucose-6-phosphate dehydrogenase. *Brit. Med. J.*, ii: 246 and 895.

Woolf, B. 1955. On estimating the relation between blood group and disease. *Ann. Hum. Genet.*, 19: 251.

Workman, P. L., B. S. Blumberg, and A. J. Cooper. 1963. Selection, gene migration and polymorphic stability in a U. S. White and Negro population. *Amer. J. Hum. Genet.*, 15: 429.

Discussion

BEUTLER: A very interesting suggestion has been made by Dr. Chaim Sheba to account for loss of the sex-linked gene for G-6-PD deficiency from Ashkenazy Jews. He pointed out that when Jewish slaves were brought to Rome, the religion and culture was carried on only by the male line. The males frequently married non-Jewish women. Thus, if a population with a high incidence of G-6-PD activity were transported to Rome, the following consequences might be anticipated: (1) the X chromosomes of the women are lost to the Jewish group. (2) The X chromosomes of the Jewish males are passed on to daughters, who are also culturally lost to the Jewish group. (3) The sons, who carry on Judaism receive X chromosomes from non-Jewish mothers, and thus lose the gene for G-6-PD deficiency.

If adhered to, such a mechanism would eliminate all of the G-6-PD deficiency genes in one generation. It serves as an example of what a powerful influence cultural patterns may have on distribution of a genetic trait in man.

ALLISON: As Dr. Slatis has pointed out, the nonreciprocal crosses between Whites and Negroes in this country imply that very few X-genes in the Negro population come from Whites. Hence, to apply dilution factors estimated from autosomal genes is invalid.

POST: Dr. Allison's description of the gradual decrease in frequency of the sickle-cell gene among Negro populations living in areas free of malaria during the past 200 years suggests another case of selection relaxation—viz. the situation of high frequencies of sex-linked "colorblindness" in Europe, the Middle East, and Far East, contrasted with low frequencies among a number of populations having more "primitive" culture. Over forty samples of populations with a fairly long history of civilization have gene frequencies estimated from about 8% in Europe to 6% in China, pooling protans and deuterans. Samples of aboriginal Australians, Melanesians, and American Indians have from zero to 2%. One explanation of this contrast is that selection against sex-linked color deficiencies has relaxed under the different ecological conditions produced by cultural growth.

12. On the Selective Advantage of Cystic Fibrosis Heterozygotes

ALFRED G. KNUDSON, JR., LOWELL WAYNE,

and

WILBUR Y. HALLETT

The high frequency of the recessive gene for cystic fibrosis of the pancreas (CF) estimated for Caucasian populations is generally attributed to a selective advantage of the heterozygous carrier. The magnitude of this advantage would need to be only about 2% in order to maintain at equilibrium a gene frequency of approximately 0.02. In our own recent study (Hallett et al., 1965), we included a survey for evidence of disease resistance among CF heterozygotes but were not able to find a difference between subjects and controls. We did notice a difference between the groups with respect to the average sizes of their sibships but were reluctant to draw attention to what could be a sampling error. We now feel that the problem should be considered further in view of the report of a similar finding from Australia (Danks et al., 1965).

Material and Methods

The selection of subjects and procedure for study have been described (Hallett et al., 1965). Although the purpose of the previous study was to establish whether heterozygous carriers might sustain some physiological disadvantage, inquiry into family histories provided the data necessary for the present report. The study group consisted of the parents of documented cases of CF, while the controls were for the

SOURCE: Alfred G. Knudson, L. Wayne, and W. Hallett, "On the Selective Advantage of Cystic Fibrosis Heterozygotes," *American Journal of Human Genetics*, **19**: 388-392 (Chicago, Illinois: The University of Chicago Press, 1967).

Supported by U. S. Public Health Service research grant AM 05989 from the National Institute of Arthritis and Metabolic Diseases, National Institutes of Health.

191

most part friends and neighbors of the study group subjects. The close similarities of the two groups have been emphasized previously.

The information regarding families was obtained by a combination of questionnaires and interviews, in order to maximize accuracy. Data pertinent to the present study included information on the parents of the subjects (Generation I), the subjects and their sibs (Generation II), and the children of the subjects (Generation III). The sib data included information on abortions, stillbirths, and postnatal deaths. All data were coded, recorded on punch cards, and originally examined by two of us with Dr. Frank Massey (Department of Biostatistics, School of Public Health, University of California at Los Angeles). The later scrutiny of data, which forms the basis for the present report, was performed by one of the authors (L. W.), using the facilities of the Computer Sciences Laboratory at the University of Southern California.

Results

The numbers of live offspring of persons in Generation I are compared in Table 1. The mean numbers of 4.34 and 3.43 for CF and control families, respectively, are significantly different ($P < 0.01$) by variance analysis.

These means are calculated for offspring at birth. We do have the additional observation that 19 of the CF offspring died during the first nine months of life, while only three of the control offspring died by then. The mean numbers of children surviving to this age are, therefore, 4.21 for CF and 3.41 for control. The difference of the means, 0.80, is still significant at the one per cent level. That this trend does not continue into adult life is shown by the observation that at the time of the study 506 of the offspring of the CF group and 356 of the offspring of the control group were still living; the means are 3.86 and 3.02, respectively, and the difference is 0.84.

We had already reported that in Generation III (eliminating of course those control families with no members in Generation III) the study families had more liveborn offspring (3.0 per family) than the control families (2.7 per family). Of course, this number could be influenced by the possibly greater willingness to participate on the part of CF families with more than one affected child, and so on the average more total children. We did wonder, however, whether the average family size in Generation II might be correlated with that in Generation III, so that the Generation II data might be biased. Evidence against this explanation is provided by the correlation coefficients for the two groups, $r = .049$ for CF, and $r = .209$ for controls.

Table 1. Comparison of Numbers of Live
Offspring of Grandparents

Number of Offspring per Family	CF Families		Control Families	
	Number	Total Sibs	Number	Total Sibs
1	9	9	15	15
2	20	40	37	74
3	33	99	23	69
4	15	60	15	60
5	21	105	11	55
6	10	60	7	42
7	10	70	2	14
8	6	48	2	16
9	–	–	3	27
10	3	30	2	20
11	1	11	–	–
12	2	24	–	–
13	1	13	–	–
14	–	–	1	14
Total	131	569	118	406

Mean number of offspring	4.34	3.43
Variance	6.07	5.12
Variance of mean	0.0463	0.0434
Difference of means	0.91	
Sum of variances of means	0.0897	
Standard error of difference of means	0.30	
$\dfrac{\text{Difference of means}}{\text{Standard error}}$	$3.0\,(P < 0.01)$	

Discussion

In agreement with the Australian workers, we find that parents of children with CF come themselves from larger sibships than do control persons. The difference is an index of relative fertility, and the difference in survivors is an index of relative survival values. The magnitude of the relative disadvantage, s, of the normal homozygote compared with the heterozygote may be obtained from the following relationship:

$$\frac{1-s}{1} = \frac{\text{average sibship size for normals}}{\text{average sibship size for heterozygotes}} \tag{1}$$

The values of s obtained from the Australian data and from our data are 0.10 and 0.21, respectively. Attention should be called to the fact that the incidence of childlessness in each group is assumed to be similar, although we have no way to check this assumption at present.

The difference is very large and would have an enormous effect on the frequency of this recessive lethal allele. At equilibrium, the relationship $q = s/(1 + s)$ would apply, and even a value of $s = 0.1$ would give a value of $q = 0.09$ and a birth incidence of cystic fibrosis of about 0.8%. Obviously then, if this value of s is correct, the Caucasian population is not in equilibrium with respect to this gene. On the other hand, the observed disease incidence and presently estimated gene frequency could be attained if the advantage is of relatively recent origin and equilibrium not yet attained. Assuming that the coefficient of selection for CF homozygotes is 1.0, for heterozygotes 0.0, and for normal homozygotes s, then the gene frequency in a new generation (q') bears the following relationship to that in the preceding generation (q):

$$q' = \frac{q-q^2}{1-s(1-q)^2-q^2} = \frac{q}{1-s+(1+s)q} \tag{2}$$

$$\triangle q = q'-q = \frac{sq-(1+s)q^2}{1-s+(1+s)q} \tag{3}$$

The number of generations (n) required for the gene frequency to change from q_o to q_n is derived as follows: Setting $dq/dn = \triangle q$, then

$$dn = \frac{1-s+(1+s)q}{sq-(1+s)q^2} \, dq$$

$$= \left(\frac{1-s}{s}\right) \frac{dq}{q} + \left(\frac{1+s}{s}\right) \left(\frac{dq}{s-(1+s)q}\right)$$

and

$$n = \int_o^n dn = \frac{1-s}{s} \ln(q_n/q_o) + \frac{1}{s} \ln \left(\frac{s-(1+s)q_o}{s-(1+s)q_n}\right) \tag{4}$$

If we estimate $q_n = 0.02$ and q_o to be approximately one-tenth of q_n, or 0.002, not an unreasonable estimate incidentally for the frequency among Oriental and Negro populations, then, for $s = 0.10$,

$$n = \frac{0.9}{0.1} \ln 10 + \frac{1}{0.1} \ln \left(\frac{0.1-(1+0.1)0.002}{0.1-(1+0.1)0.02}\right)$$

$$= 23 \text{ generations}$$

Therefore, a heterozygous carrier advantage which suddenly began operating 23 generations ago with a coefficient of selection of 0.1 could explain the presently observed CF incidence.

If this relative advantage of the carrier is real and still operating, then the gene frequency will continue to rise. The new frequency after one generation, given by equation (2), would, for $q = 0.02$ and $s = 0.1$, yield $q' = 0.022$. The new disease incidence would be q'^2, or 4.8 per 10,000, compared with the present estimate of 4.0 per 10,000. It will be observed that the *difference* (0.8 per 10,000) is of the order of magnitude of the incidence of such disorders as phenylketonuria and galactosemia.

The physiological basis for such an advantage of carriers would be of extreme interest of course. Of prime importance, however, is the necessity for establishing whether these observations are valid. Hopefully, other workers will report their findings.

Summary

Grandparents of cystic fibrosis patients were found to have an average of 4.34 offspring compared to 3.43 for controls ($P<0.01$). Among survivors at the time of the study the difference was still significant despite a higher infant mortality in the cystic fibrosis group.

The possibility that the frequency of the gene for cystic fibrosis is increasing through selection is discussed.

References

Danks, D. M., Allan, J., and Anderson, C. M. 1965. A genetic study of fibrocystic disease of the pancreas. *Ann. Hum. Genet.* (Lond.), **28**: 323-356.

Hallett, W. Y., Knudson, A. G., and Massey, F. J. 1965. Absence of detrimental effect of the carrier state of the cystic fibrosis gene. *Ann. Rev. Resp. Dis.*, **92**: 714-724.

13. Natural Selection and Degenerative Cardiovascular Disease

E. R. NYE

Considerable research is now devoted to the elucidation of the problems of the degenerative diseases. Huge sums are spent annually in many countries on the subject of degenerative cardiovascular disease and research programs range from the investigation of details of lipids metabolism to the possibility of grafting whole hearts or implanting artificial ones.

The impetus for all the effort devoted to research on coronary heart disease stems from the view that a man in his fifties, or forties, who is the victim of a coronary artery occlusion, is a needless loss, complete or temporary, to society. If nobody had a coronary attack before his sixtieth year there would be much less interest in the disease.

Mortality and morbidity statistics show that coronary artery disease is increasing and it seems likely that not all of the increase is due to the effect of antibiotics in providing an increase in the number of people in the "coronary prone" age groups. Support for the belief that coronary atheroma is a preventable disorder comes from the knowledge that a proportion of sufferers come from families with known defects of lipid metabolism and that diabetics, for whom a genetical basis is postulated for most cases, are also particularly afflicted. Progess in the correction of metabolic defects, such as diabetes, encourages the view that the same may be done for "coronary prone" individuals once the nature of the defect can be defined. It seems not unreasonable to suppose therefore that future research may show that a genetical basis of coronary artery disease may be commoner than we now suspect and acting in more subtle ways than we now conceive, possibly for example in association with other factors such as obesity, inactivity, heavy smoking, hypertension, and occult diabetic tendency (Stamler et al., 1966).

It is interesting to note at this point that the atheromatous process is widespread in the animal kingdom (Finlayson, 1962; Gresham and Howard, 1964; Roberts and Strauss, 1965), at least among the vertebrates, and man is clearly not unique in

SOURCE: E. R. Nye, "Natural Selection and Degenerative Cardiovascular Disease," *Eugenics Quarterly*, 14: 127-131 (Chicago, Illinois: The University of Chicago Press, 1966).

seeding his own destruction in his vascular tree but he appears to suffer more from coronary atheroma than other species. The question is posed therefore, has degenerative vascular disease, and possibly other degenerative disease, arisen as a result of natural selection? In other words does a time come for the individual when his death, given that he has escaped the hazards of accident and infection, may in some way be advantageous to the rest of the species? This question can only be answered in the affirmative if it can be shown that some advantage follows from the existence of "built-in obsolescence." The writer takes the view that there are grounds for believing such an advantage existed formerly even if not now for man.

It is clear that in changing environment a steady turnover of the genetic pool represented by a population provides the most rapid way of adapting the population to new circumstances. After the age of about 30-40 years primitive man had fulfilled his function to his species and was expendable as his genetic material was represented in his offspring. If by "mischance" he lived, say into his forties, he was perhaps becoming, in the evolutionary sense, a liability, largely by being past his reproductive peak, possibly by competing with younger males for females and also for available food. After passing the peak reproductive period an individual would assume the increasing role of a parasite in his community. It seems probable that in very primitive societies disease or accident would take an increasing toll of the individuals past their prime, but in the absence of such factors some built-in "safety mechanism" could ensure that a community did not become overburdened with relatively less productive individuals. The speculation that this "safety mechanism" could be the degenerative diseases, including coronary artery disease, follows.

The operation of a mechanism culling the excess of aged individuals from a population would probably be appreciated best, as a desirable evolution process, in consideration of small, primitive human groups or tribes perhaps living near subsistence level, or at least subject to periods of food scarcity at intervals, since, as pointed out by Boule and Vallois (1956), a population is limited by conditions in times of scarcity, not by available food in times of plenty. The maintenance of the status quo would depend on a rapidly reproducing population, ensuring greatest adaptability, with minimum competition from within from individuals past reproductive or food gathering prime. It is easily seen that such a group that became overburdened with healthy, probably aggressive, but relatively unproductive individuals could have been at a disadvantage compared with a group not so constituted, particularly during periods of food shortage.

If coronary artery disease evolved as a factor in population control then the question may be posed as to when in human history it assumed its particular importance in our species, since, as previously noted, the disorder is unusual in other animals. The consideration of this point and others is continued below.

Coronary Artery Disease in Women

If degenerative cardiovascular disease confers some ultimate advantage, then, unless some alternative process is operating, it is difficult to see why in the case of the human species there is a marked sex differential in incidence, men being much more affected than woman before the age of 50 years. The mortality data for men and women from affluent societies (United Nations, 1955) shows lower mortalities for women in almost all quinquennia of life for recent years. In contrast, however, countries characterized by a lower degree of overall prosperity are frequently seen to have higher female mortality than male, particularly during the childbearing years. In India for example during the decade 1891-1901 the mortality among females averaged about 14 per 1000 more than males from 10 to 14 years of age until 30 to 34. This sex differential had largely disappeared by 1941-1950 with some shift of the increased female mortality into the later quinquennia.

If the same trend of increased female mortality during the childbearing years occurred in primitive human society, it can be seen that a population "culling" process, capable of exerting the check which degenerative disease of the vascular system may have provided in the case of the male, already existed.

The above considerations are unquestionably a simplification of the effect of natural selection on the appearance of coronary artery disease in the human species. The question of the interplay between clotting and arterial lesions has not been discussed but comparative studies (Finlayson et al., 1962) do seem to indicate that the coronary artery lesion may be more a feature of human pathology than it is in other species; thus alteration in coagulability, also perhaps ultimately influenced by natural selection, could not have their often fatal effects in the absence of the coronary lesion.

It may also be argued that forces of natural selection cannot operate on a process that tends to appear after the later phases of an individual's reproductive life. Put another way, if the appearance of degenerative vascular disease is part of an undirected, fortuitous "ageing" process, then, as it comes after peak reproduction is over (or probably would be in primitive human society), the tendency cannot come under the influence of eliminative forces and will thus persist. It might also be argued that the late appearance of coronary artery disease in life is simply the consequence of the fact that any inborn tendency favoring early appearance of the disease will operate against successful reproduction by the affected individual. The same argument applies, of course, to malignant disease. These objections do not take account, however, of the fact that the life span of our species is extremely well defined and it seems not unreasonable to suppose that evolution is responsible for our expectancy of "three score years and ten." The objection is also met by considering evolutionary forces acting on communities of individuals rather than, in the present context, on the individuals themselves. As pointed out by Simpson (1958),

"Selection favours successful reproduction of the population, not of any or all individuals in it." Thus a small human community living under marginal conditions of existence is clearly at a disadvantage if it carries an excess of relatively nonproductive individuals, compared with a community not so burdened.

Coronary Heart Disease in Primitive Societies Today

The chief difficulty in accepting any such simplified concept as that outlined above, and one which will occur to any worker in the field of coronary heart disease, is that in primitive societies today coronary heart disease (CHD) is generally uncommon. Thus in diabetic Yemenite Jews, who live under relatively primitive conditions in Israel, the incidence of vascular disease, including coronary disease, is less than in their diabetic compatriots living under more affluent circumstances (Brunner et al., 1964). Similarly, the South African Bantu suffers less from atherosclerosis than affluent Caucasians. Therefore the thesis that coronary heart disease was a killing disease of early man, and that it reduced the likelihood of his living to old age, seems untenable in the light of present knowledge. One would have to postulate that the environment of our primitive ancestors was different from that of primitive societies today or that they were perhaps more prone to coronary disease and that the incidence of the disorder has decreased since then. The last possibility is not remote since modern studies on the incidence of atherosclerosis in primitive people (Higginson and Pepler, 1954; Walker and Arvidsson, 1954) emphasize that the reduction in atheroma may be associated with a diet low in animal fat content. Can it be therefore that CHD developed as a peculiarity of human pathology, and perhaps as a population limiting process at a time in human history when man was more carnivorous in his habits than at present? It seems more possible that this may have happened since anthropologists favour the view that early man went through a long term period of meat eating.

Pattern of Human Social Development and Possible Course of the Evolution of Coronary Heart Disease

If we accept the possibility that coronary artery disease is a mechanism of population control which has arisen as a result of evolutionary processes, then we may speculate as to how this arose in association with changes in human society.

Anthropologists have maintained that early human society was based on hunting, thus Bartholomew and Birdsell (1962), drawing on evidence from bone middens associated with remains of the australopithecines, express the view that this early toolmaker was a "generalized carnivorous animal."

It will be noted that this carnivorous habit is in contrast to the vegetarian diet of the manlike apes who seldom or rarely eat meat (Goodall, 1963; Howells, 1960). This change in dietary emphasis was probably a consequence of the use of tools and of fire. A rich source of previously almost unobtainable food became available; it is even possible that hunting was easier in the early days of palaeolithic man since prey species were not accustomed to regarding man as a potential predator and flight reactions to anthropoids may have taken quite some time to become established. The relatively sudden alteration in dietary pattern in response to changing circumstances is not unknown in other species, note for example the recent appearance of carnivorous tendencies in the New Zealand Kea, a formerly vegetarian parrot that is now often responsible for attacking and eating sheep. If CHD developed during the palaeolithic age and was in some way connected with the biochemical consequences of a change in dietary pattern the next question that must be posed is whether or not the duration of the stone age period allowed a sufficiently long time for the processes of natural selection to operate. Kroeber (1948) quotes estimates of the stone age lasting, according to various authorities, between 200,000 years and 1,000,000 years; this is in marked contrast to the age of metals which has not lasted longer than 4,000 years. Assuming an intermediate period of about 500,000 years, this would allow about 20,000 generations on which natural selections would operate. In any event it seems likely that after a long period of meat eating, augmented doubtless by eating obtainable fruits and roots, man in more recent times entered upon an agrarian phase with a consequent shift of dietary emphasis towards a more "primitive" style of feeding. It may be significant today that essentially stone age cultures, such as the Australian aborigine, depend on hunting to a great extent for food, whereas primitive communities of a more advanced order and skilled in the use of metals have become farmers. Examples are the Bantu of South Africa and many Indian communities; it may be noted that shortage of dietary protein and probably animal fat is not uncommon in these people.

Our present dietary position seems to be therefore that after the wheel having swung the full circle, from partial or complete vegetarianism back to the same state, we are now moving back into stone age feeding habits, the rancher and the dairy farmer having in the meantime relieved us of the need to kill our own meat.

Mechanisms which may have evolved in early man associated with the need to maintain the youthful vigour and genetic adaptability of stone age tribes may now be operating again since, biologically speaking, the palaeolithic period is the yesterday of human evolution.

Acknowledgment

My thanks are due to Dr. A. M. O. Veale for his comments and his criticisms of my arguments.

References

Bartholomew, G. A., and J. B. Birdsell, 1962. In *Culture and Evolution of Man*, Oxford University Press, London, p. 29.

Boule, M., and H. V. Vallois, 1956. *Fossil Men*. Thames and Hudson, London.

Brunner, D., S. Altman, L. Nelken, and J. Reider, 1964. The relative absence of vascular disease in diabetic Yemenite Jews. *Diabetes*, 13: 268-272.

Finlayson, R., C. Symons, R. N. T-W-Fiennes, 1962. Atherosclerosis: A comparative study. *Brit. Med. J.*, i: 501-507.

Goodall, J., 1963. My life among wild chimpanzees. *Nat. Geogr. Mag.*, 124: 272-308.

Gresham, G. A., and A. N. Howard, 1964. *Bull. Soc. Roy. Zool. d'Anvers.*, Bull. No. 34.

Higginson, J., and W. J. Pepler, 1964. Fat intake serum cholesterol concentration and atherosclerosis in the South African Bantu. Part II. Atherosclerosis and coronary artery disease. *J. Clin. Invest.*, 33: 1366.

Howells, W. W., 1960. *Mankind in the Making*, Secher and Warburg, London, pp. 84-86.

Kroeber, A. L., 1948. *Anthropology*. Harcourt, Brace, New York, p. 651.

Roberts, J. C., and R. Straus, 1965. *Comparative Atherosclerosis*, Harper and Row, New York.

Simpson, G. G., 1958. *The Study of Evolution*. Yale, New Haven.

Stamler, J., D. M. Berkson, H. A. Lindberg, Y. Hall, W. Miller, L. Mojonnier, M. Levinson, D. B. Cohen, Q. D. Young, 1966. Coronary risk factors. *Medical Clinics of North America*, 50: 229-254.

United Nations, 1955. Age and sex patterns of mortality, model life tables for under-developed countries. Population Studies No. 22.

Walker, A. R. P., and U. B. Arvidsson, 1954. Fat intake, serum cholesterol concentration and atherosclerosis in the South African Bantu. Part I. Low fat intake and the age trend of serum cholesterol concentration in the South African Bantu. *J. Clin. Invest.*, 33: 1358-1365.

Selected Bibliography Including References Cited

PART III

Allison, A. 1954. Protection Afforded by Sickle-Cell Trait Against Subtertian Malarial Infection. *British Medical Journal*, 1: 290-294.

Allison, A. 1954. Notes on Sickle-Cell Polymorphism. *Annals of Human Genetics*, 19: 39-57.

Allison, A. C. 1956. The Sickle and Haemoglobin C Genes in African Populations. *Annals of Human Genetics*, 21: 67-89.

Anderson, C., Allan, J., and Johansen, P. 1967. Comments on the Possible Existence and Nature of a Heterogygote Advantage in Cystic Fibrosis. Pp. 381-387 in Rossi, E. and Stoll, E. (eds.) 1967. *Cystic Fibrosis*. (Bibliotheca Pediatrica #86) Karger, N. Y.

Bianco, I., Montalenti, G., Silvestroni, E., and Siniscalo, M. 1952. Further Data on the Genetics of Microcythaemia or Thalassemia minor and Cooley's Disease or Thalassemia major. *Annals of Eugenics, London*, 16: 299-315.

Black, F. 1966. Measles Endemicity in Insular Populations: Critical Community Size and Its Evolutionary Implications. *Journal of Theoretical Biology*, 11: 207-211.

Ceppelini, R. 1955. [Thalassemia and Malaria] Discussion of A. C. Allison's Paper: Aspects of Polymorphism in Man. *Cold Spring Harbor Symposia on Quantitative Biology*, 252-255.

Chakravartti, M., Verma, B., Hanurow, T., and Vogel, F. 1966. Relation Between Smallpox and the ABO Blood Groups in a Rural Population of West Bengal. *Humangenetik*, 2: 78-80.

Damon, A. 1969. Race, Ethnic Group and Disease. *Social Biology*, 16: 69-80.

Dubos, R. 1965. *Man Adapting*. Yale Univ. Press, New Haven.

Dubos, R. 1968. *Man, Medicine and Environment*. Praeger, N. Y.

Guthrie, R. 1969. Senescence as an Adaptive Trait. *Perspectives in Biology and Medicine*, 12: 313-324.

Haldane, J. 1956/1957. Natural Selection in Man. *Acta Genetica*, 6: 321-332.

Hamilton, W. 1966. The Moulding of Senescence by Natural Selection. *Journal of Theoretical Biology*, 12: 12-45.

Hertzog, K., and Johnston, F. 1968. Selection and the Rh Polymorphism. *Human Biology*, 40: 87-97.

Hiernaux, J. 1952. La genetique de la sicklemia et l'interet anthropologie noire. *Ann. Mus. Congo belge Anthropologie*, 2: 1-42.

Kaplan, E., et al. 1968. Reproductive Failure in Males with Cystic Fibrosis. *New England Journal of Medicine*, 279: 65-69.

Livingstone, F. 1958. Anthropological Implications of Sickle Cell Gene Distribution in West Africa. *American Anthropologist*, 60: 533-560.

Livingstone, F. 1967. *Abnormal Hemoglobins in Human Populations. A Summary and Interpretation.* Aldine, Chicago. 482 pp.

Motulsky, A. 1960. Metabolic Polymorphisms and the Role of Infectious Diseases in Human Evolution. *Human Biology*, 32: 28-62.

Montalenti, G. 1965. Infectious Diseases as Selective Agents. pp. 135-151 in Meade, J. and Parkes, A. (eds.) 1965. *Biological Aspects of Social Problems.* Plenum Press, N. Y., 226 pp.

Peritz, E. 1967. A Statistical Study of Intrauterine Selection Factors Related to the ABO System. I. The Analysis of Data on Liveborn Children. *Annals of Human Genetics*, 30: 259-272.

Post, R. 1966. Breast Cancer, Lactation and Genetics. *Eugenics Quarterly*, 13: 1-29.

Post, R. 1966. Deformed Nasal Septa and Relaxed Selection. *Eugenics Quarterly*, 13: 101-112.

Sever, L. 1969. ABO Hemolytic Disease as a Selection Mechanism at the ABO Locus. *American Journal of Physical Anthropology*, 21: 177-186.

Siniscalo, M., Bernini, L., Latte, B., and Motulsky, A. 1961. Favism and Thalassemia in Sardinia and Their Relationship to Malaria. *Nature*, 190: 1179.

Stearn, E., and Stearn, A. 1945. *The Effect of Smallpox on the Destiny of the Amerindian.* Bruce Humphries, Boston.

Turner, J. 1969. Epistatic Selection in the Rhesus and MNS Blood Groups. *Annals of Human Genetics*, 33: 197-206.

Turner, J. 1968. How Does Treating Congenital Diseases Affect the Genetic Load? *Eugenics Quarterly*, 15: 191-197.

Turner, J., and Williamson, M. 1968. Population Size, Natural Selection and Genetic Load. *Nature*, 218: 700.

Vogel, F. 1970. ABO Blood Groups and Disease. *American Journal of Human Genetics*, 22: 464-475.

Wiesenfeld, S. 1967. Sickle-Cell Trait in Biological and Cultural Evolution. *Science*, 157: 1134-1140.

Wiener, A. 1970. Blood Groups and Disease. *American Journal of Human Genetics*, 22: 476-483.

Williams, G. 1957. Pleiotropy, Natural Selection, and the Evolution of Senescence. *Evolution*, 11: 398-411.

Woolf, C. 1962. Medical Dilemma: Is Insulin Therapy Increasing the Frequency of the Gene for Diabetes Mellitus? *Eugenics Quarterly*, 9: 228-230.

Zinsser, H. 1935. *Rats, Lice and History.* Bantam Books, N. Y.

PART IV

Natural Selection in Relation to Human Behavior

It is very difficult to quantitatively measure human behavior. Yet how human beings behave—how they react to their environment—determines whether they survive; if they survive, whether they reproduce; and if they reproduce, how many offspring they have. Thus the study of natural selection in relation to human behavior is crucial if man is to attain an adequate understanding of his past and ongoing evolution.

Natural selection has favored educability throughout the history of mankind because of the selective advantage that cultural behavior gives a population. Campbell (1966) summarized the advantages of culture when he stated that

" . . . culture is an evolutionary development of the greatest importance, since a new adaptive behavior pattern developed by one member of a group may be quickly taken up by others through imitation, and such a behavioral response to the environment on the part of the deme [population] is much faster than that made possible by gene mutation."

We must remember that while cultural change is undoubtedly the most important evolutionary change taking place today, behavioral adaptation to the environment is also taking place via genetic change. The occurrence of one type of change does not preclude the other. Both cultural and genetic adaptation to the environment are going on simultaneously (see diagram p. 207).

Man's attempts to culturally adapt to the environment and to adapt the environment to himself have drastically altered the direction and intensity of natural selection, particularly with respect to behavioral patterns. In the past, natural selection has favored aggressive behavior patterns (both genetic and cultural) in man because of the advantages that are attained by the aggressive individuals within a population and by aggressive populations in competition with other populations (see Tinbergen,

1968; Wynne-Edwards, 1962; Gilula and Daniels, 1968; Montagu, 1968; and Daniels, Gilula, and Ochberg, 1970 for a review of the hypotheses concerning the origin and maintenance of aggressive behavior patterns in populations). Violence as a form of aggressive behavior to solve disagreements among populations appears to have become maladaptive in the nuclear age. It will probably take a nuclear war to prove this contention.

No major study of the operation of natural selection in relation to human personality patterns within a human population is available. A number of individuals (Alland, 1966; Ardry, 1966; Emlen, 1966; Hamburg, 1963; Morris, 1966, 1968; Reynolds, 1966; Fox, 1967; Tinbergen, 1968) have made speculations concerning the genetic and cultural evolution of man's personality patterns in the past. Two very large follow-up research studies of populations where the participants took personality tests as school children are underway and the first findings relating human personality patterns to fertility in contemporary society should be published during the 1970's.

There is some evidence that in agricultural societies, such as India and China, selection has favored those social classes who were better educated or better off with respect to various socioeconomic criteria implying that the genes favoring intelligence were being selected for (Osborn, 1968). A number of studies have suggested that urbanization and industrialization have reversed the direction of natural selection to the point where the genes favoring high intelligence are being selected against in such Western societies as Great Britain and the United States (Cattell, 1937; Burt, 1952; Cook, 1951). Several recent studies seem to indicate that the inverse relationship between fertility and such factors as educational attainment, income, and occupational status that was observed in studies conducted during the first half of this century represents a transitional period. These recent studies have suggested that within the urban population the classic pattern of an inverse relationship between fertility and socioeconomic status or educational attainment is primarily confined to urban couples in which one or both spouses grew up on the farm (Goldberg, 1959, 1960, 1965; Freedman and Slesinger, 1961; Duncan, 1965). The Kalamazoo Fertility Study, reported on by Bajema, found a positive relationship between fertility and intelligence. This study was based on a population who spent most or all of their lives in an urban environment (as was the case for most of their spouses also).

There is also some evidence to support the contention that a postive relationship exists between socioeconomic (occupational) status and fertility among couples that plan their families. Therefore, one might expect that the relationship between intelligence and fertility will become more positive as the proportion of the population that is urban in origin increases and as the socioeconomic differentials with respect to the availability of contraceptive services and family planning attitudes decrease.

Methodologically, the papers in this section, which estimate the direction and intensity of natural selection in relation to human behavior patterns, represent the best attempts to measure selection in human populations. For instance, the paper that estimates the direction and intensity of natural selection in relation to human intelligence is the only study of natural selection in relation to human characteristics that takes all three of the factors affecting population growth (differentials in mortality, fertility, and generation length) into account simultaneously. By considering demographic factors in addition to fertility, the study of natural selection in relation to schizophrenia was able to show how a demographic trend such as an earlier average age of childbearing associated with an earlier average age at marriage can affect the reproductive fitness of schizophrenics.

The articles included in this section report on the research concerning the operation of natural selection in relation to behavior via.

1. A rare autosomal recessive gene which is typically lethal in the homozygous state before the homozygote reaches puberty,

2. A rare autosomal dominant gene which exerts its effect on behavior after the afflicted individual is typically past his peak reproductive years,

3. Schizophrenia, a mental disorder that approximately 1% of the population is afflicted with at sometime during their lifetime,

4. Intelligence, the behavioral patterns associated with educability.

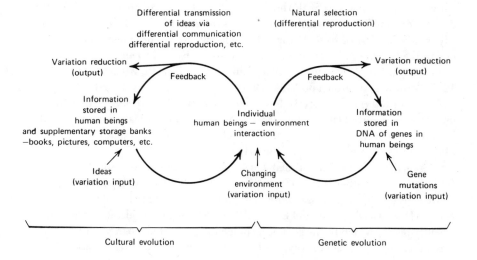

14. Population Dynamics of Tay-Sachs Disease.
I. Reproductive Fitness and Selection

NTINOS C. MYRIANTHOPOULOS

and

STANLEY M. ARONSON

Introduction

Tay-Sachs disease (TSD) or infantile amaurotic familial idiocy is a hereditary disease of lipid storage in which the sphingolipid ganglioside accumulates in the cytoplasm of the neurons of the brain. It is characterized by progressive degeneration of cerebral function which commences soon after birth and ends in death usually within the first or second year of life. The disease has been demonstrated to be due to the homozygous condition of an autosomal recessive gene, apparently with complete penetrance (Slome, 1933; Ktenidés, 1954; Kozinn, Wiener, and Cohen, 1957; Aronson, Aronson, and Volk, 1959). When the disease was first recognized, it was thought to be an exclusively Jewish disorder, but verified cases in non-Jewish infants were reported later. These cases have become increasingly numerous in the literature in recent years, and Myrianthopoulos (1962) showed that fully one-third of cases in the United States are of non-Jewish origin.

It is further well established that TSD is more frequent among the Ashkenazi Jews than among other Jewish (Sephardi) groups and non-Jewish populations. (See above investigators as well as Goldschmidt *et al.*, 1956; Goldschmidt, Ronen, and Ronen, 1960; Goldschmidt and Cohen, 1964). Although considerable literature has recently been devoted to the genetics, epidemiology, and demography of the disease, the major problem of why the TSD gene, despite its mass elimination, is

SOURCE. Ntinos Myrianthopoulos and S. Aronson, "Population Dynamics of Tay-Sachs Disease. I. Reproductive Fitness and Selection," *American Journal of Human Genetics*, 18: 313-327 (Chicago, Illinois: The University of Chicago Press, 1966).

found in such high frequency among the Ashkenazi Jews is still unresolved. The purpose of this paper is to examine the mechanisms by which the frequency of so lethal a gene could have become elevated and then maintained at the present high level and to present evidence which suggests that one such mechanism may have been responsible for this phenomenon.

Estimates of the frequency of the TSD gene among Jews and among non-Jews are in rather close agreement (Table 1). In the United States, the disease occurs

Table 1. Estimates of the Frequency of the TSD Gene
Among Jews and Non-Jews

Source	Region	Years Surveyed	Birth Incidence of TSD	TSD Gene Frequency	Carrier Frequency
Among Jews					
Goldschmidt *et al.* (1956)	Israel	1948-1952	0.00020	0.014	0.028
Kozinn, Wiener, and Cohen (1957)	New York City	1944-1955	0.00012	0.011	0.022
Myrianthopoulos (1962)	United States	1954-1957	0.00016	0.013	0.025
Aronson (1964)	New York City	1951-1962	0.00023	0.015	0.030
Among Non-Jews					
Kozinn, Wiener, and Cohen (1957)	New York City	1944-1955	0.0000022	0.0015	0.003
Myrianthopoulos (1962)	United States	1954-1957	0.0000017	0.0013	0.0026
Aronson (1964)	New York City	1951-1960	0.0000026	0.0016	0.0032

once in approximately 6,000 Jewish births and once in approximately 500,000 non-Jewish births. On the basis of this birth incidence, it is estimated that one of 40 Jewish persons and one of 380 non-Jewish persons is heterozygous for the TSD gene. Thus, the gene frequency is about ten times higher and the birth incidence 100 times higher among Ashkenazi Jews than among non-Jews. A difference of the same magnitude apparently exists between the Ashkenazi Jews and the other Jewish groups in Israel and the Eastern Mediterranean. The birth incidence of TSD among the Sephardic and Oriental communities of the Middle East and North Africa appears to be even lower than that found in non-Jewish Europeans and Americans.

The following mechanisms can be invoked to explain the differential frequency of TSD and the gene responsible for it in the two population groups: (1) differential breeding pattern; (2) genetic drift; (3) differential mutation rate; (4) differential fertility of the heterozygote, and (5) a combination of any of these mechanisms.

Differential breeding pattern would involve such socially conditioned forces as intermarriage between genetically different groups and close inbreeding within a group. It may be argued that the high frequency of TSD among the Ashkenazi Jews represents a repeated introduction of genes through local intermarriage with non-Jews which was subsequently abetted by inbreeding and genetic drift. This is improbable since we have no evidence that TSD was present in any appreciable frequency in the non-Jewish Polish, Lithuanian, and Russian populations among whom the Jews settled after the diaspora. It is from these areas that most of the ancestors of the Jewish TSD cases in the United States appear to have come (Aronson and Volk, 1962).

Consanguinity data are not helpful and merely reflect gene frequency levels in the two groups. In general, with rare recessive genes, the consanguinity rate among parents of affected children bears an inverse relationship to the gene frequency. This is precisely what we find in our data (Myrianthopoulos, 1962). In a series of 83 TSD families, the first-cousin marriage rate among parents of Jewish infants was 1.78%, while among the parents of non-Jewish infants it was 7.70%. Consanguinity of varying degrees was found in 5.26% of parents of Jewish infants and in 11.12% of parents of non-Jewish infants. First-cousin marriages among the Ashkenazi population of Israel range between 1% and 2% which is not significantly different than the rate for the Jewish population in the United States (Goldschmidt, Ronen, and Ronen, 1960), while that of the Sephardic and Oriental communities is extraordinarily high, ranging from 7% to 29% (Goldschmidt and Cohen, 1964). Yet, the birth incidence of TSD in these latter communities is very low, perhaps even lower than that found in non-Jewish European populations. It might be argued that the high consanguinity rate of the Sephardic Jews has served to deplete the gene from this population rather rapidly and the process must be continuing. However, this cannot account for the low frequency of the gene in non-Jewish populations.

The argument for genetic drift is weak enough at the outset, for drift, in the strict sense, is supposed to have its random effect on the frequencies of near neutral genes or slightly unfavorable ones. It is, perhaps, permissible to stretch the operational definition of drift to include lethal genes, such as the TSD gene, although we do not know of a precedent.

Genetic drift could be held partly accountable for the rise of the TSD gene at a very high frequency if it could be shown that the Jewish isolates of Europe, especially of northeastern Poland and the surrounding areas, were composed of very small marriageable populations without social contact with neighboring communities. There is sufficient historical commentary, however, to indicate a fertile intercommunication between these religious-cultural communities. In a review of the places of birth of the foreign-born Jewish grandparents and great-grandparents of a large number of TSD cases, Aronson (1964) found that about 40% of the grandparent marriages were between coreligionists born in different countries, 49% were between individuals born in the same East European country but in separate cities,

and only 11% of the marriages were between individuals born in the same community. These figures indicate that the grandparents and great-grandparents of children with TSD, at least as far back as 1850, were sufficiently mobile to choose
marriage partners in centers beyond their own immediate communities. Further,
from our demographic studies of these areas, it appears that the Jewish communities were large enough to support synagogues and schools and to engage in active
social life. Many thousands of Jews lived in northeastern Europe for many generations. We can find no evidence for circumstances which theoretically might favor
drift, such as migration of small groups, famine, disease, or war, affecting all or a
large number of these Jewish communities simultaneously. No doubt, such circumstances existed at one time or another, within one community or another. But even
under conditions of complete genetic isolation, random fluctuation of the TSD
gene in some communities must have been balanced in other communities. And our
demographic data show that the Jews of the United States, among whom our observations were made, emigrated not from a few selected communities but from hundreds of cities, towns, and villages of northeastern Europe (S. M. Aronson and N. C.
Myrianthopoulos, unpublished data). On the basis, then, of what is known, it is
unreasonable to hold genetic drift as the predominant factor responsible for the ele-
vation of the TSD gene frequency among the Ashkenazi Jews.

The possibility of differential mutation rate is equally unlikely. The mutation
rate needed to account for the estimated frequencies of the TSD gene would be
equal to the frequency of the trait in the population, i.e., about one mutation per
6,000 gametes per generation among Jews and one mutation per 500,000 gametes
per generation among non-Jews, which is absurd. All available evidence indicates
that mutation rates of specific loci tend to be rather constant within species.

The Possibility of Heterozygote Advantage

The possibility of heterozygote advantage was considered earlier (Myrianthopoulos, 1962) but not pursued because of lack of adequate control data to test it. Heterozygote advantage would provide an adequate explanation for the differential
increase of the TSD gene if it could be shown that Jewish heterozygotes were more
fertile than the Jewish homozygous normal and thus were able to transmit the mutant gene to the next generation differentially.

The magnitude of the selective advantage required by heterozygotes in order to
maintain the frequency of the lethal TSD gene at such a high level among the
Ashkenazi Jewish population is given by

$$S = q/(1-q)$$

where S is the selection coefficient against the normal homozygote. Substituting
the estimated frequency of the TSD gene among the Ashkenazi Jews of 0.0126,

$$S = 0.0126/(1-0.0126) = 0.0128$$

and the fitness of the three genotypes is

$$TT = 1-S = 0.9872$$
$$Tt = 1$$
$$tt = 0$$

which means that a selective advantage of about 1 1/4% on the part of heterozygotes is sufficient to maintain the gene at equilibrium despite its mass elimination via the TSD homozygotes. In order to determine if this is the case in the TSD population, it is necessary to collect and examine information on the fertility of Jewish women, particularly those known to be carriers of the TSD gene.

Our study population could, of course, be the parents of children with TSD, since these are both known to be heterozygous. But their total reproductive performance is almost certain to be biased by the birth of a child with a lethal hereditary disease and by the knowledge that there is a high probability of repeating this misfortune in subsequent pregnancies. This bias is quite evident from birth order analysis of 188 Jewish TSD cases in 150 sibships of two or more, shown in Table 2, calculated by the method of Haldane and Smith (1948). In theory, the test makes use of A, the sum of birth ranks of all affected sibs, and compares it with its theoretical mean value, calculated on the assumption that there is no birth rank effect.

Table 2. Effect of Birth Rank Among TSD Jewish Cases

	Birth Rank								
Sibship Size	1	2	3	4	5	6	7	8	Total
2	28	65							93
3	16	15	35						66
4	4	2	4	7					17
5	1	1			3				5
6									
7	1				1		1	1	4
8						1	1	1	3
Total	50	83	39	8	3	2	2	1	188

$6A = 2646; m = 2283; 6A - m = 363;$
$s^2 = 3721; s = 61.0;$ number of standard deviations exceeding mean $= 363/61 = 5.95.$

In practice, the arithmetic is simplified if $6A$ is tested in place of A. Both $6A$ and its variance are integers and are given in table form in the article by Haldane and Smith. Among the Jewish TSD cases, $6A$ exceeds its mean by almost six standard deviations. A similar birth rank effect, although not as striking, is found among non-Jewish TSD cases, where $6A$ exceeds its mean by almost three standard deviations. This finding is almost predictable and does not really indicate that birth rank effect is a phenomenon of biological significance. It merely reflects the bias which is introduced by a voluntary truncation of family size when a child with TSD has been produced. This effect is much more pronounced among the Jewish cases, since Jewish parents who have a TSD child are more aware of the social and medical consequences of such an occurrence than are non-Jews generally, and the eugenic problem concerns them more acutely.

An unbiased estimate of the fertility of the unsuspecting heterozygotes, who unknowingly perpetuate the gene, can be obtained by assessing the fertility of the grandparents of the affected children. It may be assumed that at least one maternal and one paternal grandparent of an affected child is a heterozygote. It can also be assumed that by the time the TSD homozygote (grandchild) has been identified, the fertility of the grandparents will have been completed. The fertility can be estimated by determining the number of surviving offspring, that is, the sibships containing the parents of the affected, and comparing with an appropriate control group.

Source of Data and Methodology

The data for this study came from two sources: from screening by one of us (N. C. M.) of death certificates from the Bureau of Vital Statistics assigned to rubric 325.5 (mental deficiency, other and unspecified types) of the International List of Causes of Death, for deaths which occurred in the United States during the years 1954-1957, inclusive, and from a case registry of cerebral sphingolipidoses which was begun by one of us (S. M. A.) in 1952.

The first source, which constitutes as complete ascertainment of cases as possible for the four year period, yielded 89 cases of TSD, 58 in children of Jewish parents, 29 in children of non-Jewish parents and two cases in children with one Jewish and one non-Jewish parent. The second source provided 296 cases in 242 families, mostly from the New York City area, and contains a much larger proportion of Jewish cases. From this source, 226 cases were in children of Jewish parents, 35 in children of non-Jewish parents, and 35 in children with mixed or doubtful parentage. Over 85% of cases recorded as having died from 1954 through 1957 were picked up independently through the second source. (For details about the method and criteria for selection of cases, see Myrianthopoulos, 1962, and Aronson, 1964.)

The unit of this investigation is not the affected individual but the family, particularly the sibship of the parents of the affected child. By personal contact and mailed questionnaire, we sought to obtain from the parents precise and detailed information concerning dates of birth and death and neurological conditions of siblings, parents, aunts and uncles, granparents, and first cousins of these cases, as well as other demographic data. We were able to collect all the required information for 194 families of Jewish cases and 47 families of non-Jewish cases.

In this paper, we are concerned exclusively with the sibships of the Jewish TSD parents, comprising 388 sibships with 1,244 total siblings. The distribution of the sibships of the parents of Jewish cases, which include the parents themselves, and the distribution of the control sibships, to be described below, is given by place of birth in Table 3. These are separated into U. S. born and non-U. S. born for purposes of analysis, since it is conceivable that a heterozygote effect might exist in the one group and not in the other.

Table 3. TSD and Control Sibships

	Number of Sibships	Total Number of Siblings
TSD		
U. S. born	322	1008
Non-U. S. born	66	236
Total	388	1244
Controls		
U. S. born	713	2436
Non-U. S. born	99	412
Total	812	2848

The ordinarily difficult task of selecting the proper control proved to be straightforward in our case. The lack of adequate demographic and fertility data for the Jewish population of the United States dictated the only logical alternative: to obtain a population sample which would be representative of the United States Jewry and comparable to our own TSD sample with regard to those variable which were required for comparison and control.

The sources of the control sibships were seven synagogues (Philadelphia, Pennsylvania; White Plains, Brooklyn, and Lawrence, New York; Bridgeport and Trumbull, Connecticut; and Boston, Massachusetts) as well as some fraternal societies (not affiliated with synagogues) from the New York City area. The selected controls were married couples with children among whom Tay-Sachs disease had not occurred. Thus, the claim can be made that the control sample is representative of the urban and suburban Jewish population of the northeastern United States, corresponding approximately with the TSD sample in country of birth, age, number of children, religious-cultural background, socioeconomic level, and geographic distribution. The co-operation of these people was entirely voluntary, and it was secured after an appeal and an explanation of the general scope of the study. The specific aim, i. e., the comparison of fertility and test for heterozygote advantage, was not discussed with them. Each control was asked to complete a confidential questionnaire, giving much the same information as that obtained from the parents of the TSD cases. This information was collected from 406 couples comprising 812 sibships with 2,848 total siblings. Their distribution is shown in Table 3.

One possible bias which could weaken the power of the comparison stems from the well known correlation of family size of closely related individuals. The family size of a propositus group could be greater than that of the control group to the extent that there is a correlation between the propositus sibship size and the sibship size of the parental generation. Unfortunately, it is impossible to demonstrate presence or absence of correlation between sibship size of our index cases and that of the parents because the parents have not completed their reproduction in most instances and also because Jewish parents of TSD children tend to curb their reproduction, as was demonstrated by the birth rank test (Table 2). Information about the siblings of the grandparents, almost all of whom were born in Europe and most of them now dead, is very scanty; therefore, a correlation at this level cannot be attempted either. The only positive statement that can be made is that the selected controls had, on the average, no fewer children than the TSD parents and that both groups must have come from relatively large families. We feel that this bias, if it has entered at all in the selection of the control group, is not of sufficient magnitude to influence the results.

Results

Selective advantage in modern genetics is expressed as a function of relative reproductive fitness. The concept of relative reproductive fitness of individuals with specific traits, or carriers of specific genes, although simple enough, has proven surprisingly difficult in practical application because the genetic situations on which the concept of fitness has a deciding bearing are often variable and subtle. Relative fitness in its simplest terms can be expressed as the ratio of the mean number of children of two groups, an affected group and a control group. In practice, such a ratio is not easy to derive. The affected group, for example, may not have completed their reproduction. There is no universal definition of the unit of fertility, neither need there be; further, there is the proverbial problem of what constitutes an adequate control group for a particular genetic situation and under a particular set of circumstances. These problems and their many ramifications have been dealt with extensively in methodological papers (see Krooth, 1955; Reed, 1959).

In this study, the definiton of relative fitness of the heterozygous carriers for the TSD gene is reduced to its simplest terms because the grandparents whose fertility is assessed have completed their reproduction at the time of the study. The unit of fertility is defined as a livebirth who has survived to reproductive age, for which 21 years is considered a reasonable lower limit; the analysis, however, will also include all livebirths.

The completed reproductive performance of the TSD heterozygotes, represented by the sibships of parents of TSD children and that of the controls, is shown in Table 4. The number of sibships, total number of siblings, number of siblings surviving through age 21, and average number of total and surviving siblings are given by decade of birth of each "proband" TSD and control parent. These are further subdivided according to the U. S. born versus non-U. S. born dichotomy. The "proband" parents are distributed by decade of birth so that the effects of fertility trends over a span of half a century would not obscure the over-all heterozygote effect, if any, and the contribution of each decade may be properly evaluated and weighted.

From a cursory inspection of Table 4, it becomes evident that there are some rather consistent differences between the TSD and control sibships, especially among the non-U. S. born, favoring the TSD sibships. These differences are clearly seen when the mean number of total and surviving siblings is adjusted for sample size. Table 5 shows the ratio of adjusted reproductive performance of TSD heterozygotes and controls. The adjustment was made by multiplying the number of control sibships by the average number of TSD siblings in each decade, and vice versa, and summing up over all decades. This represents a simple but true measure of fertility, and it shows that in all four categories the ratio is in favor of the TSD heterozygote. Its highest deviation from unity, about 16%, is in the category of non-U. S. born who survive to age 21; it decreases in the two following categories and becomes negligible among the total U. S. born siblings.

The differences which resulted after adjustment for sample size are, as expected, small and not statistically significant. The significance test which was considered as most appropriate for these data is a combination t test which combines separate t values for each decade, comparing the mean total siblings and those surviving beyond 21 years of the TSD sibships with the corresponding categories of the controls. The individual t values, shown in Table 6, were weighted by the inverse of their variances and summed over all decades, and a mean t was computed for each of the four categories: total U. S. born siblings, U. S. born surviving age 21, total non-U. S. born siblings, and non-U. S. born surviving age 21. The variance of t is given by

$$V_t = n/(n-2)$$

where n is the number of degrees of freedom. The weight of t is the inverse of its variance,

$$W_t = 1/V_t$$

Population Dynamics of Tay-Sachs Disease **217**

Table 4. Fertility of Jewish TSD and Control Families

Decade of Birth	1890-1899 TSD	1890-1899 Control	1900-1909 TSD	1900-1909 Control	1910-1919 TSD	1910-1919 Control	1920-1929 TSD	1920-1929 Control	1930-1939 TSD	1930-1939 Control	1940-1949 TSD	1940-1949 Control
U. S. born												
Number of sibships	2	45	13	71	81	210	136	249	84	127	6	11
Total siblings	7	217	58	338	326	803	382	732	215	322	20	24
Number dying before age 21	0	25	2	31	11	43	4	37	2	15	0	0
Number surviving to age 21	7	192	56	307	315	760	378	695	213	307	20	24
Average total	3.50	4.82	4.46	4.76	4.02	3.82	2.81	2.94	2.56	2.54	3.33	2.18
Average surviving to age 21	3.50	4.26	4.31	4.32	3.88	3.62	2.78	2.79	2.54	2.42	3.33	2.18
Non-U. S. born												
Number of sibships	2	20	5	22	14	25	30	23	15	8	0	1
Total siblings	13	121	27	118	49	84	103	66	44	21		2
Number dying before age 21	0	5	2	14	5	4	3	11	4	0		0
Number surviving to age 21	13	116	25	104	44	80	100	55	40	21		2
Average total	6.50	6.05	5.40	5.36	3.50	3.36	3.43	2.87	2.93	2.62		
Average surviving to age 21	6.50	5.80	5.00	4.72	3.14	3.20	3.33	2.39	2.67	2.62		

Table 5. Ratio of Adjusted Fertilities of TSD
Grandparents and Controls

U. S. born						
Total siblings	$\dfrac{\text{TSD}}{\text{Controls}}$	=	$\dfrac{1010.30}{1007.22}$	=	1.0031
Siblings surviving to age 21	$\dfrac{\text{TSD}}{\text{Controls}}$	=	$\dfrac{991.25}{953.70}$	=	1.0394
Non-U. S. born						
Total siblings	$\dfrac{\text{TSD}}{\text{Controls}}$	=	$\dfrac{235.85}{211.34}$	=	1.1160
Siblings surviving to age 21	$\dfrac{\text{TSD}}{\text{Controls}}$	=	$\dfrac{221.91}{191.00}$	=	1.1618

Table 6. Comparison of Mean Fertility of TSD Heterozygotes
and Controls. (Values of t for each of
four categories in each decade)

	U. S. Born		Non-U. S. Born	
Decade of Birth	Total Siblings	Surviving to Age 21	Total Siblings	Surviving to Age 21
1890-1899	−0.87	−0.56	0.22	0.43
1900-1909	−0.50	−0.02	0.03	0.25
1910-1919	0.85	1.13	0.24	−0.01
1920-1929	0.87	−0.07	1.00	2.02
1930-1939	0.30	1.00	0.46	0.08
1940-1949	1.16	1.16		

Table 7 shows the weighted mean of t, its standard error, and the standardized $\bar{t}$ in each of the four categories, where

$$\bar{t}_w = \Sigma W_t t / \Sigma W_t$$

and

$$SE_{\bar{t}_w} = 1/\sqrt{\Sigma W_t}$$

Table 7. Comparison of Mean Fertility of TSD Heterozygotes
and Controls. (Weighted mean of t, its standard error,
and standardized $\bar{t}$ in each of four categories)

Categories	$\dfrac{\bar{t}}{w}$	$\dfrac{SE-}{t}{w}$	$\dfrac{\bar{t}}{w}\Big/\dfrac{SE-}{t}{w}$	P
U. S. born				
Total Siblings	−0.01	0.42	−0.02	0.9840
Siblings surviving to age 21	0.43	0.41	1.05	0.2937
Non-U. S. born				
Total siblings	0.40	0.46	0.87	0.3843
Siblings surviving to age 21	0.53	0.45	1.18	0.2380

For n greater than 30, as is the case here, the distribution of $\bar{t}$ becomes very nearly normal, and the P values correspond to those of the standard normal variable. (For the best description of combination of tests of significance, see Hald, 1952.)

Although the differences in fertility are not significant, the apparent definite fertility gradient in the four categories suggests that differential survival might be an important component. Indeed, when the percentage loss in each decade is calculated (Table 8), the loss among controls, with only two exceptions, is higher—and survival to age 21 lower—than that among the TSD sibships.

Table 8. Loss of Siblings Not Surviving to Age 21, per Decade

Decade of Birth	1890-1899	1900-1909	1910-1919	1920-1929	1930-1939	1940-1949
U. S. born						
TSD	0.00	0.03	0.03	0.01	0.01	0.00
Controls	0.11	0.09	0.05	0.05	0.05	0.00
Non-U. S. born						
TSD	0.00	0.07	0.10	0.03	0.09	0.00
Controls	0.04	0.12	0.05	0.17	0.00	0.00

The over-all difference in survival to age 21 between the TSD and control sibships is statistically significant (Table 9). The test procedure employed here is that of combining 2 X 2 tables for each decade in each category, for which chi squares were computed in the usual way. These were then added up to command n degrees of freedom. By this test, the differences in all three categories are highly significant.

Table 9. Chi Square Test of Differences in Siblings Surviving to Age 21
Between TSD and Control Sibships

Decade of Birth	U. S. Born χ^2	Non-U. S. Born		Total Siblings χ^2
		χ^2	χ	
1890-1899	0.9	0.6	+0.78	2.0
1900-1909	2.0	0.5	+0.71	2.3
1910-1919	2.0	1.4	−1.18	0.6
1920-1929	11.5	9.9	+3.15	15.3
1930-1939	5.8	1.9	−1.38	1.9
	$\chi_5^2 = 22.2$	$\chi_5^2 = 14.3$	$\Sigma\,\chi = 2.08$	$\chi_5^2 = 22.1$
	$P < 0.001$	$P = 0.01$	$\Sigma\chi / \sqrt{n} = 0.93$	$P < 0.001$
			$P = 0.35$	

The procedure is legitimate in the categories U. S. born and total siblings because
the sign of the difference between observed and expected is the same in all decades. In the non-U. S. born category, however, the percentage loss in the decades
1910-19 and 1930-39 is higher among the TSD siblings than in the controls. In
order to account for the sign of the difference, values of χ instead of χ^2 were computed for each decade (Cochran, 1954). Since χ and, therefore, the sum of its independent values are normally distributed, the test statistic $\Sigma\chi/\sqrt{n}$ can be used and
referred to the normal curve tables. When computed in this way, the value of the
normal deviate is not significant. This, of course, is a test for survival apart from
that of fertility and measures the survival component in siblings of heterozygous
parents, half of whom are expected to be heterozygous normal.

Discussion

The failure to find statistically significant differences in fertility between TSD
heterozygotes and controls does not invalidate the argument of heterozygote advantage. The predicament here is that in order to be statistically significant, the differences in fertility would have to be huge and thus entirely out of line with the
hypothesis which requires that only a small increase be present and operating in
order to maintain the gene frequencies of a pair of alleles in polymorphic balance.
The observed differences are of the right magnitude and all in the same direction
and therefore compatible with the hypothesis that the TSD heterozygote has a

selective advantage over the presumed homozygous normal. But, unless either the fitness differential or the study sample is large enough to produce statistically significant differences, the probability of sampling variation must be kept in mind.

By this approach, then, the evidence cannot be considered as decisive and at best is only suggestive. But, since no reasonable support has been found for all the other logical hypotheses, including that of genetic drift, it is, perhaps, not idle to examine how ancillary and indirect evidence may bear on the hypothesis of heterozygote advantage.

Some evidence in favor of the heterozygote advantage theory is offered by the demonstration that, on the whole, the TSD sibships have significantly better survival than the control sibships. It appears that this advantage is largely due to the contribution of the U. S. born. This is surprising, especially since the fertility differential, although not significant, showed a definite gradient (Table 5), being highest for siblings born outside the United States who survived to reproductive age, and becoming negligible when total siblings born in the United States were considered. One possible explanation for this is that the data for the non-U. S. born group are not as reliable as those for the U. S. born, or perhaps a number of unknown factors are at play. Be that as it may, it is not unreasonable to attribute differential survival to resistance to some adverse situation, e.g. disease, conferred by the TSD gene. Such an explanation is compatible with the hypothesis of heterozygote advantage.

Historical perspective may also furnish some helpful information. At least three culturally and ethnically distinct Jewish groups are recognized in our times. These date back to the first century A.D., if not earlier. One group, known as the Ashkenazi, are those Jews who, after having left Palestine, dispersed in central, eastern, and western Europe, and includes their descendants who emigrated from there to North and South America, South Africa, and Australia. Another group, the Sephardi, lived in the countries around the Mediterranean, including the European part of Turkey. A third group, the Oriental Jews, live in Asia Minor, Iraq, Iran, and Yemen. It is generally agreed that the three Jewish groups achieved cultural and perhaps genetic individuality during the two thousand years of the diaspora. The rise of the frequency of the TSD gene in one of these groups can also be assumed to have been a part of the same process.

The conquest of Jerusalem by Titus during the first century A.D. is an historical landmark which altered fundamentally the course of life and cultural activity of the Jewish people. Although there were large numbers of dispersed Jews before the Roman conquest of Jerusalem, it is only since then that the mass exodus began. One may argue that the Jewish people who immigrated to central, eastern, and western Europe, especially those who ultimately settled in Poland and the Baltic States, developed a way of life subject to certain selective pressures which favored the rise of the TSD gene, while those who stayed around the Mediterranean and in the Oriental part of Palestine pursued a life which did not favor any appreciable change of the status quo. Post (1965) suggests that the continuous habitat of most

Jewish populations for several millennia in cities which have greater exposure to infectious and contagious diseases than rural populations, along with greater tolerance and concern of Jews for illness, may have provided an ecological environment different enough from that of the Gentiles to result in differential rates of natural selection between Jews and Gentiles.

It should be noted that TSD is not the only genetic disorder which has a uniquely high frequency among the Ashkenazi Jews. Gaucher's disease, Niemann-Pick disease, and possibly other rare metabolic disorders are known also to have a very high frequency among them and a much lower frequency among the other Jewish groups and non-Jewish populations. It is interesting that the antecedents of the majority of Jewish cases of both Gaucher's disease and Niemann-Pick disease in the United States are also traced to the northeastern provinces of Poland and the Baltic States. This may be explained on the grounds that all three lipid storage diseases are subject to the same unknown selective force and share the same polymorphic properties.

The historical argument can be carried one step further by estimating the magnitude of a presumed selective advantage required to raise the TSD gene frequency from 0.0013 at the end of the first century A.D., when the mass emigration of the Jews began, through 50 generations to the late nineteenth century, when TSD was recognized as occurring chiefly among the Ashkenazi Jews with a gene frequency of 0.0126. The following model was suggested by Dr. Alfred Naylor.

If the frequency of the TSD gene is q and the fitness S, then

	TT	Tt	tt
Frequency	p^2	$2pq$	q^2
Fitness	$1-S$	1	0

After selection

$$q' = \frac{(\frac{1}{2})\,2pq}{1-Sp^2-q^2} \sim \frac{q}{1-S} \ (\text{if } p \sim 1)$$

$$\triangle q' = \frac{Sq}{1-S} \sim \frac{dp}{dt}$$

and

$$\frac{dp}{q} = \left(\frac{S}{1-S}\right) dt = d\log_e q$$

Integrating both sides with respect to time and gene frequency, we have

$$\frac{S}{S-1}\,t \int_{t_1(\text{present time})}^{t(50 \text{ generations})} = \log_e q \int_{q_1(\text{non-Jewish frequency})}^{q_2(\text{Ashkenazi frequency})}$$

Substituting,

$$50 \, \frac{S}{S-1} = \log_e \frac{0.0126}{0.0013} = 2.3$$

$$\frac{S}{S-1} = 0.046$$

$$S = \frac{0.046}{1 + 0.046} = 0.044$$

It appears, then, that a selective advantage of about 4½% would suffice, under the assumptions, to raise the gene frequency to its present level among the Ashkenazi Jews. This is not too different from the over-all advantage of about 6% estimated from our data and, therefore, compatible with the hypothesis of heterozygote advantage.

The most promising approach is, of course, through demographic studies of the east European Ashkenazi communities in the areas where the antecedents of the TSD cases lived for many generations. If a selective agent is involved, its identification may possibly be achieved through an analysis of the political, social, ecologic, and epidemiologic forces uniquely influencing these communities. The recent genocide of the Jewish peoples during World War II, the destruction of records, and the rapidly changing traditional Jewish way of life make this undertaking extremely difficult. But in another generation, whatever evidence still remains will disappear altogether. Demographic studies of the type mentioned above have been initiated by us and are now in progress.

Summary

It is now well established that the birth incidence of Tay-Sachs disease (TSD) is a hundred times higher (and the gene frequency ten times higher) among the Ashkenazi Jews than among other Jewish groups and non-Jewish populations. There is no evidence that differential breeding pattern, genetic drift, or differential mutation rate can explain the difference in gene frequency distribution. The possibility of differential fertility of the heterozygote is examined at length. The reproductive performance of the grandparents of Jewish infants affected with TSD is compared with that of an appropriate control group. Although the differences in reproductive performance between the two groups are not statistically significant, the results suggest, but do not prove, that the Jewish heterozygote enjoys an overall reproductive advantage of about 6% over the presumed homozygous normal. This advantage appears to be greatest for offspring born outside the United States and surviving to reproductive age, diminishes for offspring born in the United States and surviving to reproductive age, and becomes negligible when total offspring born in the United States is considered. Survival to age 21, however, is significantly higher among TSD sibships than among control sibships, and this finding corroborates the heterozygote advantage hypothesis. Historical evidence also appears to corroborate the hypothesis and to place the rise of the TSD gene among the Ashkenazi Jews in historical times, perhaps during the early centuries of the diaspora.

Acknowledgement

We would like to express our thanks to Dr. Alfred Naylor for his advice and help in the analysis of the material and to Dr. Jerry D. Niswander for reading and criticizing the manuscript.

References

Aronson, S. M. 1964. Epidemiology. In *Tay-Sachs' Disease*. B. W. Volk, ed. New York: Grune and Stratton, pp. 118-153.

Aronson, S. M., Aronson, B. E., and Volk, B. W. 1959. A genetic profile of infantile amaurotic family idiocy. *Amer. Med. Assoc. J. Dis. Child.*, 98: 50-65.

Aronson, S. M., and Volk. B. W. 1962. Genetic and demographic considerations concerning Tay-Sachs' disease. In *Cerebral Sphingolipidoses: A Symposium on Tay-Sachs' Disease and Allied Disorders*, S. M. Aronson and B. W. Volk, eds. New York: Academic Press, pp. 375-394.

Cochran, W. G. 1954. Some methods for strengthening the common χ^2 test. *Biometric*, 10: 417-451.

Goldschmidt, E., and Cohen, T. 1964. Inter-ethnic mixture among the communities of Israel. *Cold Spring Harbor Symp. Quant. Biol.*, 24: 115-120.

Goldschmidt, E., Lenz, R., Merin, S., Ronen, A., and Ronen, I. 1956. The frequency of the Tay-Sachs gene in the Jewish communities of Israel. Presented at 25th Annual Meeting of Genetics Society of America, Storrs, Connecticut.

Goldschmidt, E., Ronen, A., and Ronen, I. 1960. Changing marriage systems in the Jewish communities of Israel. *Ann. Hum. Genet.* (Lond.), 24: 191-204.

Hald, A. 1952. *Statistical Theory with Engineering Applications*. New York: Wiley and Sons, pp. 407-408.

Haldane, J. B. S., and Smith, C. A. B. 1948. A simple exact test for birth-order effect. *Ann. Eugen.* (Lond.), 14: 117-124.

Kozinn, P. J., Wiener, H., and Cohen, P. 1957. Infantile amaurotic family idiocy. *J. Pediat.*, 51: 58-64.

Krooth, R. S. 1955. The use of the fertilities of affected individuals and their unaffected sibs in the estimation of fitness. *Amer. J. Hum. Genet.*, 7: 325-360.

Ktenidés, M. 1954. Au subjet de l'héredité de l'idiotie amaurotique infantile (Tay-Sachs'). Thesis no. 2264, University of Geneva.

Myrianthopoulos, N. C. 1962. Some epidemiologic and genetic aspects of Tay-Sachs' disease. In *Cerebral Sphingolipidoses: A Symposium on Tay-Sachs' Disease and Allied Disorders,* S. M. Aronson and B. W. Volk, eds. New York: Academic Press, pp. 359-374.

Post, R. H. 1965. Jews, genetics and disease. *Eugen. Quart.,*12:162-164.

Reed, T. E. 1959. The definition of relative fitness of individuals with specific genetic traits. *Amer. J. Hum. Genet.,*11: 137-155.

Slome, D. 1933. The genetic basis of amaurotic family idiocy. *J. Genet.,* 27: 363-376.

15. Huntington's Chorea in Michigan[1,2]
2. Selection and Mutation

T. EDWARD REED and JAMES V. NEEL

Introduction

This paper is the second in a series of three describing the results of a comprehensive study of Huntington's Chorea in the state of Michigan. The first paper was concerned with the demography and certain aspects of the genetics of this disease (Reed and Chandler, 1958). This presentation will be concerned with selection and mutation in relation to this condition. The third and final communication will be of a clinical nature.

Previous data have suggested that the relative fitness of choreics, compared to normal persons, is in the neighborhood of unity or above (Panse, 1942; S. Reed and Palm, 1951). If this were in fact the case, it would be a rare-not to say unique-situation in human genetics: individuals affected with a severely debilitating disease whose onset is often during the reproductive period nevertheless actually achieving a greater-than-normal fertility. Also, if these studies are correct, it then becomes necessary to explain why Huntington's chorea is a rare disease today, having a frequency in populations of European ancestry of about 4×10^{-5}. This study was undertaken primarily to analyze the population "dynamics" of the gene responsible for this disease in the state of Michigan. It will be shown that not only are individuals heterozygous for this gene at a reproductive disadvantage as compared with normal, but that there are probably at least three and possibly four different ways in which this disadvantage is brought about. The unreliability of sibling controls in studies of genetically determined fertility differentials will be demonstrated for the

SOURCE: T. Edward Reed and James Neel, "Huntington's Chorea in Michigan. 2 Selection and Mutation," *American Journal of Human Genetics*, 11: 107-136 (Chicago, Illinois: The University of Chicago Press, 1959).

[1]This study was supported in part by contract AT(11-1)-405 of the United States Atomic Energy Commission.

[2]We wish to acknowledge the assistance on clinical aspects of this study given by Dr. Joseph H. Chandler and the excellent field work done by Mrs. Estella Hughes and Mrs. Ruth Davidson.

case of Huntington's chorea. Finally, the rate with which mutation resulting in this phenotype occurs will be calculated.

Selection

The essence of this paper is an attempt to estimate the fitness of individuals with Huntington's chorea and to utilize this estimate in several calculations. Estimating fitness is always difficult in man and is especially so for Huntington's chorea. For this reason it is necessary to give considerable attention to the methods which were used.

A. Methodology

The fitness of one class of humans, say population 1 (P_1), relative to that of another, P_2 has usually been defined as $\bar{B}_1/\bar{B}_2$, where the $\bar{B}$'s refer to mean fertility. There are a variety of ways in which the mean fertility may be estimated and a number have been used, some without apparent genetic justification. Reed (1959) has recently discussed the definitions of relative fitness (W) for human populations and proposed the following definition as best: Assume that for P_1 and P_2 we have complete information on the survival and reproduction of a large random sample of liveborn individuals who have been followed from birth to death. Let

x = age in years at last birthday

N_i = original number of newborn liveborn individuals of population i

$B_{i \cdot x}$ = number of livebirths born to the survivors of the N_i individuals during their xth year of life.

$\bar{B}_{i \cdot o}$ = mean number of livebirths ever born per newborn liveborn individual of P_i

$B_{i \cdot y}$ = mean number of livebirths ever born per newborn liveborn individual of P_i per year, based on $\bar{B}_{i \cdot o}$ livebirths ever born

$P_{i \cdot x}$ = parental age frequency distribution of the N_i individuals (the proportion of livebirths, out of $\bar{B}_{i \cdot o}$ livebirths, which is born to the survivors of these individuals at age x).

Then,

$$\bar{B}_{i \cdot o} = \frac{1}{N_i} \sum_x B_{i \cdot x} \, , P_{i \cdot x} = \frac{B_{i \cdot x}}{\sum_x B_{i \cdot x}} \, , B_{i \cdot y} = \frac{1}{N_i} \sum_x \frac{B_{i \cdot x}}{x} \, ,$$

and the proposed definition of W for populations 1 and 2, is

$$W = \frac{B_{1 \cdot y}}{B_{2 \cdot y}} \tag{1}$$

or its equivalent

$$W = \frac{\bar{B}_{1 \cdot o} \sum_x \dfrac{P_{i \cdot x}}{x}}{\bar{B}_{2 \cdot o} \sum_x \dfrac{P_{2 \cdot x}}{x}}.$$

The number of livebirths ever born per newborn is used in order to include all factors affecting genetic fitness: viability, marriage, adult fertility. $B_{i \cdot y}$ is used instead of $B_{i \cdot o}$ alone because the latter is in reality a mean number of livebirths *per generation* and we have no assurance that the generation lengths of P_1 and P_2 are equal. Even if equal, the mean number of livebirths ever (to be) born per newborn *per year*, which gives an exact rate of increase of the population due to births, may differ since this is proportional to $\sum_x \dfrac{P_{i \cdot x}}{x}$. When the parental age distributions are the same, then, as (2) shows, (1) reduces to the usual ratio of means, $\bar{B}_{1 \cdot o}/\bar{B}_{2 \cdot o}$.

One further point should be noted about what may be called the reference population, P_2. Unless otherwise specified, the fitness of a group or genotype should always be related to the general population. The use of other groups, such as normal sibs, for P_2 may be allowed for convenience or necessity but it then becomes necessary to show that P_2 is representative of the general population in its fertility. As will be seen, the non-choreic sibs of the choreic individuals of the present study differ significantly in their completed fertility from the general population of Michigan. Consequently, a direct comparison with sibs would give a misleading estimate. It was therefore necessary to use several different approaches to obtain valid comparisons of choreics (or heterozygotes) with the general population.

Bias due to ascertainment of the kindreds (i.e., obtaining an unrepresentative sample) is believed to be negligible in this study since ascertainment is almost complete, sibships being used as of 1940 and ascertainment having continued until 1956.

In all the calculations to follow, the fitnesses of males and females were calculated separately to test whether there are significant differences. Such a difference has already been found for one dominant trait, namely neurofibromatosis (Crowe, Schull, and Neel, 1956).

In addition to the difficulties in estimating fitness discussed above, there is another peculiar to Huntington's chorea and other dominant traits with delayed onset, namely, the inability, at any specified age, to recognize a certain fraction of heterozygotes. Since the "normal" non-choreic sibs of choreics may still, at any age, in fact be heterozygotes who have failed to manifest the disease, and their children may receive the gene for Huntington's chorea and develop the disease themselves, it is necessary to estimate the number of non-choreic heterozygotes by the use of age-onset curves. This number, added to the number of choreics, gives an estimate of the total number of heterozygotes. It is the fitness of heterozygotes, not choreics, which is of *genetic* importance. The fitness of choreics is, of course, of interest for other reasons. The necessity for this correction, i.e., estimating the number of heterozygotes from the number of trait-bearers, has sometimes been overlooked.

Possible differences in the survival from birth to age five years, between children of choreics and children of non-choreics, were also investigated, since Panse (1942) has found evidence for increased mortality among children of choreics.

B. Results

1. *Sibship comparisons of mean number of children ever born*

For a series of choreics and their non-choreic siblings the conventional ratio of mean numbers of children was calculated, subject to the conditions discussed above. The sibships used are those containing a Michigan choreic living on April 1, 1940, all members of the sibship being used. The status of each member is taken as of the time of death or, if living, the time of last investigation (1954 to 1956). The classification of "choreic" means that either a) a person has been medically diagnosed as having Huntington's chorea (see Part 1 of this study) and/or b) he has been so classifed by reliable, non-contradictory lay reports. If only b), he is also a near biological relative (sib, son, etc.) of a medically diagnosed choreic. It is not desirable to restrict "choreic" to medically diagnosed cases since this introduces selection for severity of the disease and increased age of the individual. "Non-choreic" means that there is reliable information that the individual in question, at the time of investigation, lacked signs of the trait. The three sibships containing "Negro" choreics were excluded as well as all sibships in which information on fertility or choreic status was faulty or lacking. Sibships in which a choreic was ascertained only through an affected descendant, were omitted. The need to estimate the number of non-choreic heterozygotes imposed further restrictions. This was met by calculating the probability C_x that a non-choreic sib of a choreic has the gene for Huntington's chorea but does not show the trait at age x, given that his parent has the gene. Therefore sibships for which one parent could not be considered to be heterozygous for the gene leading to Huntington's chorea (Hh) were omitted.

In order to ensure that the reproduction of individuals was nearly, or completely, terminated, only persons living at age 45 or over, or deceased at age 15 or over, were used. Therefore choreics and non-choreics who die relatively young are included in the survey. The age 45 is a compromise between the desire to be certain that reproduction of the still-living is completed and the desire not to select for mildly affected choreics who can live to advanced ages. Age 15 was chosen because no chorea occurred earlier in this sample. Therefore the asumption that the gene has no effect on the fitness of its bearer before age 15 seems plausible, although, of course, this is not certain. This lower age limit enables us to neglect, for this calcula- tion, the many infant and juvenile deaths which, since we cannot determine geno- type, contribute no information. After making these omissions, 120 choreic males, 137 choreic females, 97 non-choreic males and 113 non-choreic females were avail- able for study. Five individuals could not be classified either as "choreic" or "non- choreic" from available information. They were omitted from the calculations since preliminary work (Reed, 1957) showed that this omission had a negligible effect.

In order to estimate C_x, we may note that if P_x is the probability that an indi- vidual who is Hh develops chorea by age x, then

$$C_x = \frac{1-P_x}{2-P_x}.$$

P_x may be estimated from the age of onset distribution as was done in Part 1 of this study. Since the values of C_x may be of interest elsewhere, they are presented, with the values for P_x, in Table 1 [omitted here]. It may be noted that at age 40, C_x is 0.240 so that the assumption that non-choreic sibs of this age are hh (homozygous normal) would lead to gross error. At age 50, C_x becomes 0.063, and at 60, 0.006. C_x at advanced ages is perhaps unreliable because the apparently normal distribu- tion of ages of onset may not, in fact, obtain for extreme deviations from the mean.

In Tables 2 and 3 the distribution by age at death or the age at time of last inves- tigation of the above-described choreic individuals and their non-choreic sibs are presented, together with the number of liveborn children ever born to them. Tables 4 and 5 give the distribution by marital status of the number of liveborn children ever born to these individuals. From the distributions in Tables 2 and 3 the number N of heterozygotes among the non-choreic sibs was calculated from the relation

$$N = \sum_x N_x C_x$$

where N_x is the number of non-choreics age x at time of death or last investigation. The number B of liveborn children ever born to these N non-choreic heterozygotes is

$$B = \sum_x B'_x C_x$$

where B'_x is the number of livebirths born to non-choreics who were age x at time of death or last investigation, assuming that the fertility of non-choreic heterozy- gotes is the same as that of all non-choreics. If one makes the more extreme as- sumption that the fertility of these non-choreic heterozygotes is the same as that of all choreics, the mean is only slightly changed, dropping about 0.005 in the males and increasing about 0.026 in the females. In this situation it seems adequate to make the former assumption. The values of N and B for all groups, together with

Table 2. Distribution by Age at Death of the Number of Deceased Choreics and Deceased Non-Choreic Sibs, and the Number of Liveborn Children Ever Born to Them. Sibships Containing a Michigan Choreic on April 1, 1940 and Selected, as Described in Text, for Estimating Relative Fitness.

N_x = Number of Individuals Age x;

B'_x = Number of Livebirths to Individuals Age x.

Age x	Males Choreic		Males Non-Choreic		Females Choreic		Females Non-Choreic		Age x	Males Choreic		Males Non-Choreic		Females Choreic		Females Non-Choreic	
	N_x	B'_x	N_x	B'_x	N_x	B'_x	N_x	B'_x		N_x	B'_x	N_x	B'_x	N_x	B'_x	N_x	B'_x
15	0	0	1	0	0	0	0	0	50	6	9	0	0	0	0	1	10
16	0	0	0	0	0	0	0	0	51	3	7	2	4	9	40	1	6
17	0	0	0	0	0	0	2	0	52	4	8	0	0	4	12	0	0
18	0	0	1	0	0	0	0	0	53	5	12	1	3	7	15	1	0
19	0	0	1	0	0	0	0	0	54	2	6	0	0	4	11	0	0
20	0	0	0	0	0	0	0	0	55	3	2	0	0	7	29	1	4
21	0	0	0	0	0	0	1	0	56	3	7	0	0	2	0	0	0
22	1	0	1	0	0	0	1	0	57	2	7	3	14	1	2	0	0
23	0	0	2	0	1	1	1	0	58	1	2	0	0	2	4	3	5
24	0	0	3	1	0	0	2	0	59	1	7	1	4	1	0	1	2
25	0	0	0	0	0	0	0	0	60	1	3	0	0	3	14	1	0
26	1	0	0	0	1	0	1	1	61	5	9	1	0	0	0	1	0
27	1	0	0	0	0	0	2	1	62	0	0	2	10	5	23	1	0
28	0	0	0	0	0	0	1	0	63	3	6	1	0	3	4	0	0
29	0	0	3	2	0	0	1	0	64	5	19	1	2	3	11	0	0
30	1	0	1	0	2	1	3	6	65	2	3	3	15	0	0	0	0
31	2	0	0	0	2	2	1	3	66	0	0	0	0	1	4	0	0
32	0	0	0	0	0	0	1	2	67	3	4	1	1	3	3	0	0
33	1	0	0	0	0	0	1	5	68	0	0	0	0	1	6	1	2
34	0	0	2	4	0	0	1	0	69	1	0	2	7	2	9	1	0
35	1	2	0	0	1	2	0	0	70	2	2	1	0	0	0	2	3
36	1	5	0	0	1	2	0	0	71	0	0	1	0	1	0	0	0
37	5	9	0	0	2	4	0	0	72	0	0	1	2	0	0	0	0
38	1	0	0	0	0	0	0	0	73	2	6	3	5	2	2	1	2
39	1	4	1	0	2	0	3	3	74	0	0	0	0	2	4	0	0
40	2	6	1	0	3	5	2	5	75	1	0	2	4	1	4	1	0
41	2	0	0	0	1	9	0	0	76	0	0	1	7	1	12	1	3
42	2	3	0	0	1	0	0	0	77	1	3	1	1	0	0	1	8
43	1	0	0	0	2	11	1	0	78	1	1	1	0	2	9	0	0
44	3	5	0	0	2	3	1	4	79	0	0	0	0	0	0	1	4
45	6	5	1	0	4	9	0	0	80	1	0	0	0	0	0	0	0
46	1	0	1	0	4	12	2	4	81	0	0	0	0	0	0	0	0
47	3	2	0	0	1	3	0	0	82	0	0	0	0	0	0	1	6
48	1	0	2	11	5	13	0	0	83	0	0	0	0	0	0	0	0
49	2	1	0	0	5	17	0	0	84	0	0	1	1	0	0	0	0
									85	0	0	0	0	0	0	1	4
									Total	97	165	51	98	107	312	50	93

Table 3. Distribution by Age at Time of Last Investigation of the Number of Living Choreics and Living Non-Choreic Sibs and the Number of Liveborn Children Ever Born to Them. Sibships Containing a Michigan Choreic on April 1, 1940 and Selected, as Described in Text, for Estimating Relative Fitness. N_x and B'_x as in Table 2.

Age x	Males				Females			
	Choreic		Non-Choreic		Choreic		Non-Choreic	
	N_x	B'_x	N_x	B'_x	N_x	B'_x	N_x	B'_x
45	0	0	3	7	1	2	1	0
46	0	0	1	1	2	6	7	23
47	0	0	2	0	0	0	0	0
48	2	4	1	1	3	9	1	4
49	2	3	0	0	1	7	3	4
50	0	0	3	2	0	0	3	4
51	3	4	2	5	1	1	2	6
52	0	0	2	2	2	5	1	1
53	1	3	0	0	1	2	2	6
54	2	7	2	2	1	0	0	0
55	0	0	2	6	0	0	2	7
56	1	1	0	0	2	9	2	6
57	0	0	0	0	1	2	2	0
58	2	4	1	4	3	1	3	4
59	0	0	0	0	1	0	2	7
60	1	0	1	0	0	0	1	2
61	1	0	4	15	2	4	1	3
62	2	12	2	10	1	4	1	0
63	1	2	0	0	1	4	1	1
64	1	8	1	0	0	0	5	14
65	1	0	4	8	0	0	0	0
66	1	2	0	0	0	0	2	4
67	0	0	1	1	0	0	2	0
68	0	0	1	0	0	0	5	9
69	0	0	1	0	0	0	1	5
70	1	0	1	2	2	11	1	3
71	0	0	3	6	2	2	1	4
72	0	0	0	0	0	0	1	2
73	0	0	2	5	0	0	2	5
74	0	0	2	6	0	0	2	4
75	0	0	0	0	1	5	0	0
76	0	0	2	10	0	0	3	1
77	1	7	0	0	0	0	0	0
78	0	0	0	0	1	0	1	4
79	0	0	0	0	0	0	0	0
80	0	0	1	2	1	0	0	0
81	0	0	0	0	0	0	1	0
82	0	0	1	8	0	0	1	3
Total	23	57	46	103	30	74	63	136

Table 4. Distribution of Number of Liveborn Children Ever Born to the Males Tabulated in Tables 2 and 3.
S = Single (Never Married); M = Married (Ever Married)

Number of Livebirths	Choreic									Non-Choreic								
	Deceased ≥ 15			Living ≥ 45			Total			Deceased ≥ 15			Living ≥ 45			Total		
	S	M	Σ	S	M	Σ	S	M	Σ	S	M	Σ	S	M	Σ	S	M	Σ
0	27	13	40	3	4	7	30	17	47	15	8	23	6	11	17	21	19	40
1	1	12	13	0	3	3	1	15	16	0	7	7	0	3	3	0	10	10
2	0	14	14	0	5	5	0	19	19	0	7	7	0	6	6	0	13	13
3	0	17	17	0	2	2	0	19	19	0	4	4	0	10	10	0	14	14
4	0	4	4	0	2	2	0	6	6	0	5	5	0	2	2	0	7	7
5	0	2	2	0	1	1	0	3	3	0	0	0	0	4	4	0	4	4
6	0	4	4	0	0	0	0	4	4	0	0	0	0	1	1	0	1	1
7	0	2	2	0	1	1	0	3	3	0	1	1	0	1	1	0	2	2
8	0	0	0	0	1	1	0	1	1	0	1	1	0	1	1	0	2	2
9	0	1	1	0	0	0	0	1	1	0	1	1	0	1	1	0	2	2
10	0	0	0	0	1	1	0	1	1	0	1	1	0	0	0	0	1	1
11	0	0	0	0	0	0	0	0	0	0	1	1	0	0	0	0	1	1
12	0	0	0	0	0	0	0	0	0	0	0	0	0	0	0	0	0	0
Total no. of livebirths	1	164	165	0	57	57	1	221	222	0	98	98	0	103	103	0	201	201
Total no. of individuals	28	69	97	3	20	23	31	89	120	15	36	51	6	40	46	21	76	97
Mean no. of livebirths	0.036	2.377	1.701	0.000	2.850	2.478	0.032	2.483	1.850	0.000	2.722	1.922	0.000	2.575	2.239	0.000	2.645	2.072
Variance	0.036	3.944	3.941	0.000	7.818	7.715	0.032	4.775	4.700	0.000	8.549	7.554	0.000	5.481	5.519	0.000	6.845	6.547
S.E. of mean	0.036	0.239	0.202	0.000	0.625	0.579	0.032	0.232	0.198	0.000	0.487	0.385	0.000	0.370	0.346	0.000	0.300	0.260

Table 5. Distribution of Number of Liveborn Children Ever Born to the Females Tabulated in Tables 2 and 3.
S = Single (Never Married); M = Married (Ever Married)

Number of Livebirths	Choreic Deceased ≥15 S	M	Σ	Choreic Living ≥45 S	M	Σ	Choreic Total S	M	Σ	Non-Choreic Deceased ≥15 S	M	Σ	Non-Choreic Living ≥45 S	M	Σ	Non-Choreic Total S	M	Σ
0	9	13	22	0	8	8	9	21	30	11	12	23	9	8	17	20	20	40
1	1	17	18	1	1	2	2	18	20	0	3	3	0	9	9	0	12	12
2	0	22	22	0	7	7	0	29	29	0	8	8	0	14	14	0	22	22
3	0	13	13	0	3	3	0	16	16	0	6	6	0	9	9	0	15	15
4	0	8	8	0	6	6	0	14	14	0	4	4	0	7	7	0	11	11
5	0	2	2	0	2	2	0	4	4	0	2	2	0	2	2	0	4	4
6	0	8	8	0	0	0	0	8	8	0	2	2	0	2	2	0	4	4
7	0	5	5	0	1	1	0	6	6	0	0	0	0	2	2	0	2	2
8	0	4	4	0	1	1	0	5	5	0	1	1	0	1	1	0	2	2
9	0	1	1	0	0	0	0	1	1	0	0	0	0	0	0	0	0	0
10	0	1	1	0	0	0	0	1	1	0	1	1	0	0	0	0	1	1
11	0	1	1	0	0	0	0	1	1	0	0	0	0	0	0	0	0	0
12	0	2	2	0	0	0	0	2	2	0	0	0	0	0	0	0	0	0
Total no. of livebirths	1	311	312	1	73	74	2	384	386	0	93	93	0	136	136	0	229	229
Total no. of individuals	10	97	107	1	29	30	11	126	137	11	39	50	9	54	63	20	93	113
Mean no. of livebirths	0.100	3.206	2.916	1.000	2.517	2.467	0.182	3.048	2.818	0.000	2.385	1.860	0.000	2.519	2.159	0.000	2.462	2.027
Variance	0.100	8.145	8.210	0.000	4.687	4.602	0.164	7.390	7.415	0.000	5.717	5.429	0.000	3.764	4.007	0.000	4.534	4.615
S.E. of mean	0.100	0.290	0.277	0.000	0.402	0.392	0.140	0.242	0.233	0.000	0.383	0.330	0.000	0.264	0.252	0.000	0.221	0.202

the number of individuals and livebirths and the means for all groups, are given in Table 6.

Table 6. Number of Individuals, Number of Livebirths, and Mean Number of Livebirths. Data of Tables 2, 3, 4, and 5. Numbers for Specified Genotypes are Estimated. See Text.

Hh = Heterozygote for gene for Huntington's chorea
$(Hh)_n$ = Non-choreic Hh individual
hh = Homozygous normal individual
B = Observed number of liveborn children ever born to the N individuals (age ≥ 15 years)
$\bar{B}_0{}^*$ = Estimated mean number of liveborn children ever born per *newborn* individual
N = Number of individuals

Status		Males					Females				
		Cho-reics	Non-Cho-reics	$(Hh)_n$	Hh	hh	Cho-reics	Non-Cho-reics	$(Hh)_n$	Hh	hh
Deceased	N	97	51	7.839	104.839	43.161	107	50	10.369	117.369	39.631
≥ 15	B	165	98	4.297	169.297	93.703	312	93	11.412	323.412	81.588
	$\bar{B}$	1.701	1.922	0.548	1.615	2.171	2.916	1.860	1.101	2.756	2.059
	S.E.	0.202	0.385	–	–	–	0.277	0.330	–	–	–
Living	N	23	46	1.358	24.358	44.642	30	63	1.884	31.884	61.116
≥ 45	B	57	103	1.979	58.979	101.021	74	136	4.656	78.656	131.344
	$\bar{B}$	2.478	2.239	1.457	2.421	2.263	2.467	2.159	2.471	2.467	2.149
	S.E.	0.579	0.346	–	–	–	0.392	0.252	–	–	–
Total	N	120	97	9.197	129.197	87.803	137	113	12.253	149.253	100.747
	B	222	201	6.276	228.276	194.724	386	229	16.068	402.068	212.932
	$\bar{B}$	1.850	2.072	0.682	1.767	2.218	2.818	2.027	1.311	2.694	2.114
	S.E.	0.198	0.260	–	–	–	0.233	0.202	–	–	–
Total	$\bar{B}_0$	1.388	1.554	–	1.325	1.664	2.170	1.561	–	2.074	1.628

* $\bar{B}_0 = \bar{B}l_{15}$ where $l_{15} = 0.75$ for males and 0.77 for females. These values may be in error by about two per cent. See text.

Mean of males plus females:

	$\bar{B}$	S.E.	$\bar{B}_0$
Non-choreic	2.048	0.162	1.556
hh	2.162	–	1.643

About ten percent of the non-choreic sibs in this sample are calculated to be heterozygotes. The number of these heterozygotes plus the number of choreics give the estimated number of heterozygotes in each category. We note that the mean for all non-choreic males, 2.072 ± 0.260, does not differ from the mean for all non-choreic females, 2.027 ± 0.202, so that males and females may be pooled to give a better estimate of the fertility of non-choreics. There is, however, a marked and significant difference in the fertility of male choreics and female choreics, the former being 1.850 ± 0.198, the latter 2.818 ± 0.233. A large fraction of this difference is explainable by the relatively high proportion of male choreics who fail to marry. In Tables 4 and 5 we see that the proportion of "never married" male choreics is $31/120 = 0.258$, while the corresponding proportion for female choreics is $11/137 = 0.080$. The probability that this difference is due to chance is less than .001. The proportions for non-choreic males and for non-choreic females are 0.216 and 0.177, respectively. This increased marriage rate of female choreics has already been noted in Part 1 of this study. That this sex difference in proportion of married choreics may be only one factor in the difference in mean numbers of children is suggested by the fact that the mean for all married male choreics is 2.483 ± 0.232 while for all female married choreics it is 3.048 ± 0.242. These means do not differ significantly but the difference is suggestive. It is worth emphasizing here that these means are the fertilities of individuals who have survived to the age of 15 years. To obtain the mean fertility per (liveborn) newborn individual $(\bar{B}_o)$ we must multiply each of the above-calculated means by the proportion of individuals of the specified category who survive to age 15, i. e. by l_{15}. Values of l_{15} are not known precisely since a direct count of deaths in these sibships between birth and age 15 is subject to some error and the division between heterozygotes and homozygotes cannot be made, while estimating l_{15} from life tables is made difficult by the wide range of birth years and geographical origins of individuals in these sibships. However, it should still be possible to estimate l_{15} for the present study from the life tables in the *United States Life Tables*, 1890, 1901, 1910, and 1901-1910 (U. S. Bureau of the Census, 1921) since these tables are based on data from ten north-eastern states, including Michigan. Since the living members of these sibships in 1940 were about 48 years of age, the l_{15} values for 1890 will be appropriate for individuals of this mean age and are an approximation for all individuals in this sample. For 1890 the l_{15} values are given only for Massachusetts (whites plus non-whites), being 0.706 for males and 0.723 for females. For 1901 and 1910 they are available for the ten states and can be further limited to whites only (native plus foreign-born). The values for these males in 1901 and 1910 are 0.780 and 0.805 respectively, while for females they are 0.807 and 0.831. It therefore seems adequate to consider that for the fertility sample composed of native and foreign-born whites, l_{15} is about 0.75 for males and 0.77 for females. The error in these estimates should not exceed two or three per cent for the general population of Michigan in 1890. Since the earliest observed age of onset was 15 years, it is believed that there is no difference in the l_{15} values of choreics, their non-choreic sibs, and the general population of Michigan. Using these values we note, in Table 6, that now, *measuring from birth*, only choreic females and *Hh* females have a $\bar{B}_o$ greater than two, while individuals in other categories have $\bar{B}_o$ values ranging from 1.325 to 1.664. Since the mean number of children required for exact replacement is two, these results indicate that only these two classes of females, on the average, are replacing themselves in the course of one complete life cycle.

Using the mean numbers of children ever born given in Table 6 we can calculate the relative fitness from ratios of means as is usually done. We use the values actually obtained in our sample since correcting to $\bar{B}_o$ involves multiplying both numerator and denominator by l_{15}. We first calculate two relative fitnesses, $W_{c \cdot n}$, defined as the ratio of the mean for choreics, of specified sex, to the mean for non-choreics of both sexes (since the males and females don't differ), and $W_{H \cdot h}$, the ratio of the estimated mean for Hh individuals, of specified sex, to the estimated mean for hh individuals of both sexes. [Because of the many comparisons of relative fitness to be made, the following terminology is adopted. The subscripts of W indicate the two groups whose fitnesses are being compared, the first subscript indicating the numerator and the second the denominator. The subscripts are c (choreic), n (non-choreic sib), H (estimated Hh), h (estimated hh sibs), and p (general population). It should be noted at the outset that $W_{H \cdot p}$ (for the mean of males and females) is the relative fitness of genetic importance. The other relative fitnesses are interesting for other reasons or were used in obtaining the mean $W_{H \cdot p}$.] We find the following values, uncorrected for parental age distribution:

	$W_{c \cdot n} \pm$ S.E.	$W_{H \cdot h} \pm$ S.E.
Males	0.903 ± 0.120	0.817 ± 0.110
Females	1.376 ± 0.157	1.246 ± 0.142
Mean of males and females	1.140 ± 0.117	1.032 ± 0.105

The basis of these estimates is negligible compared with the standard error since the standard errors of the mean number of livebirths are about 0.1 of the mean (Cochran, 1953). The standard error of a ratio of means requires knowledge of the correlation between pairs of observations used in estimating the numerator and denominator. This correlation is assumed to be zero here since, in the present data, there is no definite mechanism operating to produce a correlation between the fertility of a choreic, taken at random, with the fertility of a non-choreic, taken at random. A positive correlation would reduce the above standard errors. In testing the significance of the W's the differences between means were used. Only $W_{c \cdot n}$ for females is significantly different from unity ($P < 0.01$). For both sexes $W_{H \cdot h}$ is about ten per cent less than that of $W_{c \cdot n}$ so that the relative fitness of heterozygotes (relative to sibs, not to the general population) is less than a simple comparison of affected with normal sibs would indicate. A fairly striking difference between males and females shown by both $W_{c \cdot n}$ and $W_{H \cdot h}$ is apparent, in each case significant at the 0.01 level.

As shown previously (Eq. 2) the definition of W should include the parental age distribution if these distributions differ in the two groups being studied. In Table 7 the distribution is given of individuals by exact age at the birth of a liveborn child.

Table 7. Distribution of Individuals in Tables 2 and 3 by Age at Birth of Liveborn Child. Only Individuals All of Whose Ages at Birth Are Known. Deceased $\geq$ 15 Years Plus Living $\geq$ 45 Years.

x = Age at birth of child
N = Total number of individuals
B = Total number of children born to N individuals

Age x	Males Choreic	Males Non-Choreic	Females Choreic	Females Non-Choreic	Age x	Males Choreic	Males Non-Choreic	Females Choreic	Females Non-Choreic
15	0	0	0	1	35	7	4	10	4
16	0	0	1	1	36	4	7	10	3
17	1	2	4	2	37	7	5	6	3
18	1	0	7	5	38	3	2	4	2
19	1	4	9	7	39	3	4	7	4
20	1	2	19	10	40	6	2	9	1
21	5	3	23	5	41	3	2	2	0
22	7	5	17	9	42	2	3	2	0
23	8	2	13	12	43	2	2	3	0
24	8	9	20	9	44	3	2	3	0
25	17	6	16	7	45	1	1	1	0
26	11	10	18	7	46	2	1	0	0
27	15	3	14	7	47	2	2	0	0
28	13	6	14	10	48	3	0	1	0
29	8	10	10	9	49	0	2	0	0
30	9	5	14	4	50	1	0	0	0
31	9	8	4	11	51	2	1	0	0
32	9	4	13	4	53	0	0	0	1
33	12	6	10	6	54	1	0	0	0
34	8	7	9	3	56	1	0	0	0
					N	65	41	86	51
					B	196	132	293	147
					$\bar{x}$	31.158	31.068	27.843	26.823
					S.E.$\bar{x}$	0.539	0.649	0.401	0.515

In Table 8 the distribution of these parental age groups, by five year intervals, is given, together with the parental age (P_x) for males and females in Michigan in 1935 (the earliest year for these data). Using expected values from these Michigan P_x values, the significance of departures of the observed values from the state (testing within sexes) was calculated. As is shown, only female choreics differ significantly $(P < 0.01)$, having an excess of births at high ages. The parental age distributions within sexes, cannot be shown to differ between choreics and non-choreics. Since the $W_{c \cdot n}$ and $W_{H \cdot h}$ values obtained must later be used in estimating fitness relative to the general population, the values based on female choreics must be weighted by

$$\sum_x \frac{P_{1 \cdot x}}{x} \bigg/ \sum_x \frac{P_{2 \cdot x}}{x}$$

Table 8. Comparison of Observed Parental Age Distributions with the Distribution in Michigan in 1935. Data of Table 7 and of U. S. Bureau of the Census (Vital Statistics–Special Reports)

Age Group	Number of Livebirths								Proportion of Livebirths, Michigan, 1935	
	Males				Females					
	Choreics		Non-Choreics		Choreics		Non-Choreics		Males	Females
	Obs.	Exp.	Obs.	Exp.	Obs.	Exp.	Obs.	Exp.		
15-19	3	2.35	6	1.58	21	32.82	16	16.46	.012	.112
20-24	29	35.67	21	24.02	92	92.88	45	46.60	.182	.317
25-29	64	56.25	35	37.88	72	77.06	40	38.66	.287	.263
30-34	47	44.69	30	30.10	50	50.40	28	25.28	.228	.172
35-39	24	30.18	22	20.33	37	28.42	16	14.26	.154	.097
40-44	16	16.66	11	11.22	19	10.26	1	5.15	.085	.035
45+	13	10.00	7	6.73	2	0.88	1	0.44	.051	.003
Total	196	195.80	132	131.86	293	292.72	147	146.85	.999	.999
χ^2_5*		4.53		0.58		16.22		3.07	–	–
P		> 0.3		> 0.9		< 0.01		> 0.5	–	–

* Ages 15-24 pooled for males, 40+ for females.

(see Eq. 2) to correct for the different parental age distribution, where subscript 1 refers to female choreics and 2 to the females of Michigan. From Table 8, using the mid-ages, 17.5, 22.5, etc., this ratio is calculated to be 0.969. The corrected values of $W_{c \cdot n}$ and $W_{H \cdot h}$ are then:

	$W_{c \cdot n}$	$W_{H \cdot h}$
Females	1.333	1.207
Mean of males and females	1.118	1.012

2. Comparison of fertility of choreic and non-choreic sibs with 1940 census fertility data

The 1940 Census (April 1, 1940) obtained information on the number of (live-born) children ever born to white women of specified ages in Michigan. This permits direct comparisons to be made with the fertility of choreics and their non-choreic sibs who were living in Michigan on April 1, 1940. These data are presented in Table 9. It is important to note several characteristics of these data. They give mean number of livebirths per woman *surviving* to the specified age and therefore are not the desired mean number of livebirths per newborn individual. This latter statistic is not available for any representative American population. This limitation does not, however, invalidate direct comparison between the means of the census and those of non-choreic males and non-choreic females living at the time of the census because the viability of these normal sibs should not differ appreciably from that of the general population. There are no census data for males but it is obvious that, in the absence of migration, the mean number of livebirths ever to be born to a newborn male must be very nearly equal to that of a newborn female. For ages before the end of the reproductive period the mean age differentials between spouses will make the age-specific means differ, but for ages of 40 or over this difference should be negligible. The non-choreic sibs may therefore be directly compared to the general population of females of the state in terms of fertility of surviving individuals, but this is not the case for the choreics. In this latter case a bias arises as a consequence of using choreics who have reached a specified age. This produces a selection for choreics who are, on the average, less severely affected and may, therefore, differ in fertility from the mean for all choreics. One might expect that they would be more fertile because their chorea is milder. Because of this bias it was not thought worthwhile to make the additional corrections for a) the choreics who may have been selected, wholly or in part, because of having had children, and b) for non-choreics who are in fact heterozygotes and should be added to the choreics to obtain the total heterozygotes. Because of these biases it is very likely that the means for choreics are really an upper limit of the true mean for all choreics and for all heterozygotes. Since 22 individuals living in Michigan in 1940 and non-choreic at that time later developed Huntington's chorea, they are included in Table 9 in the category "Pre-choreic." The sum of choreics and "pre-choreics" should be a good approximation to the actual number of heterozygotes in 1940. Conversely, by not classifying the "pre-choreics" as non-choreics, the latter should be a good approximation to the homozygous normals.

Table 9 shows that, except for choreic females and the few "pre-choreics," individuals in almost every age group are less fertile, but usually not significantly so, than white Michigan females of the corresponding age groups. To obtain an approximation to the number of children ever born and increase the numbers of individuals in a specified category, the fertility of individuals age 40 and over, classified by whether ever married or not, was analyzed. These data are given in Table 10. It is seen that (total) male choreics are much less fertile than the state population of the same age distribution as those choreics (mean 1.867 ± 0.289; expected mean 2.836). This difference is significant at the .001 level. For total male choreics plus male pre-choreics the difference is still significant (2.047 ± 0.317; expected 2.824) at the .02 level. The difference between observed and expected for male choreics (with or without pre-choreics) who were ever married is not significant but in each case the fertility of choreics or choreics plus pre-choreics is less. The only other significant difference for separate sexes is that between total non-choreic females and the expected (2.255 ± 0.292; expected 2.902), significant at the .02 level. However, the mean for total non-choreic males, 2.224 ± 0.373, differs from the expected 2.933, at $P = .07$, and if one makes a similar comparison for total non-choreic males age 30 or over, the difference is significant ($P = .02$), the mean again being lower than expected. In each of these cases where a significant difference is found using total individuals, the mean based on "ever married" individuals is also lower than the corresponding expected mean, but not significantly so. The mean fertilities of non-chorecis do not differ significantly between sexes and therefore the fertilities of males plus females, for "ever married" and for total individuals, were calculated. For the former category the observed mean is 2.589 ± 0.249, expected 3.086, $P < .05$, and for the latter 2.240 ± 0.232, expected 2.917, $P < .01$. Since the mean number of livebirths ever born per non-choreic age 40 or more ($\bar{B}_{n \cdot 40}$) is less than that of the general population of Michigan females, ($\bar{B}_{m \cdot 40}$) it is very likely that the mean for newborn non-choreics ($\bar{B}_{n \cdot o}$) is also significantly less than the corresponding population mean ($\bar{B}_{m \cdot o}$). It is therefore inadmissible to consider the non-choreics as representative of the general population in any estimation of the relative fitness of choreics or heterozygotes. If one can assume that $\bar{B}_{n \cdot 40}/\bar{B}_{n \cdot o}$ is very near to $\bar{B}_{m \cdot 40}/\bar{B}_{m \cdot o}$ (this appears reasonable, see Reed [1959]), one can estimate the fitness ($W_{n \cdot p}$) of non-choreic sibs relative to that of the general population. We estimate this as

$$W_{n \cdot p} = \frac{\bar{B}_{n \cdot 40}}{\bar{B}_{m \cdot 40}} = \frac{2.240 \pm 0.232}{2.917} = 0.768 \pm 0.080,$$

the mean value for males plus females.

3. Estimate of the fitness of Hh individuals relative to the general population

Using normal sibs and the 1940 census: If we multiply the estimates of $W_{c \cdot n}$ (corrected for parental age distribution) of the preceding section by $W_{n \cdot p}$ we will

Table 9. Fertility of Choreics, Their Non-Choreic Sibs, and White Females of the General Population. All Individuals Living in Michigan on April 1, 1940. Data as of April 1, 1940. See Text Concerning Biases in the Means of Choreics Relative to Other Means

N = No. of individuals B = No. of liveborn children (Liveborn in "1940" are counted 1/4)

Age	Ever Married	Item	Males				Females				White Females, Michigan, 1940**
			Choreic*	Pre-Choreic*	Choreic + Pre-Choreic	Non-Choreic	Choreic	Pre-Choreic*	Choreic + Pre-Choreic	Non-Choreic	
15-29	No	N	3	2	5	2	3	2	5	3	—‡
		B	0	0	0	0	1	0	1	0	—
		$\bar{B}$	0	0	0	0	0.333	0	0.200	0	—
	Yes	N	2	2	4	10	3	1	4	10	277,020
		B	2.50	2.50	5.00	6.50	2	1	3	10	342,716
		$\bar{B}$	1.25	1.25	1.25	0.650	0.667	1.000	0.750	1.000	1.237
		S.E.	0.625	—	0.313	0.256	—	—	—	0.258	—
	Total	N	5	4	9	12	6	3	9	13	622,240
		B	2.50	2.50	5.00	6.50	3	1	4	10	342,716
		$\bar{B}$	0.500	0.625	0.556	0.542	0.500	0.333	0.444	0.769	0.551
		S.E.	0.313	—	0.139	0.224	0.224	—	0.176	0.231	—
30-39	No	N	2	0	2	2	0	0	0	1	—
		B	0	0	0	0	0	0	0	0	—
		$\bar{B}$	0	—	0	0	—	—	—	0	
	Yes	N	11	4	15	11	17	3	20	20	307,140
		B	21	5	26	16	50	5	55	35	668,540
		$\bar{B}$	1.909	1.250	1.733	1.455	2.941	1.667	2.750	1.750	2.177
		S.E.	0.368	—	0.316	0.312	0.539	—	0.481	0.428	—
	Total	N	13	4	17	13	17	3	20	21	343,380
		B	21	5	26	16	50	5	55	35	668,540
		$\bar{B}$	1.615	1.250	1.529	1.231	2.941	1.667	2.750	1.667	1.947
		S.E.	0.368	—	0.311	0.303	0.539	—	0.481	0.416	—
40-49	No	N	5	0	5	1	2	0	2	2	—
		B	1	0	1	0	0	0	0	0	—
		$\bar{B}$	0.200	—	0.200	0	0	—	0	0	

Age	Married										
40-49	Yes	N	17	3	20	10	29	5	34	15	269,180
		B	37	14.25	51.25	27	90	22	112	33	766,772
		$\bar{B}$	2.177	4.75	2.563	2.700	3.103	4.400	3.294	2.200	2.849
		S.E.	0.472	2.529	0.552	0.817	0.531	1.123	0.482	0.536	—
	Total	N	22	3	25	11	31	5	36	17	289,720
		B	38	14.25	52.25	27	90	22	112	33	766,772
		$\bar{B}$	1.727	4.750	2.090	2.455	2.903	4.400	3.111	1.941	2.647
		S.E.	0.407	2.529	0.481	0.779	0.515	1.123	0.473	0.504	—
50+	No	N	6	0	6	5	4	0	4	6	—
		B	0	0	0	0	0	0	0	0	—
		$\bar{B}$	0	—	0	0	—	—	—	0	—
		S.E.	0	—	0	—	—	—	—	0	—
	Yes	N	17	0	17	33	42	0	42	32	373,520
		B	46	0	46	82	135	0	135	91	1,186,644
		$\bar{B}$	2.706	—	2.706	2.485	3.214	—	3.214	2.844	3.177
		S.E.	0.452	—	0.452	0.469	0.475	—	0.475	0.376	—
	Total	N	23	0	23	38	46	0	46	38	393,420
		B	46	0	46	82	135	0	135	91	1,186,644
		$\bar{B}$	2.000	—	2.000	2.158	2.935	—	2.935	2.395	3.016
		S.E.	0.417	—	0.417	0.429	0.454	—	0.454	0.359	—
Total‡‡	No	N	16	2	18	10	9	2	11	12	—
		B	1	0	1	0	1	0	1	0	—
		$\bar{B}$	0.063	0	0.056	0	0.111	—	0.091	0	—
		S.E.	0.063	—	0.056	0	0.111	—	0.091	0	—
	Yes	N	47	9	56	64	91	9	100	77	1,226,860
		B	104	20.50	124.50	131.50	277	28	305	169	2,964,672
		$\bar{B}$	2.213	2.278	2.223	2.055	3.044	3.111	3.050	2.195	2.416
		S.E.	0.260	0.999	0.266	0.292	0.296	0.824	0.278	0.229	—
	Total	N	63	11	74	74	100	11	111	89	1,648,760
		B	105	20.50	125.50	131.50	278	28	306	169	2,964,672
		$\bar{B}$	1.667	1.864	1.696	1.777	2.780	2.545	2.757	1.899	1.798
		S.E.	0.227	0.855	0.299	0.265	0.282	0.767	0.264	0.214	—

* Non-choreic on April 1, 1940, choreic later.
** U. S. Census. Data on total white women, except ever married and not reporting on children.
‡ Single women were not asked about children.
‡‡ Means for total choreics and total non-choreics sibs are not comparable with the total state population since the age distributions are very different.

Table 10. Fertility of Choreics and Their Non-Choreic Sibs and the Expected Fertility $(\bar{B}_m)$ of White Females in Michigan with Similar Age Distribution. Only Individuals Age 40 Years or Over. Data and Symbols of Table 9

Ever Married	Item	Males				Females			
		Choreic	Pre-Choreic	Choreic + Pre-Choreic	Non-Choreic	Choreic	Pre-Choreic	Choreic + Pre-Choreic	Non-Choreic
No	N	11	0	11	6	6	0	6	8
	B	1	0	1	0	0	0	0	0
	$\bar{B}$	0.091	—	0.091	0	0	—	0	0
Yes	N	34	3	37	43	71	5	76	47
	B	83	14.25	97.25	109	225	22	247	124
	$\bar{B}$	2.441	4.750	2.628	2.535	3.169	4.400	3.250	2.638
	S.E.	0.325	2.528	0.359	0.403	0.353	1.123	0.338	0.308
	$\bar{B}_m$	3.013	2.849	3.000	3.101	3.043	2.849	3.030	3.072
Total	N	45	3	48	49	77	5	82	55
	B	84	14.25	98.25	109	225	22	247	124
	$\bar{B}$	1.867*	4.750	2.047*	2.224	2.922	4.400	3.012	2.255*
	S.E.	0.289	2.528	0.317	0.373	0.339	1.123	0.327	0.292
	$\bar{B}_m$	2.836	2.647	2.824	2.933	2.867	2.647	2.854	2.902

$\bar{B}$ for non-choreic males plus non-choreic females:

	Observed	Expected	P
Ever married	2.589 ± 0.249	3.086	$<.05$
Total	2.240 ± 0.232	2.917	$<.01$

* Different from expected at 5 per cent level of significance.

obtain estimates of fitness of choreics relative to the general population itself. We may call these estimates $W_{c \cdot p}$. To obtain an estimate of the fitness of heterozygotes relative to the general population $(W_{H \cdot p})$ we need to multiply the fitness of heterozygotes relative to non-choreic sibs $(W_{H \cdot n})$ by $W_{n \cdot p}$. The values of $W_{H \cdot n}$ obtained from Table 6, and including the previous parental age correction, are 0.863, 1.315, and 1.089 for males, females, and their mean, respectively. These estimates are:

	$W_{c \cdot p} \pm$ S.E.	$W_{H \cdot p} \pm$ S.E.
Males	0.694 ± 0.117	0.663 ± 0.107
Females	1.024 ± 0.164	0.979 ± 0.148
Mean of males and females	0.859 ± 0.129	0.821 ± 0.116

The variance of the product of two means, $\bar{x}\bar{y}$, is, in part, a function of the correlation between x and y, increasing with positive correlation and vice versa. There appears to be no obvious cause for positive correlation between $W_{c \cdot n}$ and $W_{n \cdot p}$ or between $W_{H \cdot n}$ and $W_{n \cdot p}$. There is some reason, however, to believe that any correlation might be negative, since the mean fertility of non-choreics occurs in the denominator of $W_{c \cdot n}$ and $W_{H \cdot n}$ and in the numerator of $W_{n \cdot p}$. If this were so, the above standard errors, which assume no correlation, would be reduced slightly. Using the standard errors as calculated (which are based on the standard errors for $W_{c \cdot n}$ and $W_{H \cdot h}$, believed to be slightly too large) $W_{c \cdot p}$ and $W_{H \cdot p}$ for males are seen to be significantly different from unity, indicating that male choreics and male Hh are less fit than the general population. Females, and the mean of males and females, however, do not differ significantly. However, since the value for the mean $W_{H \cdot p}$, 0.821, (which is *the* relative fitness of genetic importance) is based half on the fitness of male Hh which do differ significanlty ($P < .01$) from unity, it really is different from unity. This comparison therefore indicates that heterozygotes are less fit genetically than the general population, having a relative fitness of about 0.82 of normal.

Using estimate of $\bar{B}_o$ obtained from Cohort Fertility: The preceding estimate of the mean $W_{H \cdot p}$ was partly based on the mean fertilities of living normal sibs and the 1940 census fertility data because a direct comparison of choreics or heterozygotes with census data is not valid. An estimate of $W_{H \cdot p}$ not dependent on the non-choreic sibs is desirable, however. The fertility data assembled by Whelpton (1954) can be used for such an estimate. Using all available United States census data on fertility and mortality, Whelpton (1954) has calculated, among other quantities, the mean number of liveborn children ever born ($\bar{B}_s$) to U. S. native-born white women, of certain specified cohorts, who *survive* to specified age x. ("Cohort" here is all of the above-specified liveborn females who are born in a particular year, say 1890). The mean number of liveborn children per original member of the cohort ($\bar{B}_o$) is not given here, nor is it available elsewhere. $\bar{B}_s$ and $\bar{B}_o$ differ appreciably for high x, $\bar{B}_s$ exceeding $\bar{B}_o$, because $\bar{B}_s$ is based only on women who survive to, say, 40 years, neglecting those who die below this age. If one knew the age-specific cumulative fertility of women dying at age x, it would be possible to calculate $\bar{B}_o$ exactly. This fertility is unknown but, as a first approximation, one can assume it is the same as that of women living at age x. It is possible to think of reasons why it should be lower; it is also possible to find reasons why it may be higher. Our interest is in the cohort of 1890 since this approximates the present data well. By multiplying the number of women dying at age x (obtained by subtraction from Table C, Whelpton, 1954) by the cumulative fertility at age x (mean value for age x in Table A, *ibid.*), and summing these products from birth to age 47, the number of births to women dying before age 47, $\bar{B}_d$, is obtained. $\bar{B}_d$ plus the total births from women reaching age 47 or over gives the total births from the original cohort. For the cohort of 1890 these data are not given before age 30 so that it is necessary to estimate these early births from data on the cohort of 1900. (The authors are indebted to P. K. Whelpton, Scripps Foundation for Research in Population Problems, for suggesting this general procedure. The calculations are our own.) If the cohort of 1890 numbered 10,000 liveborn females, it may be calculated that 536 children are born to women dying under 30, 1,538 to women dying between 30 and 46 inclusive, and 17,753 to women surviving to 47 or over. The total of 19,827 children is equivalent to 1.9827 livebirths ever born per original member of the cohort. The accuracy of this estimate is somewhat uncertain but it is clear that the true value is appreciably greater than the minimum estimate of 1.7753, the value which would result if women dying under 47 had no children. An absolute maximum, clearly too high, is given by assuming that all women reaching age 15 survive through age 47. This maximum is 2.340 children ever born. The estimate of 1.9827 is below replacement value, but this is believed to have been characteristic for several native white American cohorts of this period (P. K. Whelpton, personal communication). Since this estimate is for native-born white women of all the United States, we may approximate the corresponding value for Michigan white (native-born plus foreign-born) women by multiplying 1.9827 by the ratio, for women age 45-49 years in 1940, of a) number of livebirths ever born per Michigan white women to b) number of livebirths ever born per U. S. native white woman. This ratio is 2.730/2.602 = 1.049. The estimate for a Michigan cohort of 1890 is then 2.080, slightly over replacement. This estimate of $W_{H \cdot p}$ requires an estimate of $\bar{B}_o$ for all Hh individuals. From Table 6 and from the weighting factor of 0.969 to correct for the parental age distribution of female choreics, this is $[1/2]$ $[1.325 + 0.969 \times 2.074] = 1.668$. The mean $W_{H \cdot p}$ is then $1.668/2.080 = 0.802$.

Using the estimate of $\bar{B}_o$ for the general population derived by Reed (1959): Reed (1959) calculated $\bar{B}_o$ for the normal sibs of sporadic (parents normal) propositi of individuals affected with multiple neurofibromatosis, using published and unpublished data of Crowe *et al.* (1956). These propositi were ascertained in Michigan between 1934 and 1953. Because their parents and ancestors were normal these sibs are believed to be representative of the general population. A comparison of their completed fertility with data of the 1950 census confirms this belief; no significant differences were found. For 107 such individuals who had lived to age 40 years or over, or who died at any age, $\bar{B}_o$ was found to be 2.037 ± 0.240 livebirths ever born. Using this value, the mean $W_{H \cdot p}$ is estimated to be $1.668/2.037 = 0.819$. The relatively large standard error of $\bar{B}_o$ makes this estimate less reliable than the preceding two but it is useful as a check

The "best" estimate of $W_{H \cdot p}$: Since the estimates vary only from 0.802 to 0.821, the range being less than the smallest standard error, the question of which is "best" is rather academic. It seems reasonable to use the mean of the three estimates as the "best" estimate. This is 0.81, to two decimals.

4. Infant mortality among the children of choreics and non-choreic sibs

Panse (1942) reported an increase in infant mortality among children of choreics, (28.64 ± 1.8 per cent dying in the first 10 years of life for children born 1880-
1899) relative to that in children of non-choreic sibs of choreics (22.64 ± 2.08 per
cent in the first 10 years). He does not distinguish the sex of the choreic parent.
This possible mode of selection was investigated in the present study by considering
mortality between birth (livebirth) and the fifth birthday. These data are presented
in Table 11. The proportion of deaths among children of male non-choreics does
not differ significantly from that of female non-choreics so we may pool these
groups to obtain an estimate for non-choreics, in general, of the proportion of liveborn children dying before their fifth birthday, 0.043 ± 0.008. The proportion for
male choreics does not differ significantly, being 0.048 ± 0.012, but that of females
does, being 0.076 ± 0.011 (χ^2_1 = 4.958, P = .025). If this difference between male
and female choreics is real it is not surprising since the mother is probably more important to an infant's survival than is the father. These data indicate that the real
fitness of married female choreics should be decreased by about 0.033 and that of
all choreics by about 0.015.

Table 11. Proportion of Liveborn Children of Choreics and of Non-Choreic Sibs
Dying Under Five Years of Age. Only Children Born Five or More Years
Before the Time of Last Investigation to White, Married Parents

	Male Parent		Female Parent	
	Choreic	Non-Choreic	Choreic	Non-Choreic
Total number of children	314	266	529	360
Number dying under 5 years	15	10	40	17
Proportion dying	0.048 ± 0.012	0.038 ± 0.012	0.076 ± 0.011	0.047 ± 0.011

Proportion dying under five, non-choreic parents, males plus females: 0.043 ± 0.008

Mutation

Almost all rare dominant traits in man with high penetrance, when intensively
studied, give examples of mutation. Dentinogenesis imperfecta (Witkop, 1957) is
perhaps the only such trait which, after a large-scale investigation, yielded no evidence for mutation. It is therefore appropriate to look for mutation in Huntington's
chorea even though, because of its high relative fitness and late onset, good examples are expected to be rare.

The possibility of very late onset of the symptoms of chorea in individuals who are *Hh* is a serious handicap to any attempt to demonstrate mutation. In fact, we consider it impossible, with the present diagnostic facilities, to pick out individual families in which mutation has occurred. At the same time, if one studies a large number of choreic families it may be possible to demonstrate with considerable reliability that mutation is occurring in some of the families. The latest onset of choreiform movements among 204 choreic individuals in the present study was 65 years, this onset age being found in two individuals. The next highest age was 54 years, again occurring in two individuals. These ages indicate the difficulty of proving mutation in specific cases in the face of the alternative of late onset in a parent. An instructive example of this difficulty can be given. Through correspondence we learned of a patient, male, age 44 years, with Huntington's chorea whose parents were reported normal at ages 77 years (father) and 68 years (mother). The patient's wife, the family physician, and the superintendent of the nearby mental institution to which the patient was committed, all stated that the parents and all other relatives were normal. Dr. J. H. Chandler, then of this Department, visited the family (Kindred #4655) (which lives outside of Michigan), and made neurological examinations on the patient, his brother (age 41 years), and both parents. Typical Huntington's chorea was found in both the patient and his brother. Definite, but mild, symptoms of this disease were also observed in the father (e.g. flexion-extension movements at fingers and wrists, jerking movements of shoulders, torsions of the trunk, fleeting universal flexion of lower extremities). The mother appeared entirely normal. Since the family (whose members were intelligent) did not recognize any abnormality in the father, his age at onset is unknown. The important point is that without careful examination this family would have been thought to offer good evidence for mutation, considering the ages of the parents. Other workers have also noted very late onset, 70 years or later (e.g., Entres, 1921; Bell, 1934; Brothers, 1949).

Although we do not believe that specific instances of mutation in Huntington's chorea can be demonstrated, we have recorded, as a matter of interest, the number of instances where a) both parents of a *single* case of Huntington's chorea reached the age of 60 or over and b) were reported to be normal. The kindreds containing parents meeting both requirements are therefore possible examples of mutation. This procedure, however, cannot be used to obtain an estimate of the mutation rate because, obviously, mutation can occur in young parents as well as old and these young parents are excluded. In 196 apparently separate kindreds (groups of biologically related individuals) of Huntington's chorea, one or more of whose choreic members (medically diagnosed) lived in Michigan at some time, there was sufficient information (although usually second-hand) to classify the parents. In eight of the 196 kindreds both were at least 60 and had only one choreic child. Only four of these eight kindreds could be thoroughly investigated and of these four, the father in one was found to have committed suicide at age 62, raising the question of his being in the early stages of the disease. In the kindred presenting the strongest case for mutation (Kindred #4455) the propositus, who was the only case of Huntington's chorea in the kindred, died at age 38 years. Autopsy findings, including brain sections (reviewed by Dr. J. H. Chandler), was typical of the disease. Four sibs, ranging in age from twenty to forty-four years, were reported to be normal. The parents, both age 65 years, were visited by a trained fieldworker, familiar with early and late stages of the disease, and were found to be normal. The parents had no knowledge of the disease in their own ancestors. These facts definitely suggest that mutation has occurred within the last several generations of this kindred but, in view of the fact that this kindred is one of 196 kindreds, we may have merely selected for late onset in the heterozygous ancestors of the patient, all of whom died before showing symptoms of the disease. It is therefore not justifiable to conclude that this example demonstrates mutation. The possibility of diagnostic error, when there is only one choreic in a kindred, is further reason to be cautious.

If, as seems probable, mutation from the normal allele to an allele causing Huntington's chorea does occur, it is obvious that an upper limit of the frequency of such mutation, μ [mutations/locus (loci, if there are several producing the phenotype of Huntington's chorea)/generation], can be found. If we classify each individual in Michigan on April 1, 1940 who was known to be Hh, (either being choreic at that time or developing chorea at a later date) according to whether or not one of his parents appeared to be Hh, being classified Hh because of being choreic or, if not choreic, having two or more choreic children and/or a collateral or antecedent choreic relative, we find the following: Of 231 Hh individuals known, 206 had an Hh parent, 7 were not known to have an Hh parent, and 18 could not be classified, usually because they had no known relatives in Michigan. A maximum estimate of the proportion of Hh individuals lacking an Hh parent, p, is clearly given by 25/231 = 0.108. The frequency of Hh individuals in Michigan, f, was estimated in the first part of this study as 1.01×10^{-4}. Therefore, whether or not the population is in equilibrium with respect to origin and loss of H genes, $\mu_{max} = pf/2 = 5.4 \times 10^{-6}$. This is clearly a maximum estimate and probably is several times too high.

No useful estimate of the minimum mutation rate can be derived from the present data. The fact that of 231 Hh individuals only 7 are reported to have had normal parents (neither known to be Hh) would seem reasonable even if we knew that there were no mutation. If, in fact, there were no mutation this small proportion of Hh parents might be expected not to show chorea because of failure to reach the age of onset. Examination of the ages at death or time of last investigation of these seven pairs of parents is not very informative. These ages are, giving the father's age first: (75, 21), (70, 25), (63, 40), (48, 71), (80, 55), (62, 70), and (69, 87). The P_x and C_x statistics, previously derived from the age of onset distribution, if reliable at advanced ages, could be used to calculate the probability that both members of some of these pairs of parents are hh and, therefore, demonstrate mutation. Unfortunately, such reliability cannot be assumed, but, even if it existed, the problem of sampling error would be very difficult. For these reasons no estimate of the minimum mutation rate is attempted.

If the population is in genetic equilibrium with respect to Huntington's chorea, which is quite possibly not the case, we may estimate μ from $\mu = (1 - W)f/2$, where W is the relative fitness estimated previously, 0.81. Using this value of W, μ is estimated to be 9.6×10^{-6}, almost twice the previously calculated μ_{max}. Since the 95 per cent confidence interval of μ_{max}, based on the estimation of p, does not include 9.6×10^{-6}, and the true value of μ is believed to be at least several times less than μ_{max}, there is reasonably good evidence that either the estimated value of W is in error or that the assumption of genetic equilibrium is incorrect. f is used in both estimates and is therefore not implicated here. It is quite possible, of course, as a result of sampling error or some unknown bias, that W is in error. For example, if the true μ is 2×10^{-6}, the observed f is correct, and genetic equilibrium exists, then W should be 0.96. On the other hand, the assumption of genetic equilibrium for Huntington's chorea seems inherently dubious. The possibility of recent change in the social and psychological characteristics of our society, which can affect the reproduction of Hh individuals, is obvious. Although no decision between these two alternatives, or a combination of them, can be made, lack of genetic equilibrium seems most likely to be the explanation. It may be noted in passing that these data demonstrate the uncertainty of indirect estimates of μ when W is near unity since small changes in W may produce large changes in $1 - W$.

In summary, several examples of possible mutation were observed but, because of the possibility of late onset of chorea definite instances of mutation could not be demonstrated. A direct *maximum* estimate of the mutation rate was calculated: 5.4 $\times 10^{-6}$. If mutation occurs it probably does so at an appreciably lower rate than this.

Discussion

The general difficulties in estimating relative fitness for specific traits, as well as the additional difficulties in the particular case of Huntington's chorea, have been discussed at some length by Reed (1959). The present study has shown how these difficulties are made acute when the relative fitness is high, around 0.8 (mean $W_{H \cdot p}$) in this case, instead of being under 0.5 as is the situation for most genetic traits for which estimates of W have been obtained. Biases which are completely negligible when W is near zero may be very important when W approaches one, thus making the definition of W more difficult than has usually been the case. Examples of such biases are a) age differences between affected and normal sibs (since fertility is age-dependent) when incompleted fertilities are used, b) fertility differences between the normal sibs of the affected and the fertility of the general population, and c) differences in the parental age distribution ($P_{i \cdot x}$ as defined on page 108).

The definition of W proposed in equations (1) and (2) is designed to eliminate a) and c). Bias a) is obvious but bias b) and bias c) apparently have not been recognized previous to the findings of this study. The demonstration that b) occurs made it necessary to make all estimates with reference to the general population of Michigan ($W_{H \cdot p}$) and not with respect to the unaffect sibs ($W_{c \cdot n}$) as is usually done. The increasingly available age-specific census data on number of children ever born makes this procedure practical in a number of countries.

Bias in W as a consequence of its usual definition of the ratio of mean numbers of children uncorrected for parental age distributions, has not usually been recognized. Reed (1959) discussed this and derived equations (1) and (2) to correct for it. The finding that the parental age distribution of female choreics differs significantly from the females of Michigan provided an example of this type of bias.

The estimate of 0.81 for Huntington's chorea and the estimate of 0.78 for multiple polyposis of the colon, (Reed and Neel, 1955), are among the highest relative fitnesses determined for rare dominant or sex-linked recessive traits, clearly a consequence of the late age of onset in these traits. It is probable that there are rare gengenetic traits, for example dentinogenesis inperfecta, with higher W's but such estimates do not seem to have been made.

It is not known why the normal non-choreic sibs were less fertile than the females of Michigan. Reed and Chandler (1958) found no significant differences between these sibs and the general population with regard to occupation and marital status and the general impression gained was that there were no remarkable differences. An obvious possibility, for which we have no evidence, is that in kindreds having a number of affected persons some of the normal members, or their spouses, limit reproduction because of the fear of developing the disease or of transmitting it to their children. Since the fertility of "ever married" non-choreics as well as total non-choreics is decreased, another possibility is that non-choreics marry later, on the average, than the general population. Unfortunately, we do not have data on age at marriage for non-choreics. Although the mechanism causing this reduction in fertility is unknown, it is an interesting example of "gene action" in man. No other published examples of decreased fertility of normal sibs of individuals with genetic diseases are known to the authors. Data on the fertility of non-choreic sibs given by Panse (1942), S. C. Reed and Palm (1951), and Kishimoto et al. (in press) do not permit a comparison with census fertility data since distributions by age and year are not presented. It is worth noting that a tabulation by Reed (1959) of the original data of Crowe, Schull, and Neel (1956) on neurofibromatosis, obtained in a partial survey in the state of Michigan, revealed no indication of a difference in fertility between normal sibs of sporadic propositi and the female population of Michigan. For the normal sibs of familial (one parent affected) propositi, however, the mean fertility was appreciably, but not significantly, lower than the females of Michigan. Since most normal sibs of choreics have an affected parent, the question seems raised whether the presence of an affected parent, or other ancestor, depresses the fertility of unaffected individuals. Further data are clearly required.

The reduced fitness of *Hh* individuals (i.e., the mean for males and females) may have several explanations. The proportion of choreics who marry could not be shown by Reed and Chandler (1958) to differ significantly from the general population but the proportion may, in fact, be lower because male choreics appear to have a lower marriage rate. A more definite factor is the termination of reproduction because the disease is sufficiently advanced to require either institutionalization or other segregation. To estimate the magnitude of this factor, use can be made of the "Relative Reproductive Span" (*RRS*) (Reed and Neel, 1955), assuming (for this calculation only) that the only cause of reduced fitness is institutionalization of some choreics before the end of their potential reproductive period. The age at time of first institutionalization from Table 16 of Reed and Chandler (1958) was used, equating this age to the termination of actual reproduction, and the distribution of parental ages for Michigan in 1935 (Reed and Neel, 1955) was also used. Under these assumptions *W* is about 0.93. If, as is probable, fertility is reduced before this age, *W* would be less. Since the estimate of 0.81 doesn't differ significantly from 0.93 this factor alone might explain all of the reduction in fitness, but this does not seem likely.

The significant difference in fitness observed between male and female choreics, and also *Hh* individuals, is similar to that noted by Panse (1942). He found (see his Table 20) that male choreics had a mean of 3.13 children while female choreics had a mean of 3.56, the ratio of these means being 0.88. The significance of this difference cannot be determined from his data but it probably does not quite reach significance. Kishimoto *et al.* (in press), using data from Japanese pedigrees, report a mean fertility of 3.53 for male choreics and 3.88 for female choreics. These differences are very probably not significant. Other data on fertility, by sex, of choreics do not appear to be published and, in fact, the only other dominant trait similarly analyzed seems to be neurofibromatosis (Crowe, Schull, and Neel, 1956). [The calculation of Vogel (1957) on retinoblastoma, which found no difference between males and females in fertility, employed unilaterally affected adults of varying ages.] Using the method advocated by Krooth (1955), the relative fitness of affected males was found to be 0.413, while that of affected females was 0.748. The significance of these differences was not determined. Calculations by Reed (1959), using the original data of Crowe *et al.*, of the mean completed fertility gave values of 0.451 ± 0.208 livebirths for affected males and 1.292 ± 0.383 livebirths for affected females. These means do not quite differ significantly (*P* = .07) but the difference is suggestive. Although, except for the present study, we lack strong evidence for the reality of these differences, it is interesting that in each study the males appear less fertile. In each case (except for the data of Kishimoto *et al.*, which do not mention the marriage frequency) this seems to be in part, a result of a lower marriage rate among affected males. A reasonable explanation for much, if not all, of the decrease in the proportion of male choreics who marry, relative to that of female choreics, is given by the higher age at time of first marriage of males compared to females. If m_x is the proportion, out of all first marriages in the general population, which occur at age x, and P_x, as defined previously, is the probability that an Hh individual will develop Huntington's chorea by age x, then the probability that an Hh person marries at age x is approximately proportional to $m_x(1-P_x)$. $M = \sum_x m_x(1-P_x)$ should be approximately proportional to the probability that an Hh person ever marries. Data to estimate m_x are not available for Michigan before 1950, but data for the neighboring state of Wisconsin for 1922 (when many of the choreics in our study were marrying) should be suitable. M_δ is found to be 0.752 and M_φ is 0.834; their ratio is 0.902. The mean age at marriage is 28.18 years for males and 24.62 years for females; these are slightly high because the data are for all marriages, not first marriages. (Mean age at first marriage for Michigan in 1950 was 24.88 years for males and 21.89 year for females.) From Tables 4 and 5 the observed proportions of male choreics ever marrying is seen to be 0.742, for female choreics 0.920, giving a ratio of 0.807. Thus, the fact that, on the average, a male Hh individual is less likely to be non-choreic at the time of his (actual or potential) marriage than is a female Hh individual, seems capable of explaining much of the observed difference. This mechanism is a good example of interrelationship of genetic and social factors in determining fertility. Other possible explanations might be considered but they appear to lack reasonable support. Another example of such interrelationships is the increased mortality found among the children of female choreics but not of male choreics. This increased mortality decreases the relative fitness for all choreics by only about 1.5 per cent from the values estimated on the assumption of no difference in mortality, but is a good example of selection operating at a stage of the life cycle not usually examined.

There appear to have been only three studies, in additon to the present one, in which numerical estimates of the relative fertility of choreics have been obtained. Of these, only Panse (1942) has fertility data from a complete, or nearly complete, survey of a large population, in this case the Rhineland of Germany. In his Table 20, Panse presents data on the number of children born to 457 choreics and 505 non-choreic sibs, all age 30 or over and apparently ascertained through their parents. One can calculate from his data that the mean number of children of choreics is 3.344, that of non-choreics 2.837, giving an estimate of $W_{c \cdot n}$ of 1.18. This is an excellent agreement with the estimate of 1.14 found in the present study (uncorrected for parental age). There is a bias which will increase his value of $W_{c \cdot n}$ although it probably is not of major importance. This bias is a consequence of his minimum age of 30 since, for the age range 30-45, say, choreics will have a higher mean age than their non-choreic sibs because the probability of ascertaining choreics increases with age but the probability for ascertainment of non-choreic sibs does not. This greater mean age in turn produces a greater mean fertility. Panse did not estimate $W_{H \cdot h}$ but it seems probable that such an estimate again would not differ appreciably from our own. He did not compare the choreics' fertility with census fertility data but instead had his own data on 219 normal, more distant, relatives (often first cousins) of choreics, age 30 and over, whose fertility was known. But since the mean fertility of these relatives is 1.973 children, it is difficult to believe that they adequately represent the general population. In fact, since they are close relatives of choreics, their lower fertility raises the question of whether the same mechanism is responsible for their lowered fertiltiy as for that of the normal sibs in the present study. It is therefore not known whether the non-choreic sibs differ from the general population of the Rhineland.

S. C. Reed and Palm (1951) presented data on the fertility of an unstated number of choreics and their non-choreic sibs in the state of Minnesota, U. S. A. They reported a mean number of 6.07 ± 0.9 children born to choreics and 3.33 ± 0.5 born to non-choreic sibs. These figures are from the data of Palm (1953) who indicates that they are based on 29 choreic and 49 non-choreic individuals, respectively, who were married and had at least one livebirth. Palm (1953) also reports a significant (at the five per cent level) difference in the mean fertility of 34 choreics and 60 non-choreics unselected for having had children, the respective means being 5.15 ± 0.87 and 2.72 ± 0.46. The age distributions are not reported nor is the method of ascertaining these individuals. It is not stated that choreic parents who were ascertained only through their children were excluded from the calculations. The ratio $5.15/2.72 = 1.89$ differs markedly from the $W_{c \cdot n}$ ratios found by Panse and in the present study. It is possible that, among other factors, such as age difference between choreics and non-choreics, the fact that they ascertained only a small fraction of the Huntington's chorea kindreds in Minnesota accounts for their high value for $W_{c \cdot n}$. When ascertainment is quite incomplete the probability of ascertaining a kindred is approximately proportional to the number of choreics (or the number known to official sources). As a result there is selection for large kindreds, and consequently, fertile choreics. S. C. Reed and Palm (1951) also discuss a single large kindred, finding that the total number of descendants of a choreic man is four times that of his normal brother. It is not demonstrated that this kindred is representative of choreic kindreds in Minnesota and the mode of ascertainment of the kindred does not exclude bias for large size and hence, fertile choreics. Their data clearly show that in some kindreds choreics are very fertile, but it does not follow that the mean fertility for the state is as high as their results suggest. In our data, for example, there are many kindreds whose choreics are indeed very fertile; there are also others in which choreics are conspicuously less fertile than their normal sibs, usually because of early onset, institutionalization, and death. S. C. Reed and Palm (1951) and Palm (1953) also report that a tabulation of published pedigrees also shows greater fertility (but not significantly so) for choreics than for non-choreic sibs. Since, except for those of Panse (1942), none of these pedigrees are from complete surveys, the above criticisms concerning age distribution, method of ascertainment, and bias for large size also apply here. Because of these biases we do not believe that the data reported by S. C. Reed and by Palm are adequate for estimating the relative fitness of choreics, not to mention that of heterozygotes.

The third study is that of Kishimoto et al. (in press) and Kishimoto (personal communication) in Japan, the only data not on Caucasian populations. Fifty-five Japanese kindreds, published and unpublished, were used for the fertility calculations. The mean age of the choreics, living and dead, was about 51 years. No restrictions as to age or ascertainment are mentioned. They found that 58 male choreics had 3.53 children, 56 female choreics had 3.88 children, and 23 sibs (from Aichi Prefecture only, mean age 51 years) had 5.71 children. Using these figures they estimate relative fitness for choreics as 0.649. However, Kishimoto (personal communication) reports that if one compares only choreics and non-choreic sibs studied in his survey of Aichi Prefecture (numbers and ages not specified), the relative fertility of choreics is 0.85. An alternative estimate of the fertility of the general population of Japan can be obtained from the 10 per cent sample tabulation of the 1950 census. The mean number of children ever born per Japanese woman age 50-54 years in 1950 was 4.719. Valid comparisons are not possible since the age and year distributions of choreics are not given and their data are too heterogeneous and subject to bias for large size to yield a valid estimate of W. It does appear, however, that in Japan choreics (and therefore heterozygotes) are also less fertile than the general population, especially considering that, because of biases, the reported choreic fertilities are probably too high. It seems quite possible that the relative fitness of heterozygotes in Japan may be less than that in populations of European ancestry and it is tempting to correlate this with the higher fertility of the general population in Japan. Kishimoto et al. also reported a frequency estimate for Huntington's chorea, in Aichi Prefecture, of 3.8×10^{-6}, about one-tenth that found in the U. S. A. and in Europe. For a fixed mutation rate and genetic equilibrium, lower relative fitness should produce a lower frequency. Whether this relationship explains these differences between Japan and other countries must remain speculative.

Estimation of mutation rates in dominant traits with late onset like Huntington's chorea and multiple polyposis of the colon (Reed and Neel, 1955) is of necessity indirect and inferential. Nevertheless we believe that the procedure followed in the present study provides a firm upper limit of the mutation rate. We consider it almost certain that mutation does occur since Huntington's chorea is now, and in the past has been, a rare disease, implying a relative fitness less than one. Recurrent mutation seems to be the only way in which the present frequencies can be maintained. The inability to specify a minimum rate of mutation was a consequence of the definite possibility of non-penetrance as an alternative to mutation.

Kishimoto et al. (in press) estimate a mutation rate for Huntington's chorea by the indirect method, using their estimate of W and frequency. Their estimate is 6.7×10^{-7}. As mentioned above, this estimate of W is subject to several biases and the calculation is not made in terms of heterozygotes, but this estimate may give the order of magnitude. Several authors, e.g., Panse (1942), mention the possibility of mutation but make no estimates. Our estimate of 5.4×10^{-6} is one of the lowest human mutation estimates made, in keeping with the high estimate for relative fertility of about 0.8. It seems unlikely that accurate or even minimum estimates will be possible until diagnostic techniques improve.

Summary

This paper is the second of three papers based on data from a survey of all known cases of Huntington's chorea (H.c.) in the lower peninsula of the state of Michigan, U. S. A. The first paper (Reed and Chandler, 1958) was concerned with certain demographic and genetic aspects while the third will consider clinical features of the disease.

For the study of natural selection all sibships containing one or more individuals with H.c. living in the lower peninsula of Michigan on April 1, 1940 were considered. From these sibships choreic individuals and non-choreic sibs meeting the following requirements were chosen: 1) classification as "choreic" or "non-choreic" is adequate, 2) ascertainment was independent of fertility, 3) the individual was living at age 45 years or over, or was deceased at age 15 years or over, and 4) one parent of these sibships was very probably heterozygous for the gene for H.c. Two hundred fifty-seven choreics and 210 non-choreic sibs met these requirements. Requirement 3) ensured that only essentially completed fertilities were being measured. Requirement 4) was necessary for the estimation from the age of onset distribution, of the proportion of non-choreic sibs who were in fact unrecognized heterozygotes for the gene for H.c. This proportion was found to be about 10 per cent.

The relative fitness of choreics, when compared to their non-choreic sibs, is about 1.12 ± 0.12, while the relative fitness of heterozygotes (i.e., overt choreics plus individuals heterozygous for the gene but thus far without chorea), compared to homozygous normal sibs, is about 1.01 ± 0.11. The fertility of male choreics (or heterozygotes) was significantly less than that of female choreics (or heterozygotes).

A significant difference between the fertility of the non-choreic sibs and the general population was found. This comparison utilized non-choreic sibs living on April 1, 1940, who were 40 or more years old, and fertility data of the 1940 U. S. census on females having the same age distribution. The relative fertility of the non-choreic sibs is about 0.77 ± 0.08 of the latter. The estimate of the relative fitness of individuals heterozygous for the H.c. gene, *compared to the general population*, is about 0.81.

For the study of mutation, 196 kindreds (groups of biologically related individuals) were available. No specific instances of mutation in H.c. could be demonstrated because non-penetrance could not be definitely excluded. It is noteworthy, however, that in only eight of these kindreds were both parents of a single case of H.c. 1) 60 years of age, or older, and 2) normal. Because mutation occurs in almost all well-studied rare dominant traits in man and also because the relative fitness of H.c. heterozygotes is less than that of the general population, we believe that mutation occurs in H.c. From consideration of the age and diagnostic status of the parents of the heterozygous individuals living in Michigan on April 1, 1940, a direct estimate of the *upper limit* of the mutation rate was obtained. This estimate is 5×10^{-6} mutations per locus (loci, *if* there are several) per generation. An estimate of the lower limit could not be made.

These and other findings are discussed. The reasons for the higher estimates of relative fitness reported in other studies of H.c. are considered.

Acknowledgments

The authors wish to express their thanks to Dr. Friedrich Vogel and Dr. William J. Schull for their comments.

References

Bell, Julia. 1934. *The Treasury of Human Inheritance.* Vol. 4. Nervous diseases and muscular dystrophies. Part 1. Huntington's chorea. Cambridge: Cambridge University Press.

Brothers, C. R. D. 1949. The history and incidence of Huntington's chorea in Tasmania. *Proc. R. Australa. Coll. Physicians,* 4: 48-50.

Cochran, W. G. 1953. *Sampling Techniques.* New York: Wiley and Sons.

Crowe, F. W., Schull, W. J., and Neel, J. V. 1956. *A Clinical, Pathological, and Genetic Study of Multiple Neurofibromatosis.* Springfield: C. C Thomas.

Entres, J. L. 1921. Über Huntingtonsche Chorea. *Zschr. ges. Neur. Psychiat.,* 73: 541-551.

Kishimoto, K., Nakamura, M., and Sotokawa, Y. On population genetics of Huntington's chorea in Japan. *Proc. First Internat. Cong. Neurol. Sc.,* Brussels, 1957. (In press.)

Krooth, R. S. 1955. The use of the fertilities of affected individuals and their unaffected sibs in the estimation of fitness. *Am. J. Human Genet.,* 7(4): 325-359.

Palm, J. D. 1953. *Detection of the Gene for Huntington's Chorea.* Ph.D. Thesis. University of Minnesota.

Panse, F. 1942. *Die Erbchorea.* Leipzig: Thieme.

Reed, S. C. and Palm, J. D. 1951. Social fitness versus reproductive fitness. *Science,* 113: 294-296.

Reed, T. E. 1957. A genetic survey of Huntington's chorea in the state of Michigan. Proc. Symp. Twin Research Pop. Genetics, Tokyo, 1956. Supplement, *Jap. J. Human Genet.,* 2: 48-53.

Reed, T. E. 1959. The definition of relative fitness of individuals with specific genetic traits. *Am. J. Human Genet.,* 11: 137-155.

Reed, T. E. and Chandler, J. H. 1958. Huntington's chorea in Michigan. 1. Demography and genetics. *Am. J. Human Genet.,* 10: 201-225.

Reed, T. E. and Neel, J. V. 1955. A genetic study of multiple polyposis of the colon (with an appendix deriving a method of estimating relative fitness). *Am. J. Human Genet,.* 7: 236-263.

Vogel, F. 1957. Neue Untersuchungen zur Genetik des Retinoblastoms (Glioma retinae). *Zschr. menschl. Vererb.,* 34: 205-236.

Whelpton, P. K. 1954. *Cohort Fertility.* Princeton: Princeton University Press.

Witkop, C. J. 1957. Hereditary defects in enamel and dentin. *Acta Genet. et Stat. Med.,* 7: 236-239.

16. Selection and Schizophrenia[1,2]

L. ERLENMEYER-KIMLING[3]

and

WILLIAM PARADOWSKI

One of the troublesome problems for human behavior genetics is that of fitting grossly maladaptive behavior into evolutionary perspective. Speculations regarding evolution and the psychopathologies can be cast in either the past or future tense: (1) what adaptive mechanisms might there have been to enable a given behavioral aberration to persist over time? (2) what effects may cultural manipulation of the environment have upon the frequency of the disorder in future generations? These two avenues of inquiry will be considered here in relation to schizophrenia. Several hypotheses that have been suggested to account for the maintenance of schizophrenia in the past will be reviewed. The possible interaction between cultural and population genetic changes will be examined through the presentation of data from an ongoing study of reproductive rates in schizophrenia.

Schizophrenia Past

Although some forms of mental disease apparently existed in all populations of historic record, there is no way of determining whether they were the same behavior-genetic units as the various aberrations encountered nowadays. Nor is it

SOURCE: L. Erlenmeyer-Kimling and William Paradowski, "Selection and Schizophrenia," *American Naturalist,* 100: 651-665 (Chicago, Illinois: University of Chicago Press, 1966).

[1]Professor Franz J. Kallmann, founder of the Department of Medical Genetics and initiator of this study, died on May 12, 1965, before the work had reached completion. This paper is dedicated to him in recognition of the guidance offered by him in the conduct of research and in appreciation of his outstanding influence on the field of psychiatric genetics.

[2]This research is supported by grant MH-03532 from the National Institute of Mental Health, United States Public Health Service.

[3]Based on papers presented at the 3rd annual Congress of Scientists on Survival, New York City, June, 1964, and the 4th annual meeting of the New England Psychological Association, Symposium on "Behavioral Consequences of Genetic Differences in Man," Chicopee, Massachusetts, November, 1964.

259

certain that the frequencies of particular psychopathologies have remained unchanged. Schizophrenia, which occurs at the relatively high rate of about 1% in modern populations, could have been either more or less common in the past. The definiton of schizophrenia was introduced by Bleuler as late as 1911, while the earlier and more limited concept of *dementia praecox* was first used by Morel only in 1860. Knowledge of the disorder as a diagnostic entity is thus bound to an exceedingly minute span on the time-scale of evolution.

It is unlikely, however, that schizophrenia could have arisen *de novo* within the past two centuries or so. Indeed, schizophrenia has probably endured over many generations. Whatever may be the transmission of genes predisposing to this disorder—recessive (Garrone, 1962; Kallmann, 1938), dominant (Böök, 1953; Slater, 1958), polygenic (Edwards, 1960; Gottesman, 1963) or a mechanism involving two separate genes (Burch, 1964; Karlsson, 1964; Rüdin, 1916)—the present high frequency of occurrence could not have been sustained by recurrent mutation alone. Estimations of mutation rate depend upon the mode of transmission, and, as shown by Kishimoto (1957), upon the method of calculation, as well as the population under study. Calculations reported by Kishimoto, for four different estimation formulae and three different Japanese isolates, range from 3.35×10^{-3} to 8.0×10^{-3} on the assumption of dominance with incomplete penetrance, and from 7.5×10^{-5} to 9.3×10^{-3} on the assumption of recessivity. Six of the 12 estimates obtained are between 3 to 5×10^{-3}. Whatever method or population is used, however, the required mutation rates in conjunction with schizophrenia would be too high to be reconciled with more general estimates of human gene mutability, in the order of 10^{-5} (Haldane, 1949). Thus, if schizophrenia has persevered at a relatively constant rate of incidence, explanations for the stability of the polymorphism must be sought in terms of a balance between the patent disadvantages of the disease and less evident advantages accruing to affected individuals or the unaffected carriers (Caspari, 1961; Huxley et al., 1964).

The maladaptive aspects of schizophrenia represent enormous liabilities to be offset on the "balance sheet" of selection. Among other characteristics, the affected individual tends to be socially withdrawn and suspicious of others, to the point of being, in many cases, completely alienated from human contact. The disease usually makes its overt appearance well before completion of the reproductive period—with average age at onset in the mid-twenties—and is often of a chronic nature, displaying a course of progressive deterioration. At least in our time, the schizophrenic's future is not bright with the promise of either economic or reproductive prosperity.

Social Advantage Hypothesis

Hypotheses formulated in an attempt to assign a compensating adaptive value to schizophrenia run the gamut from possible social advantages to physiological mechanisms. According to one line of thought, for instance, schizophrenic behavior may have conferred high social status upon affected individuals in earlier periods. The hallucinatory and delusional features of the disease might previously have earned the affected individual esteem and protection by presenting him to his community as visionary, mystic, shaman, or saint. As Lemkau and Crocetti correctly point out, this is "an hypothesis entirely impossible to prove" (1958, p. 66). Even were it possible to establish that social reinforcement of this kind existed over any prolonged period of history, it is difficult to see how the hypothesis could serve as an explanation for the continued transmittal of the genes involved in schizophrenia. Saints and prophets not being renowned for high fecundity, the benefits ordinarily garnered by them or their isomorphs would scarcely have bestowed a reproductive advantage of the needed magnitude.

Sibling Advantage Hypothesis

A second suggestion often encountered is that the presumed maintenance of schizophrenia may be related to heterozygote superiority. Since heterosis is the chief means of maintaining a variety of polymorphic situations in man and other organisms (cf. Allison, 1964; Dobzhansky, 1957), this hypothesis merits serious attention. Likewise, if the major factor in schizophrenia is assumed to be a dominant gene with low penetrance, a selective advantage might rest with the carriers in whom the disease is not manifest. Difficulties arise, however, in attempting to divine the source of the postulated effect. One possibility is that the heterozygotes or unaffected carriers are endowed with favorable psychological attributes, which, in turn, are reflected in higher differential reproductivity.

Stemming perhaps from the early notion that *dementia praecox* occurs in brilliant young persons, it is sometimes suggested that the siblings may possess exceptionally high intelligence or creativity. Studies affording direct comparison of intellective differences between siblings of schizophrenics and control populations are lacking. In Lane and Albee's study of affected individuals and their siblings (1965), most of the subjects were stated to have come from culturally-deprived backgrounds, so that the data cannot be generalized to the problem at hand. Nevertheless, the average IQ of the siblings appears to be comparable to, rather than superior to, the norm for their peers in the general population. Furthermore, the relationship between intelligence and fertility appears to be a dynamic one (Bajema, 1963), contingent upon prevailing cultural practice; and it is by no means clear that superior intelligence would have predicated a reproductive advantage at all periods of history. Similar criticisms may be levelled against the supposition that a sibling advantage has its roots in high creativity.

A more plausible case could be made for some other possibilities in conjunction with a sibling advantage. For example, carrier genotypes might express themselves in favorable factors of personality, such as flexibility (Gottesman, 1965) from which an adaptive value would be derived. It is equally reasonable to conjecture that carrier fitness might be based on an unusually high sex drive or low thresholds of sexual arousal. As with all of the speculations outlined here concerning the evolutionary overview of schizophrenia, the hypotheses pertaining to carrier effects are tenuous. No data as yet affirm that the siblings of schizophrenics do indeed possess any of the behavioral properties postulated as favoring increased reproductivity. Moreover, there have been no demonstrations that the siblings have greater reproductive fitness than the population at large. Data presented in a later section bear further on this problem.

Physiological Advantage Hypothesis

While the foregoing hypotheses stressed differential reproductivity, a third approach suggests that differential viability is the selection mechanism responsible for the retention of genetic factors underlying schizophrenia. According to this hypothesis, schizophrenics, and possibly unaffected carriers, command a physiological advantage expressed as resistance to diseases, wounds, or infections (Huxley et al., 1964). Such an hypothesis is attractive on two counts: (a) it allows for the possibility that psychological maladaptativeness and physiological adaptiveness function as balanced pleiotropic effects of gene action; (b) it has at least one known analogue in the resistance of sickle-cell carriers to malaria (Allison, 1964).

Increased resistance to diseases that make their appearance after the end of the reproductive period would, of course, not enter into the picture. In fact, the selective efficacy of differential viability would be limited to the period prior to the onset of the psychosis, since it is then that 50 to 70% of schizophrenic reproductivity takes place (Erlenmeyer-Kimling, Rainer, and Kallmann, 1966; Essen-Möller, 1935; Kallmann, 1938). Critical data in support of the immunity hypothesis would be those showing that prospective schizophrenics enjoy increased resistance to diseases which appreciably affect survival in childhood through early adulthood. In addition, for the polymorphism to have been maintained solely on this basis, a substantial difference in survival between schizophrenics and unaffected persons would have had to continue over many generations. No single major disease has been shown to meet the qualifications: schizophrenics are not known to demonstrate lessened vulnerability to cancer, diabetes, syphilis, or pneumonia, for example, while they do display heightened susceptibility to tuberculosis; cardiovascular diseases primarily affect age groups older than those in question.

Remaining possibilities are a fairly generalized resistance to infections or perhaps specific resistance to one or more of the epidemic diseases of the past (e.g., smallpox) that were responsible for annihilating large portions of the population. As Huxley et al. (1964) point out in suggesting these possibilities, a selective advantage based on immunity to infection or epidemic diseases would be expected to dwindle "in the more hygienic modern world" (p. 220). Immunological effects of the gene or genes involved in schizophrenia would thus represent assets undergoing depreciation in value. An analogous hypothesis has been suggested by Neel (1962) with

respect to diabetes mellitus. Neel notes that the pre-diabetic is probably exceptionally efficient in the intake and utilization of food, and that this "thrifty" aspect of
the genotype could have conferred particular advantage in the feast-or-famine conditions of hunting and food-gathering cultures.

The existence of buffering mechanisms providing a significant selective advantage
unfortunately continues to elude identification. Critical data, contingent upon differential viability rates in childhood, are presently impossible to obtain, Methods for
identifying the pre-schizophrenic during infancy have not as yet been successfully
elaborated. Once such methods become available and lend themselves to mass
screening, as in the case of phenylketonuria—and provided that any previously
existing survival differential has not already been erased through general improvements in infant and child care—it will be possible to determine via longitudinal
studies whether mortality rates among nonschizophrenics exceed those of prospective schizophrenics. At this stage of technical development, only an approximation
could be attained by means of follow-up studies which compare infants born in
families with a high degree of pathology (e.g., the children of two schizophrenic
parents) to infants from families established as being free of pathology.

Genotypic Heterogeneity

None of the hypotheses reviewed above can be said to provide a fully satisfying
answer to the problem of the maintenance of schizophrenia. The social advantage
hypothesis and the notion of psychological superiority of the siblings, in particular,
carry little weight. It is possible that the compensatory contributions are made by
immunologic advantages still to be identified, in combination with a limited het-
erotic (or carrier) effect mediated either through physiological or through psycho-
sexual mechanisms.

Another possibility should be considered. Genotypic heterogeneity could explain why conflicitng interpretations have prevailed regarding the mode of transmission of hereditary factors, as well as hitherto disappointing attempts to replicate
biochemical data. Moreover, if "schizophrenia" is viewed, not as a single genetic
entity, but a heterogeneous collection of genotypes that produce similar phenotypes, a more significant role may be assigned to mutation. As noted earlier, it
would be necessary to propose an improbably high rate of mutation at a single locus to account for a constant frequency of schizophrenia in the face of opposing
selection. Mutations at several different loci, however, could occur at rates well
within the bounds calculated for other conditions and still substantially replenish
gene losses following from diminished reproductivity.

Parallels might be sought between schizophrenia and the pool of conditions
broadly classed as "mental deficiency"—which had at one time been thought to represent a single genetic entity. During the past three decades, research on low-grade
mental deficiency has led to the recognition of at least 38 types of gene-controlled
biochemical errors (Jervis, 1961), a number of heritable lesions of the central nervous system and various chromosomal defects. Discovery of other etiologically

distinct conditions will undoubtedly further reduce the currently undifferentiated group to a residue of nonspecific cases whose intellectual deficiencies reflect unfortunate combinations of genes involved in "normal" intelligence.

If known dominant syndromes and chromosomal anomalies are excluded, calculations for the remaining aggregate of low-grade mental deficiencies indicate a mutation rate of 1.4×10^{-3} per genome (Morton, 1960). It is estimated, however, that the load is distributed over approximately 69 independent loci, each mutating at the rate of 2.0×10^{-5} per generation. Calculations of this sort rest on the assumption that the rates of mutation and the population frequencies are the same for all of the alleles involved. If, as seems probable, mutation rates and selection coefficients vary, such calculations underestimate the number of loci (i.e., the number of genetically distinct entities).

The number of loci involved in schizophrenia would have to be relatively large to account for the maintenance of the disorder on the basis of mutation alone.[4] On the other hand, the assumption of heterogeneity does not necessarily rule out the possibility of favorable selective properties. Indeed, the indeterminacy of evidence regarding, for instance, physiological advantages, could well be due to differences occurring among various forms of schizophrenia.

Attempts to differentiate subclassifications within schizophrenia have thus far been based chiefly upon behavioral measures, and the resulting categoreis are clearly not genetically distinct. By analogy with mental deficiency where behavioral classification is of little assistance, delineation of specific entities in schizophrenia will probably best be sought at organic levels. In the meantime, large-scale family investigations, designed for modern methods of genetic analysis, are needed to provide clues toward the disentanglement of heterogeneity and toward the clarification of selection mechanisms.

Schizophrenia Present and Future

The problem of how schizophrenia has persevered for any major length of time remains open. On the other hand, the nature of the "positive feedback relation" between cultural and genic changes (Caspari, 1963), can be approached by the examination of current research data. The effect of cultural changes on the reproductivity of schizophrenics can be illustrated by data from a partially completed study of marriage and fertility trends.

The Changing Culture

Knowledge about schizophrenia in any statistical form begins at the turn of the century. In two early studies concerned with the differential reproductivity of

---

[4]Professor Ernst Mayr (personal communication) has pointed out that calculations based on up to six genes leave an excessive frequency of schizophrenia still to be explained.

schizophrenics—Kallmann's (1938) and Essen-Möller's (1935) studies of German patients hospitalized around the turn of the century—the reproductive rate for schizophrenics was far below that of the general population. Similar results were also obtained in Ødegaard's more recent Norwegian (1946, 1960) study along with the finding that schizophrenic reproductivity was lower than that of any other psychiatric category except psychosis with mental deficiency.

The marked reproductive disadvantage attached to schizophrenia in the periods previously investigated need not be a permanent state of affairs. Both the intensity and direction of natural selection depend upon the current environment, and, as Dobzhansky has pointed out, "the environment that exerts a decisive influence upon the human species is the social environment" (1962, p. 322).

There are cogent reasons for asking whether selection pressures against schizophrenia might be in the process of being reduced. Developments in recent years can be seen as directly or indirectly affecting the social expectations of the individual who becomes schizophrenic. The introduction of the new somatic therapies appears to be increasing the chances of discharge back to the community. Continued maintenance on drugs together with attendance at aftercare clinics may also help discharged patients to realize better adjustment and to remain in the community for longer periods than could previously have been expected. Follow-up data from a number of studies (cf. Epstein and Morgan, 1961) indicate that an increasingly smaller percentage of patients remain in the hospital for long durations.

There exists, at the same time, a greater readiness on the part of the community to understand, accept, and re-absorb the discharged patient. A network of factors may therefore be operating to change segments of the out-look for the schizophrenic, which might be affecting the marriage and fertility patterns of this group. These factors provided the rationale for the currently ongoing study.

Project Procedures

To evaluate possible changes in the marriage and reproductive rates of schizophrenics, samples of patients were drawn in two survey periods. One group consists of patients admitted to New York State hospitals in the years 1934-36, while the other group is composed of patients admitted in 1954-56. Between the two periods, new treatment methods were introduced and hospital policies liberalized.

The sampling procedures, described in detail elsewhere (Deming, 1962), were carried out at 11 state hospitals and involved the screening of records on about 9,800 patients. Each case drawn was re-evaluated, and only consistently diagnosed and uniformly verified cases of schizophrenia were retained as index cases to form the two samples for study. Of the total patient group, 3,354 were thus ascertained as schizophrenic index cases. Of these, 17 were later omitted because they were found to be transients, hospitalized in New York State for only a few days. The working base is thus 3,337 cases.

The initial clinical and statistical data on each index case were obtained from the hospital records. Such data included: a detailed psychiatric history, a resume of personal characteristics, and information about the patient's family, as well as marital and childbirth histories up to the time of admission.

In the second—and still ongoing—phase of the study, follow-up data are being collected on hospitalizations subsequent to the survey admission and on the updated reproductive histories of the index cases and their siblings. This part of the study usually requires an interview or other personal contact with the index case or members of the family. At the same time, psychiatric hospitalizations among the relatives are being ascertained. Diagnostic classification of the relatives from their hospital records are, of course, made according to exactly the same psychiatric criteria used in verifying the index cases.

The demographic structures of the two samples are compared in Table 1. The sex ratio is identical, with 49% male and 51% female cases in each period. Within

Table 1. Demographic Characteristics of the 1934-36 and 1954-56
Samples of Schizophrenics

Characteristic		1934-36 Sample (N = 1922)		1954-56 Sample (N = 1415)	
Color:	Nonwhite and				
	Puerto Rican	7.7%		16.8%	
	White	92.3		83.2	
Religion:	Protestant	24.9%		29.6%	
	Catholic	47.0		44.4	
	Jewish	24.0		21.5	
	Other	4.1		4.5	
Sex:		Male (48.8%)	Female (51.2%)	Male (48.9%)	Female (51.1%)
Age*:	15-24	25.2%	20.4%	19.5%	14.6%
	25-34	37.6	31.4	31.0	26.9
	35-44	23.7	28.1	25.8	33.4
	45-54	10.1	14.3	15.9	14.2
	55 and over	3.4	5.8	7.8	10.9

* Age distributions based on 2,706 white index cases.

each sex, the age distributions are roughly similar for the two samples. Females in each sample are somewhat older than males at admission, as would be expected on the basis of previous studies. The distribution of religions is nearly the same for the two periods, but there is a substantially higher proportion of nonwhite and Puerto

Rican cases in the 1954-56 sample due to changing immigration patterns in New York State. Data for this report pertain to 2,706 white patients.

Marriage and Reproductivity Trends

The majority of the marriages and childbirths for schizophrenics continue to occur prior to the onset of illness, as had been found in earlier studies (Essen-Moller, 1935; Kallmann, 1938). The marriage and reproductivity rates of the 1954-56 cases, however, exceeded those of the 1934-36 group even in the prepsychotic stage; and the intersample differences increased further in the interval between onset and admission (Goldfarb and Erlenmeyer-Kimling, 1962). Thus, by the time of admission, the proportion of married index cases was substantially greater in the 1954-56 sample. As seen in Table 2, the differences between the samples are maintained in all age-groups. The differences are especially notable in the younger age-groups that still had most of their reproductive years ahead of them at the time of survey admission.

Schizophrenics are still more prone to remain single than are members of the general population. Nevertheless, the 1954-56 sample approaches the general population's marriage rate somewhat more closely than was true in the earlier period, as seen in Table 2. During the time period under study, the population at large experienced an increase in marriage rate, while schizophrenics showed slightly larger increases. The marital frequency among schizophrenic males, for instance, rose from 26.4 to 40.8%. This 14% gain represents a *relative* change of 55%, compared to a relative increase of only 17% in the marriage rate of the general population males. Female schizophrenics showed similar, though less striking, gains.

Besides an increase in marriages, the schizophrenics are following other trends that have been noted in the general population: a decline in childless, or nonfertile marriage, a decrease in the proportion of one-child families, and a corresponding increase in two- and three-child families.

Changes in reproductivity may partly reflect changes in fertility patterns within marriage, as indicated by the child-per-fertile-marriage and child-per-marriage rates shown in Table 3. Within marriage, schizophrenic women of the 1954-56 sample had more, and schizophrenic men fewer, children than previously. Since the samples differed in marital rate, however, their total reproductivity is best compared by considering the average number of children per person. By this measure, males of the later sample show a small gain in reproductivity, amounting to ten children per 100 men. The females, on the other hand, exhibit a dramatic increase: for every 100 women there were 60 more children born in the later period than in the earlier.

Follow-up data now being collected will permit comparisons of the two samples for the years subsequent to the survey admission. Preliminary analysis of a portion of the longitudinal material (Erlenmeyer-Kimling, Rainer, and Kallmann, 1966) indicates that the intersample differences in marital and reproductive rates continued to widen during, at least, the five-to-seven year period following admission. At that time (1941 for the 1934-36 sample, 1961 for the 1954-56 sample) the proportions of index cases who had produced at least one child were: in the 1934-36 sample,

Table 2. Per Cent of Schizophrenics Ever-Married by Age at Admission

Schizophrenics by Age at Admission	Male		Female	
	1934-36 (N = 776)	1954-56 (N = 541)	1934-36 (N = 817)	1954-56 (N = 572)
15-34	12.0	26.7	40.2	47.5
35-44	46.6	50.7	70.2	79.8
45+	53.2	60.9	65.2	71.3
Total	26.4	40.8	54.2	64.3
General Population	66.8*	77.9†	74.0*	82.7†

*U.S. Bureau of the Census, *Current Population Reports*, Series P-20, No. 33. Figures apply to 1940, ages 15 and over.

†U.S. Bureau of the Census, *Current Population Reports*, Series P-20, No. 56. Figures apply to 1954, ages 15 and over.

Table 3. Reproductivity of Schizophrenics at Admission

Measures	Male		Female	
	1934-36 (N = 776)	1954-56 (N = 541)	1934-36 (N = 817)	1954-56 (N = 572)
Percent childless married	27.4%	26.2%	28.3%	19.6%
Children per fertile married	2.7	2.3	1.8	2.5
Children per married	1.9	1.7	1.3	2.0
Children per person	0.5	0.6	0.7	1.3

16% of the men and 35% of the women; in the 1954-56 sample, 32% of the men and 51% of the women.

It is not yet possible to evaluate from these data the cultural influences that may have affected the observed changes in reproductivity, especially those noted prior to admission. Of relevance for the period following admission is the preliminary finding that index cases in the 1934-36 sample were more likely to remain continuously hospitalized, or, if discharged, to be readmitted, than were the 1954-56 cases. At least part of the post-admission gain in reproductivity shown by the 1954-56 sample thus appears to be attributable to an overall change in hospitalization patterns of schizophrenics.

Assortative Mating

Although assortative mating does not *per se* effect changes in gene frequency, it may result in the increased production of affected individuals, who in turn are subject to selection pressures. If "schizophrenia" consists of heterogeneous conditions, assortative mating will be phenotypic without necessarily being genotypic; and the proportion of schizophrenic offspring deriving from such matings will be less than otherwise expected. Age-corrected risk figures vary between 40 and 68% for the children of two schizophrenic parents, compared with 9 to 16% for children of one affected parent (Kallmann et al., 1964). The data on assortative matings suggest that more than one genotype is involved in schizophrenia and that the frequencies of occurrence may not be equal.

Of the 2,706 index cases reported on here, 26 are known to have married another schizophrenic. Two per cent of all marriages are assortative matings, and their fertility rate is at least as high as that found in marriages betwen schizophrenics and unaffected individuals. There are six additional assortative matings among the remaining 631 index cases on whom admission data have not yet been completed, and another 40 assortative-mating families have been located for study outside the framework of the two samples (Kallmann et al., 1964).

There appears to be little change in the tendency toward assortative mating. The index cases who married other schizophrenics represent 0.9% of the 1934-36 sample and 1.1% of the 1954-56 sample. Not only are the two survey periods comparable with respect to the proportion of assortative matings, but the percentages are similar to those observed in Kallmann's (1938) study of schizophrenics hosptialized in Berlin between 1893 and 1902. Whatever may be the social factors that are boosting the marriage rate among present-day schizophrenics, they are not increasing the likelihood of marriage between two affected individuals.

Sibship Data

If carrier advantage is to be considered as one of the possible mechanisms through which schizophrenia is maintained in the population, study of the sibships of affected individuals is essential. Although sibship data are now available for only

659 index cases, it is expected that final analyses of the completed samples will bear out the trends emerging here.

Fig. 1 shows the family data for the 659 index cases. Sixteen per cent of the cases were only children, while the remainder had a total of 2,007 siblings. The 1934-36 group had an average of 3.3 siblings (excluding the index cases), compared to an average of 2.8 siblings for the 1954-56 group. More of the siblings of the earlier group, however, failed to survive to reproductive age: 13% opposed to 3% in the 1954-56 group. The net reproductive potential of the sibships was thus about the same for the two samples.

The siblings designated as schizophrenic in Fig. 1 are those whose hospital records have been examined and whose diagnsoses have been verified. The siblings referred to as "nonschizophrenic" are those not known to have been hospitalized or whose hospitalizations related to psychiatric conditions other than schizophrenia (e.g., mental deficiency, epilepsy, etc.). The "nonschizophrenic" siblings may, therefore, include a small portion of schizophrenics who were never hospitalized or whose hospitalization histories could not be located.

Approximately 7% of the surviving siblings in each sample had become schizophrenic by the time of observation. (Note: this is not an age-corrected morbidity risk calculation.) As might be expected, their child-per-person rates are similar to those of the index cases, and—like the index cases—the schizophrenic siblings of the later sample show an appreciable gain in reproductivity. In each survey period, the reproductive rates of the "non-schizophrenics" far exceed those of their affected brothers and sisters. Curiously, the unaffected siblings, too, have an increased birth rate in the later period, so that the reproductive potential of schizophrenics and non-schizophrenics alike is more fully used by the sibships of the 1954-56 sample.

Differential Reproductivity Comparisons

From a psychiatric or a sociologic point of view, an absolute increase in the number of children born to schizophrenics is important in itself. The children are either (a) reared in a home with an emotionally disturbed parent, (b) reared in a home with one parent absent for variable periods of hospitalization, or (c) placed in foster homes or institutions. None of these alternatives suggest optimal conditions for the developing child.

From the standpoint of population genetics, the increasing number of children born to schizophrenics has significant implications only if there is a change in the previously existing reproductive differential between schizophrenics and the general population. Similarly, the birth rates attained by the siblings can be evaluated only in comparison with those prevailing in the population at large. Differential reproductivity data for female schizophrenics, the sisters of schizophrenics, and women of the general population are presented in Table 4.

At the time of admission, schizophrenic women in the 1934-36 sample had produced 0.7 children-per-women, or only 60% as many children as produced by

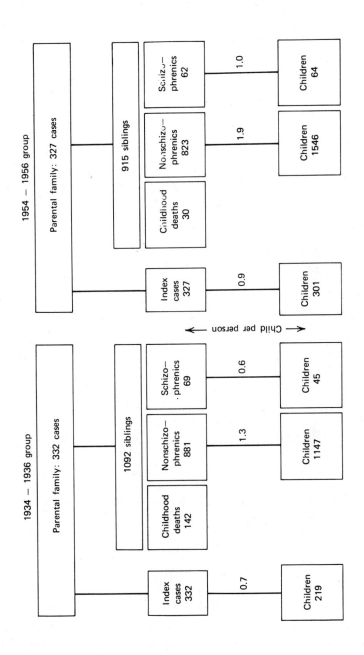

FIGURE 1. Reproductivity of the sibships for the two samples of schizophrenics.

Table 4. Child Per Woman for Schizophrenics, Their Siblings
and General Population

Measures	1934-36			1954-56		
	Siblings	Schizo-phrenics	General Population*	Siblings	Schizo-phrenics	General Population
Child per woman	1.2	0.7	1.2	2.2	1.3	1.5
Reproductivity ratios of schizophrenics	0.6		0.6	0.6		0.9

*SOURCE: U. S. Bureau of the Census, *Current Population Reports*, Series P-20, No. 108, Table 1. Figures apply to the years 1940 and 1954.

comparably-aged women in the general population. By contrast, schizophrenic women of the 1954-56 period, with 1.3 children-per-women, had reached 90% of the population's reproductive rate. With this marked gain, the reproductivity of schizophrenics—at least the females—appears to be approaching that of their general population peers more closely than had been found in any previous investigation.

The most surprising changes are those seen for the nonschizophrenic sisters. In the earlier period, the birth rate of the nonschizophrenic sisters was identical with that of the general population women. Neither a selective advantage nor disadvantage was operating at that time. During the 20-year interval under study, all three categories—schizophrenics, siblings, and general population—showed reproductivity gains. The increment was most pronounced among the siblings, however, so that by 1954-56 their reproductive rate was 140% that of the general population's rate.

It seems evident, at this stage in the collection and analysis of the data that the reproductive disadvantage earlier associated with schizophrenia is diminishing. Attempts will be made through pattern analyses of the completed material to examine specific sociocultural influences hypothesized as being responsible for the observed trend. Especially important is the indication of a newly emerging sibling advantage, which—if borne out by the further data—will present challenging problems of explanation. Finally, if etiologically distinct groups of index cases can be segregated, it will be particularly interesting to determine whether the reproductive changes (for affected individuals as well as for unaffected siblings) apply equally across the groups or concentrate within certain ones.

Summary

The merits of several current hypotheses regarding the maintenance of schizophrenia over time are considered on logical grounds, in the absence of empirical data. Viewing schizophrenia as a genotypically heterogeneous collection of conditions would allow a significant role to be assigned to recurrent mutation, while not excluding the possibility of favorable selective properties rooted in immunological advantages.

Although the balance of selective advantages and disadvantages of the disorder in the past remains a matter for speculation, changes in reproductive fitness during recent periods are demonstrated by data from an ongoing study. Interim results show (a) the reproductive rate of schizophrenics is increasing and approaching that of the general population; (b) the reproductivity of the unaffected siblings is also increasing and surpassing that of the general population.

Acknowledgments

We are grateful to Professor Ernst Caspari (Department of Biology, University of Rochester, Rochester, N. Y.) not only for his valuable suggestions upon reading the manuscript, but for his contributions to the development of many lines of thought expressed here during the course of discussions with the first author over the past several years.

We should also like to thank Professor Ernst Mayr (Museum of Comparative Zoology, Harvard University, Cambridge, Mass.) for helpful comments on the manuscript.

Literature Cited

Allison, A. C. 1964. Polymorphism and natural selection in human populations. *Cold Spring Harb. Symp. Quant. Biol.*, 24: 137-149.

Bajema, C. J. 1963. Estimation of the direction and intensity of natural selection in relation to human intelligence by means of the intrinsic rate of natural increase. *Eugen. Quart.*, 10: 175-187.

Böök, J. A. 1953. A genetic and neuropsychiatric investigation of a North-Swedish population with special regard to schizophrenia and mental deficiency. *Acta Genet. Statist. Med.*, 4: 1-100.

Burch, P. R. J. 1964. Schizophrenia: Some new aetiological considerations. *British J. Psychiat.*, 110: 818-824.

Caspari, E. 1961. Implications of genetics for psychology. Review of Behavior genetics by J. L. Fuller and W. R. Thompson. *Comtemp. Psychol.*, 6: 337-339.

——. 1963. Selective forces in the evolution of man. *Amer. Natur.*, 97: 5-14.

Deming. W. E. 1962. Some statistical principles for efficient design of surveys and experiments, p. 32-41. In F. J. Kallmann [ed.], *Expanding goals of genetics in psychiatry*. Grune and Stratton, New York.

Dobzhansky. Th. 1957. Mendelian populations as genetic systems. *Cold Spring Harb. Symp. Quant. Biol.*, 22: 385-393.

_____. 1962. Mankind evolving. Yale Univ. Press, New Haven. 381 p.

Edwards, J. H. 1960. The simulation of mendelism. *Acta Genet. Statist. Med.*, 10: 63-70.

Epstein, L. J., and R. D. Morgan. 1961. Trends in release rates of schizophrenic patients. *Compreh. Psychiat.*, 2: 196-202.

Erlenmeyer-Kimling, L., J. D. Rainer, and F. J. Kallmann. 1966. Current reproductive trends in schizophrenia, p. 252-276. *In* P. Hoch and J. Zubin [eds.], *Psychopathology of schizophrenia.* Grune and Stratton, New York.

Essen-Möller, E. 1935. Untersuchungen über die Fruchtbarkeit gewisser Gruppen von Geisteskranken. *Acta Psychiat. Neurol. Scand.* (Suppl. 8). 314 p.

Garrone, G. 1962. Étude statistique et génétique de la schizophrénie a Genève de 1901 à 1950. *J. Génét. Hum.*, 11: 89-219.

Goldfarb. C., and L. Erlenmeyer-Kimling. 1962. Mating and fertility trends in schizophrenia, p. 42-51. *In* F. J. Kallmann [ed.], *Expanding goals of genetics in psychiatry.* Grune and Stratton, New York.

Gottesman, I. I. 1965. Personality and natural selection, p. 63-80. *In* S. Vandenberg [ed.], *Methods and goals in human behavior genetics.* Academic Press, New York.

Haldane, J. B. S. 1949. The rate of mutation of human genes, p. 267-273. In *Proceedings of the seventh international congress of genetics* (Hereditas Supplement).

Huxley, J., E. Mayr, H. Osmond, and A. Hoffer. 1964. Schizophrenia as a genetic morphism. *Nature*, 204: 220-221.

Jervis, G. A. 1962. Genetic aspects of mental deficiency, p. 167-173. In F. J. Kallmann [ed.], *Expanding goals of genetics in psychiatry.* Grune and Stratton, New York.

Kallmann, F. J. 1938. *The genetics of schizophrenia.* J. J. Augustin, New York. 291 p.

Kallmann, F. J., A. Falek, M. Hurzeler, and L. Erlenmeyer-Kimling. 1964. The developmental aspects of children with two schizophrenic parents. p. 136-145. In P. Solomon and B. C. Glueck [eds.], *Recent research on schizophrenia.* Psychiatric research report no. 19, American Psychiatric Association, Washington, D. C.

Karlsson, J. L. 1964. A hereditary mechanism for schizophrenia based on two separate genes, one dominant, the other recessive. *Hereditas*, 51: 74-88.

Kishimoto, K. 1957. A study on the population genetics of schizophrenia, p. 20-28. In *Proceedings of the second international congress for psychiatry*, vol. 2. Orell Fussli Arts Graphiques, Zurich.

Lane, E. A., and G. W. Albee. 1965. Childhood intellectual differences between schizophrenic adults and their siblings. *Amer. J. Orthopsychiat.*, 4: 747-753.

Lemkau, P. V., and G. M. Crocetti. 1958. Vital statistics of schizophrenia, p. 64-81. In L. Bellak [ed.], *Schizophrenia: A review of the syndrome.* Logos Press, New York.

Morton, N. E. 1960. The mutational load due to detrimental genes in man. *Amer. J. Human Genet.*, 12: 348-364.

Neel, J. V. 1962. Diabetes mellitus: A "thrifty" genotype rendered detrimental by "progress"? *Amer. J. Human Genet.*, 14: 353-362.

Ødegaard, Ø. 1946. Marriage and mental disease: A study in social psychopathology. *J. Ment. Sci.*, 92: 35-59.

_____. 1960. Marriage rate and fertility in psychotic patients before hospital admission and after discharge. *Int. J. Soc. Psychiat.*, 6: 25-33.

Rüdin, E. 1916. *Studien uber Vererbung und Entstehung Geistiger Storungen. I. Zur Vererbung und Neuentstehung der Dementia Praecox.* Springer, Berlin.

Slater, E. 1958. The monogenic theory of schizophrenia. *Acta Genet. Statist. Med.*, 8: 50-56.

17. Estimation of the Direction and Intensity of Natural Selection in Relation to Human Intelligence by Means of the Intrinsic Rate of Natural Increase

CARL JAY BAJEMA

Introduction

The great changes now taking place in the social structure of human societies undoubtedly have an effect on the direction and intensity of natural selection. Whether a given human characteristic, such as intelligence, is favored or discriminated against in terms of reproductive performance may very well be a function of the social practices prevailing at the time. It is desirable, therefore, to investigate reproductive differentials in a variety of human societies at frequent intervals in order to assess the biological consequences of various social practices.

Cole (1954), in his discussion of the theoretical consequences of life history phenomena, has clearly demonstrated the necessity of taking into account the total life history pattern of the population being studied if an accurate estimation of the direction and intensity of natural selection in relation to a particular trait is to be made. The probability of making an erroneous conclusion concerning the direction and intensity of natural selection in relation to a behavioral trait such as intelligence is greatly increased if subtle differences in such factors as generation length, mortality rates, and the proportion of nonreproductive individuals are ignored. Anastasi (1956), in her review of the literature concerning the relationship between intelligence and fertility, has pointed to the fact that, since the observed correlations between intelligence and fertility are generally quite low, the operation of a very small selective factor could produce a completely spurious result.

SOURCE: Carl J. Bajema, "Estimation of the Direction and Intensity of Natural Selection in Relation to Human Intelligence by Means of the Intrinsic Rate of Natural Increase," *Eugenics Quarterly*, 10: 175-187 (Chicago, Illinois: The University of Chicago Press, 1963).

In the past, investigators have sometimes erred in their estimates of the direction and intensity of natural selection in relation to intelligence because they failed to consider one or more of the variables which affect population growth. Differentials in the following variables must be taken into account when measuring natural selection:

1. Number of offspring per fertile individual.
2. Proportion of nonreproductive individuals.
3. Mortality rates up to the end of the childbearing period.
4. Generation length.

The life table method, which involves computing the intrinsic rate of natural increase, provides the only means currently available whereby all of the biological variables affecting population growth can be taken into account simultaneously. During the early part of this century Lotka (1907a, 1907b, 1911, 1922, 1925) devised a statistic, r_m, called the intrinsic rate of natural increase—or the Malthusian parameter—by which differentials in fertility, mortality, and generation length can be taken into account simultaneously when determining the growth rates of various segments of a human population. Although several biologists (Cole, 1954; Crow, 1962; Fisher, 1958) have urged the use of this statistic to estimate the direction and intensity of natural selection in relation to different genotypes, it has yet to be applied to human data for this purpose.

This investigation proposes to estimate by means of the intrinsic rate of natural increase the direction and intensity of natural selection in relation to human intelligence among a group of native white individuals born in 1916 or 1917.

Methods and Materials

Type of Population Under Study

Terman group intelligence test scores were obtained on 1,144 native white individuals born in 1916 or 1917 and tested in the sixth grade by the Kalamazoo Public School System. The average age at time of testing was 11.6 years. The study was restricted to individuals born in 1916 or 1917 because this was the youngest age group for which completed fertility data were available. It was necessary to choose all individuals born in two successive years in order to obtain a sample of sufficient size.

Data Collecting Procedures

The following life history data were collected for 979 of the 1,144 individuals in the population under study: (1) date of birth; (2) number of siblings (excluding stepsiblings) who lived past the age of one; (3) marital status; (4) number of offspring produced who lived past the age of one; (5) date of death if the tested individual was deceased; and (6) place of residence.

The following sources of information were utilized in locating and obtaining information about the individuals included in the study:

1. Relatives and close friends.
2. Telephone directories.
3. City directories.
4. School records—transcripts and census records.
5. Marriage records.
6. Birth records.
7. Death records.
8. Records of funeral directors.
9. College alumni offices.
10. Present and former employers.

The individuals included in the study were interviewed personally whenever possible. An attempt was made to interview in person all individuals living within a 200 mile radius of Kalamazoo. Questionnaries were sent to 77 individuals whose life histories could not be ascertained personally by the investigator.

Definitions and Formulas

In the present study, the symbol r_m always represents the intrinsic rate of natural increase for the total sample of 979 individuals, including both males and females.

The age-specific rates of survival, l_x, and fertility, m_x, (where x = age in years) are defined as follows:

$$l_x = \frac{\text{number of individuals surviving from age of testing to age x}}{\text{number of individuals tested}} \qquad (1)$$

$$m_x = \frac{\text{1/2 times the number of offspring surviving to age one who were born to individuals of age x}}{\text{number of individuals tested who survived to age x}} \qquad (2)$$

The intrinsic rate of natural increase, r_m, is the value of r which satisfies the equation:

$$\sum_{x=0}^{x=\infty} l_x m_x e^{-r_m x} = 1 \qquad (3)$$

The average generation length, T, is obtained from the relation:

$$T = \frac{Ln(\Sigma l_x m_x)}{r_m} \qquad (4)$$

The relative fitness, W_i of subgroup i, is defined as

$$W_i = \frac{e^{r_{mi}T}}{e^{r_{mh}T}} \tag{5}$$

where e is the base of the Napierian logarithms, T is the average generation length for the total sample, Ln is the Napierian logarithm, r_{mi} is the intrinsic rate of natural increase for the ith subgroup of the sample, and r_{mh} is the intrinsic rate of natural increase for the subgroup of the sample having the fastest growth rate (i.e., the largest value of r_m).

The estimate of the population growth rate per individual, e^{r_mT}, is derived from the relationship:

$$N_T = N_o e^{r_mT}$$

where N_t is the number of individuals alive at time T, N_o is the number of individuals alive at time O, r_m is the intrinsic rate of natural increase calculated by equation (3), and T is the average generation length of the total sample calculated by equation (4). If N_o is taken as unity, then e^{r_mT} is the population growth rate per individual for a period of time equal to the average generation length of the total sample.

The intensity of phenotypic selection in relation to a trait is defined as

$$I = 1 - \frac{\overline{W}}{W_o} \tag{6}$$

where $\overline{W}$ is the fitness of the total population under study and W_o is the fitness of the optimum phenotype.

Results

Characteristics of the Sample

Life histories were compiled on 979 (85.6%) of the 1,144 native-born white individuals in the population under study. This included 72 individuals who died before reaching the age of 45 as well as 61 individuals out of 77 individuals who were contacted by mail.

The average test score of the sample was 101.46 IQ points, and the standard deviation of a random observation was 12.66 IQ points. The 979 individuals had 2,189 offspring who lived past the age of one—or an average of 2.24 offspring per individual.

Possible Biases

The sample was taken at random from the population under study only insofar as the methods used to locate individuals were unbiased. The sample was compared with the population from which it was taken to determine the presence and/or the importance of three types of biases: (1) differences in the sex ratio; (2) differences in the distribution of IQ scores; and (3) differences in place of residence (living in Kalamazoo County versus living outside of Kalamazoo County).

It was thought that females might be underrepresented in this sample because they change their last names at time of marriage. Table 1, which gives the sex ratios of the population under study and the sample, clearly indicates that the sample is not biased in relation to the sex ratio.

Table 1. Sex Ratio of the Population Under Study
and of the Sample

	Male	Female
Population Under Study	575	569
Sample	493	486

Table 2, giving the per cent of the original population contained in the sample with respect to the five IQ groups into which the sample was broken down, shows that the inclusion of an individual in the sample was not a function of his intelligence test score.

Table 2. Proportion of Individuals in Population Under Study
Included in Sample in Relation to IQ

IQ Range	No. Included in Sample/ No. in Population	Per Cent Included
$\geqslant 120$	82/91	90.1
105-119	282/327	86.2
95-104	318/377	84.4
80-94	267/312	85.6
69-79	30/37	81.1
Total Sample	979/1144	85.6

The sample is definitely biased with respect to place of residence. Since it can be safely assumed that almost all of the 165 individuals not contacted do not reside in Kalamazoo County, the bias in the average number of offspring per individual due to place of residence might adversely affect any conclusions concerning the relationship between intelligence and fertility. A 2 X 5 analysis of variance with disproportionate subclass numbers, as discussed by Snedecor (1956), was used to test the effect on fertility of the two residence categories within the five IQ groups (Table

3). This analysis indicated that the effects on fertility due to place of residence and interaction between place of residence and IQ were negligible.

Table 3. Average Number of Offspring Per Individual Surviving to Age 45
by Place of Residence in Relation to IQ

IQ Range	Living in Kalamazoo Co.		Not Living in Kalamazoo Co.	
	No.	Avg. No. of Offspring	No.	Avg. No. of Offspring
≥ 120	32	2.56	47	2.66
105-119	164	2.52	99	2.09
95-104	200	2.14	92	2.05
80-94	176	2.49	71	2.73
69-79	21	1.57	5	2.00
Total Sample	593	2.35	314	2.31

Thus it can be concluded that the sample of 979 individuals does not deviate significantly from the population in terms of sex ratio or test score distribution and that the average number of offspring per surviving individual is not significantly different for place of residence within the five IQ groups.

Average Number of Offspring in
Relation to Intelligence

The data were subdivided into five groups in relation to test scores: IQ ≥ 120; 105-119; 95-104; 80-94; and 69-79. These subgroupings were chosen so that (1) one of the groups (IQ 95-104) encompassed the average IQ of the sample; (2) the bimodal nature of the data would be apparent (IQ ≥ 120 and 80-94); and (3) to maintain sufficient numbers within each group without including too wide an IQ range.

The average number of offspring per individual in relation to IQ is given in Table 4. A one-way analysis of variance showed that the average number of offspring per individual for the five IQ groups is significantly heterogeneous at the 1% level of significance ($F = 4.136 > 3.32$).

Further statistical analyses were performed using the Duncan Multiple Range Test (Duncan, 1955) with corrected tables (Harter, 1960). This test maintains the protection level against making a type II error (asserting that a mean comes from the same population as another mean when in reality both means come from different populations) at the same level that it is in the Student's t test which protects against the type I error (asserting that the two means come from two different populations when in reality they come from the same population) at the 95 per

Table 4. Average Number of Offspring Per Individual
in Relation to IQ

IQ Range	No. Reporting	No. Offspring Per Individual
⩾ 120	82	2.598
105-119	282	2.238
95-104	318	2.019
80-94	267	2.464
69-79	30	1.500
Total Sample	979	2.236

cent level. The results of the Duncan Multiple Range Test indicate that the average number of offspring per individual for the IQ ⩾ 120 group is not significantly greater than that of the IQ 80-94 group, and that the average number of offspring per individual for the highest two reproductive groups (IQ ⩾ 120 and IQ 80-94) in the bimodal relationship between IQ and fertility are both significantly greater than the IQ 69-79 group and the average IQ group (95-104) but are not significantly greater than the IQ 105-119 group.

Table 5 compares the results of this study with those of Higgins et al. (1962) using the IQ subgroupings employed by them in their investigation. While the results of the study by Higgins et al. also indicate a bimodal relationship between IQ and fertility, it is not so pronounced as the bimodal relationship found in this study. Both studies provide strong evidence for the existence of a high reproductive rate for the IQ > 130 group which is probably a quite recent development. No other previous studies have reported a bimodal relationship between IQ and fertility, and it is

Table 5. Comparison of the Results of This Study with
Those of Higgins et. al. (1962)

IQ Range	Higgins et al (1962)		This Study	
	No.	Avg. No. Offspring	No.	Avg. No. Offspring
> 130	25	2.96	23	3.00
116-130	269	2.45	107	2.51
101-115	778	2.26	344	2.08
86-100	583	2.16	427	2.30
71-85	208	2.39	75	2.05
56-70	74	2.46	3	0.00
0-55	29	1.38	—	—

doubtful that the biased techniques employed in the past could have completely obscured a high reproductive rate of the high IQ group.

Since both studies indicate that a bimodal relationship exists between IQ and fertility, Penrose's equilibrium model (Penrose, 1948, 1950a, 1950b), which assumes a very high reproductive rate among the low-normal IQ groups and a very low reproductive rate among the high IQ groups, cannot be used to explain the current relationship between IQ and fertility nor any changes that could occur in the mean IQ of the population due to this relationship. Penrose assumed that differential fertility in relation to intelligence was a permanent phenomenon. The results of Higgins et al. (1962) and this study contradict Penrose's assumptions and support the position that the relationship between IQ and fertility is a dynamic one.

Intrinsic Rate of Natural Increase
In Relation to Intelligence

Life tables were compiled for each of the five IQ groups as well as for the total sample (See Appendix). The data were programmed, and the computer facilities at Michigan State University were utilized to find the value of the intrinsic rate of natural increase for each group that satisfied equation (3). The values of the intrinsic rate of natural increase for each of the five IQ groups and for the total sample are given in Table 6. The bimodal nature of the relationship between IQ and total reproductive performance (including the effect of generation length as well as fertility) is indicated by the r_m values for the IQ $\geqslant$ 120 and IQ 80-94 groups.

Table 6. The Intrinsic Rate of Natural Increase and the Average Generation Length in Relation to IQ

IQ Range	Intrinsic Rate of Natural Increase r_m	Average Generation Length T
$\geqslant$ 120	+0.008885	29.42 years
105-119	+0.003890	28.86 years
95-104	+0.000332	28.41 years
80-94	+0.007454	28.01 years
69-79	−0.010001	28.76 years
Total Sample	+0.003915	28.49 years

The value of r_m for the total Kalamazoo sample was only +0.004. The fact that r_m was extremely small for this study can be explained by the characteristics of the population under study which are known to affect fertility in a negative way and thus would reduce r_m. The population under study consisted of: (1) white individuals only; (2) native-born Americans only; (3) individuals with above-average educational attainments; (4) predominantly Protestant individuals; (5) individuals who spent almost all of their potentially most productive childbearing years during the Great Depression and World War II and (6) individuals who spend most or all of their lives in an urban environment. The urban nature of the population under study is probably the major reason for finding such a small intrinsic rate of natural increase. It is a well known fact that the completed fertility of the total urban population (age 45 and older) of the United States has been below replacement level in the past (Grabill, 1959).

Generation Length in Relation
To Intelligence

The almost universally held hypothesis that the generation length of the lower IQ groups is shorter than the generation length of the higher IQ groups has been supported by casual observations and by data compiled by Conrad and Jones (1932), who correlated the intelligence of the parent with the age of the parent at the birth of the first, second, third, and fourth child and found that the age of the parent at the time of birth of his children is positively correlated with IQ (the higher the IQ, the older the parent at the time of birth of his children). However, no exact estimates concerning generation length (given in years) in relation to IQ have been made.

Table 6 gives the average generation length, T, calculated from equation (4) for the five IQ groups as well as for the total sample. The results indicate a very slight positive relationship between IQ and generation length. The IQ 69-79 group is the only IQ group deviating from this positive relationship. The average generation length for the total sample approximately agrees with the estimated average generation length for the population of the United States (29-30 years).

Relative Fitness in Relation
To Intelligence

Relative fitness is defined in this paper as the ratio of population growth rate per individual of a particular phenotype (IQ group) to the population growth rate per individual of the optimum phenotype (IQ group) for the same trait. The optimum phenotype is that phenotype which has the highest population growth rate per individual (IQ $\geqslant$ 120 in this study).

The relative fitness of each of the five groups using the average number of offspring per individual as the measure of population growth is given in Table 7, while Table 8 gives the relative fitness of the five IQ groups using $e^{r_m T}$ which takes all the variables affecting population growth into account. Note that because the IQ $\geqslant$ 120 group has the longest generation length, the relative fitness of each of the other four IQ groups is increased when $e^{r_m T}$ is used to measure relative fitness instead of the average number of offspring per individual. This indicates the importance of generation length as a variable which can affect the population growth rates of several phenotypic classes of a particular behavioral trait unequally.

Table 7. Relative Fitness in Relation to IQ Using the Average Number of
Offspring Per Individual as the Measure of Population Growth

IQ Range	Relative Fitness
≧ 120	1.0000
105-119	0.8614
95-104	0.7771
80-94	0.9484
69-79	0.5774

Table 8. Relative Fitness in Relation to IQ Using $e^{r_m T}$ as the
Measure of Population Growth

IQ Range	Relative Fitness
≧ 120	1.0000
105-119	0.8674
95-104	0.7838
80-94	0.9600
69-79	0.5839

As pointed out by Dobzhansky and Allen (1956), "fitness" is meaningful only in relation to a particular environment. The relative fitness values given in this paper pertain only to the population under study and cannot be assumed to be the same for IQ groups in different environments.

Intensity of Natural Selection in Relation to Intelligence

When measuring the effect of natural selection on a trait, it is interesting to determine the population's phenotypic load or the proportion by which the population fitness is decreased in comparison with the optimum phenotype. This reduction in the fitness of the population relative to the optimum phenotype is called the intensity of natural selection (Haldane, 1954; Spiess, 1962). The intensity of natural selection in this study is measured by subtracting the relative fitness of the total sample (where fitness is measured by $e^{r_m T}$) from one (equation 6).

The intensity of natural selection in relation to intelligence in man was found to be 0.13 in this study where the optimum phenotype is the IQ ≥ 120 group. The fitness (reproductive performance) of the population under study, therefore, is 0.13 less than what it would be if all the IQ's were in the optimum range (IQ ≥ 120). That is to say, the phenotypic load due to variability in intelligence is 13 per cent.

Correlation Analyses

The negative relationship between the IQ of an individual and the size of the family from which the individual comes has been observed many times and is one major evidence used to support the hypothesis that the IQ of the population is declining. The correlation coefficient between the IQ of an individual and the size of the completed family from which he comes is—0.2599 for this study and is significantly different from 0 at the 1 per cent level. This agrees with the results of other studies on samples of a similar nature which have found the negative correlation to be between $r = -0.20$ and $r = -0.30$. However, when the IQ of an individual was correlated with his subsequent completed fertility in this study, the correlation coefficient is +0.0503, significanlty greater than 0 at the 6 per cent level by a one-tailed test. Thus the mean IQ of the population under study has probably increased slightly due to the small positive relationship between IQ and fertility.

The fact that in this study the IQ of an individual is positively correlated with the number of offspring he produces but is negatively correlated with the size of the family from which he comes appears to be paradoxical at first. However, these observed correlations are due to the operation of two factors.

First, the correlation between the IQ of an individual and the size of the family from which he comes has an inherent defect which makes any estimation concerning the relationship between intelligence and fertility based on it subject to considerable error. That part of the population which leaves no offspring (20.2 per cent in this study) is completely ignored. The results (Table 9) indicate that the relationship between the intelligence of an individual and the size family from which he comes is biased in a negative direction due to the fact that as intelligence decreases the probability of leaving no offspring increases. The results of this study show that the probability of leaving no offspring is more than two times as great for the IQ 69-79 group compared with the IQ $\geqslant 120$ group. A comparison of Table 9 with Table 10 indicates that this is primarily due to differentials in childbearing as

Table 9. Proportion of Individuals Who Left No
Offspring in Relation to IQ

IQ Range	Number	Per Cent Leaving No Offspring
$\geqslant 120$	11/82	13.41
105-119	48/282	17.02
95-104	70/318	22.01
80-94	60/267	22.47
69-79	9/30	30.00
Total Sample	198/979	20.22

Table 10. Proportion of Individuals Who Never Married
in Relation to IQ

IQ Range	Number	Per Cent Never Married
≥ 120	5/82	6.10
105-119	14/282	4.96
95-104	21/318	6.60
80-94	15/267	5.62
69-79	3/30	10.00
Total Sample	58/979	5.92

opposed to the differentials in marriage rates, which were found by Higgins et al. (1962) to be the important factor in their study. The correlation coefficient between the IQ of individuals who were not childless and the number of offspring they produced was found to be +0.0077—not significantly different from 0 at the 10 per cent level, and a shift in the negative direction from r = +0.0503. Thus it can be concluded that the relationship between the intelligence of an individual and the size of the family from which he comes has an inherent bias in the negative direction at the present time.

Secondly, the relationship between intelligence and fertility is definitely a dynamic one and appears to be changing rapidly. Higgins et al. (1962) and this study have shown that a bimodal relationship exists between IQ and fertility at the present time. It is doubtful that the biased techniques employed by previous studies could have completely obscured a high reproductive rate of the high IQ group. Therefore, it is highly probable that fertility is more positively correlated with intelligence now than at any time during the past 75 years. It is a well known fact that a tremendous change has taken place with respect to family size in the United States. There has been a decrease in the proportion of unmarried individuals, childless families, and one-child families in addition to a decrease in the proportion of families having five or more children (Grabill, 1959). In this study the correlation between the size of the family from which an individual comes and the number of offspring produced is quite low (r = +0.07). Thus a positive change in the relationship between intelligence and fertility occurring simultaneously with the great changes taking place with respect to family size could help explain the observed positive correlation of intelligence with the number of offspring produced and the negative correlation of intelligence with the size of the family that the tested individual comes from.

Estimation of the Change in the Frequency of the
Genetic Factors Favoring High Intelligence in
the Population Under Study

It is generally held that a major part of the variation in intelligence is due to genetic factors usually considered to be primarily quantitative in nature. It is highly probable that at least 50 to 60 per cent of the total variation in intelligence is due to hereditary factors. Vandenberg (1962), for instance, has found that approximately 60 per cent of the variation in numerical, verbal, spatial, and word fluency abilities in dizygous twins is due to hereditary factors. While an estimate of the amount of change in the frequency of the genetic factors favoring high intelligence must await further elucidation of the exact genetic factors involved, it is possible to estimate the direction of change in relation to these factors. A positive relationship between intelligence and fertility would indicate that an increase in the frequency of the genetic factors favoring high intelligence is taking place, while a negative relationship would indicate that a decrease in the frequency of these factors is taking place.

The observed positive relationship between the IQ of an individual and his subsequent completed fertility ($r = +0.05$) would seem to indicate that a small but positive increase in the genetic factors favoring high intelligence has taken place in the population under study. However, the negative effect of generation length (see Table 6) tends to counterbalance the positive effect due to the number of offspring produced to a certain extent. When generation length is taken into account, it has been shown that there is little difference between the relative fitness of the IQ $\geqslant$ 120 group and that of the 80-94 group. This bimodal relationship between IQ and total reproductive performance complicates any estimation of the change in the frequency of the genetic factors favoring high intelligence.

The degree of positive assortative mating in relation to intelligence was not determined for the population under study. While the direction of natural selection in relation to human intelligence is not affected by assortative mating, the rate at which the frequency of the genetic factors favoring high intelligence change is a function of the degree of positive assortative mating.

In spite of the complications due to generation length, the bimodal nature of the relationship between IQ and fertility, and the lack of assortative mating data, it can be safely concluded that the population under study has been in a state of equilibrium, or, more likely, has actually experienced a very slight increase in the genetic factors favoring high intelligence during the one generation that the population was studied. The equilibrium or slight increase in the frequency of the genetic factors favoring high intelligence is due not to the reproductive success of the average IQ group but to the counterbalancing effects of the high reproductive rates of the high IQ group ($\geqslant$ 120) and of the low-normal IQ group (80-94).

Summary

This is a follow-up study of 979 native white individuals born in 1916 or 1917 and who took the Terman Group Intelligence Test in the sixth grade while attending the Kalamazoo Public School System. It is the first study that has taken into account all of the variables that affect population growth (fertility, mortality, and generation length) by means of the intrinsic rate of natural increase when estimating the direction and intensity of natural selection in relation to human intelligence.

A bimodal relationship between intelligence and fertility was observed in this study. The high fertility of the $IQ \geqslant 120$ group relative to the other IQ groups is probably a quite recent development brought about by changes in the cultural environment during the last 30 to 40 years.

The observation that the intelligence of an individual is positively correlated with the number of offspring he produces but negatively correlated with the size of the family from which the individual comes was explained by the facts that (1) the bias inherent in the relationship between the intelligence of an individual and the size of the family from which he comes tends to produce a negative relationship by itself; and that (2) the relationship between intelligence and fertility is a dynamic one. It is highly probable that fertility is more positively correlated with intelligence now than at any time during the past 75 years.

When all variables affecting population growth have been taken into account, the population under study has probably been in equilibrium with respect to the genetic factors favoring high intelligence or, more likely, has experienced a slight increase in the frequency of the genetic factors favoring high intelligence.

The intensity of natural selection in relation to human intelligence (the phenotypic load due to the variability in human intelligence) was found to be 0.13 in this study where the $IQ \geqslant 120$ group was the optimum phenotype.

The conclusions of this paper pertain only to the population under study. Because of the small size and uniqueness of the sample it is impossible to estimate accurately the direction and intensity of natural selection in relation to human intelligence for the general population. Such an estimation must await the results of future studies which take into account all the variables affecting population growth and which are based on larger and different types of samples.

Acknowledgments

The author wishes to express his deep appreciation to Dr. Philip J. Clark, Professor of Zoology, Michigan State University, for his interest, direction, and encouragement during the course of this investigation. Special thanks are due Dr. John

Cochran, Assistant Superintendent of the Kalamazoo Public School System, and Mr. Anthony Stamm, County Clerk of Kalamazoo County, for their assistance in making available to the author many of the records upon which this investigation is based.

The author is also indebted to the American Eugenics Society for its financial support of this investigation.

References

Anastasi, A., 1956. Intelligence and family size. *Psychological Bull.*, 53: 187-209.

Cole, L. C., 1954. The Population Consequences of Life History Phenomena. *Quart. Rev. Biology*, 29: 103-137.

Conrad, H. S., and H. E. Jones, 1932. A Field Study of the Differential Birth Rate. *J. Am. Stat. Assoc.*, 27: 153-159.

Crow, J., 1962. Population Genetics: Selection. In W. J. Burdette (ed.), 1962. *Methodology in Human Genetics*. Holden-Day, San Francisco, pp. 53-75.

Dobzhansky, T., and G. Allen, 1956. Does Natural Selection Continue to Operate in Mankind? *Am. Anthrop.*, 58: 591-604.

Duncan, D. B., 1955. Multiple Range and Multiple F Tests. *Biometrics*, 11: 1-41.

Fisher, R. A., 1958. *The Genetical Theory of Natural Selection*. Second Revised Edition. Dover Publications, New York.

Grabill, W. H., 1959. Fertility and Reproduction. In D. J. Bogue, 1959. *The Population of the United States*. The Free Press, Glencoe, Illinois, pp. 288-324.

Haldane, J. B. S., 1954. The Measurement of Natural Selection. *Carylogia*, 6: 480-487.

Harter, H., 1960. Critical Values for Duncan's New Multiple Range Test. *Biometrics*, 16: 671-685.

Higgins, J., E. Reed, and S. Reed, 1962. Intelligence and Family Size: A Paradox Resolved. *Eugenics Quart.*, 9: 84-90.

Lotka, A. J., 1907a. Relation Between Birth Rates and Death Rates. *Science*, 26: 21-22.

———. 1907b. Studies on the Mode of Growth of Material Aggregates. *Amer. J. Science*, 24: 199-216.

———, and F. Sharpe, 1911. A Problem in Age-Distribution. *Philosophical Magazine*, 21: 435-438.

———, 1922. The Stability of the Normal Age Distribution. *Proc. National Acad. Sciences*, 8: 339-345.

———, 1925. Elements of Physical Biology. Williams and Wilkins, Baltimore.

Penrose, L. S., 1948. The Supposed Threat of Declining Intelligence. *Am. J. Mental Deficiency*, 58: 114-118.

———, 1950a. Genetical Influences on the Intelligence Level of the Population. *Brit. J. Psychology*, 40: 128-136.

———, 1950b. Propagation of the Unfit. *Lancet*, 2: 425-427.

Snedecor, G., 1956. *Statistical Methods*. Fifth Edition, Iowa State College Press, Ames, Iowa.

Spiess, E. B. (ed.), 1962. *Papers on Animal Genetics.* Little, Brown and Co., Boston.

Vandenberg, S. G., 1962. The Hereditary Abilities Study: Hereditary Components in a Psychological Test Battery. *Am. J. Human Genetics*, 14: 220-237.

Selected Bibliography Including Literature Cited

PART IV

Adam, A. 1969. A Further Query on Color Blindness and Natural Selection. *Social Biology*, 16: 197-202.

Akesson, H. 1967. A Study of Fertility and Mental Deficiency. *Act Genetica et Statistica Medica*, 17: 234-242.

Alland, A. 1967. *Evolution and Human Behavior.* Natural History Press, N.Y.

Alland, A. 1970. *Adaptation in Cultural Evolution: An Approach to Medical Anthropology.* Columbia University Press, New York.

Ardrey, R. 1966. *The Territorial Imperative.* Atheneum, N.Y.

Bajema, C. 1966. Relation of Fertility to Educational Attainment in a Kalamazoo Public School Population: A Follow up Study. *Eugenics Quarterly*, 13: 306-315.

Bajema, C. 1968. Relation of Fertility to Occupational Status, IQ, Educational Attainment and Size of Family of Origin: A Follow-up Study of A Male Kalamazoo Public School Population. *Eugenics Quarterly*, 15: 198-203.

Benedict, B. 1966. Co-operation in Primitive Human Societies. *Eugenics Review*, 58: 71-76.

Bodmer, W., and Cavalli-Sforza, L. 1970. Intelligence, Race and Society. *Scientific American*, 223: (Oct.): 19-29.

Brosin, H. W. 1967. Human Aggression in Psychiatric Perspective. *UCLA Forum Med. Sci.*, 7: 267-296.

Burt, C. 1952. *Intelligence and Fertility.* Cassell, London.

Campbell, B. 1966. *Human Evolution.* Aldine, Chicago.

Carter, C. 1966. Differential Fertility by Intelligence pp. 185-200 in Meade, J. and Parkes, A. (eds.) 1966. *Genetic and Environmental Factors in Human Ability.* Plenum Press, N.Y. 242 pp.

Cattell, R. 1937. *The Fight for Our National Intelligence.* King, London.

Cook, R. 1951. *Human Fertility: The Modern Dilemma.* Wm. Sloane, N.Y.

Cruz-Coke, R. 1970. *Color-Blindness: An Evolutionary Approach.* Thomas, Springfield.

Duncan, O. 1965. Farm Background and Differential Fertility. *Demography*, 2: 240-249.

Eckland, Bruce. 1967. Genetics and Sociology: A Reconsideration. *American Sociological Review*, 32: 173-194.

Erlenmeyer-Kimling, L. 1968. Mortality Rates in the Offspring of Schizophrenic Parents and a Physiological Advantage Hypothesis. *Nature*, 220: 798-800.

Emlen J. 1966. Natural Selection and Human Behavior. *Journal of Theoretical Biology*, **12**: 410-418.

Falconer, D. 1966. Genetic Consequences of Selection Pressure (with special reference to intelligence). Pp. 219-232 in Meade, J. and Parkes, A. (eds.) 1966. *Genetic and Environmental Factors in Human Ability*. Plenum Press, N.Y. 242 pp.

Fox, R. 1967. In the Beginning: Aspects of Mominid Behavioral Evolution. *Man*, **2**: 415-433.

Freedman, R., and Slesinger, D. 1961. Fertility Differentials for the Indigenous Non-Farm Population of the United States. *Population Studies*, **15**: 161-173.

Gilula, M. and Daniels, D. 1969. Violence and Man's Struggle to Adapt. *Science*, **164**: 396-405.

Goldberg, D. 1959. The Fertility of Two-Generation Urbanites. *Population Studies*, **12**: 214-222.

Goldberg, D. 1960. Another Look at the Indianapolis Fertility Data. *Milbank Memorial Fund Quarterly*, **38**: 23-36.

Goldberg, D. 1965. Fertility and Fertility Differentials: Some Observations on Recent Changes in the United States. Pp. 119-142 in Sheps, M. and Ridley, J. (eds.) 1965. *Public Health and Population Change*. Univ. of Pittsburgh Press, Pittsburgh. 557 pp.

Hamburg, David. 1963. Emotions in the Perspective of Human Evolution. Pp. 300-317 in Knapp, Peter (ed.) 1963. *Expressions of the Emotions in Man*.

Harris, M, Fried, M., and Murphy, R. (eds.) 1968. *War: The Anthropology of Armed Conflict and Aggression*. Natural History Press, N. Y.

Hockett, C., and Ascher, R. 1964. The Human Revolution. *Current Anthropologist*, **5**: 135-168.

Hopkinson, G. 1963. Celibacy and Marital Fertility in Manic-Depressive Patients. *Acta Psychiatrica Neurologica Scandinavia*, **39**: 473-476.

Jensen, A., et al. 1969. Environment, Heredity and Intelligence. *Harvard Educational Review Reprint Series No. 2*, 1-246.

Lancaster, J. 1968. On the Evolution of Tool-Using Behavior. *American Anthropologist*, **70**: 56-65.

Maxwell, J. 1969. Intelligence, Education and Fertility: A Comparison Between the 1932 and 1947 Scottish Surveys. *Journal of Biosocial Science*, **1**: 247-271.

Montagu, A. (ed.) 1968. *Man and Aggression*. Oxford Univ. Press, N. Y. 178 pp.

Morris, D. 1967. *The Naked Ape*. McGraw-Hill, N. Y.

Morris, D. 1969. *The Human Zoo*. McGraw-Hill, N. Y.

Morton, N. 1967. Population Genetics of Mental Illness. *Eugenics Quarterly*, **14**: 181-184.

Neel, J., and Post, R. 1963. Transitory "Positive" Selection for Color blindness? *Eugenics Quarterly*, **10**: 33-35.

Osborn, F. 1968. *The Future of Human Heredity. An Introduction to Eugenics in Modern Society*. Weybright and Tally, N. Y.

Pickford, R. 1963. Natural Selection and Colour Blindness. *Eugenics Review*, **55**: 97-101.

Post, R. 1962. Population Differences in Red and Green Color Vision Deficiency: A Review and a Query on Selection Relaxation. *Eugenics Quarterly*, **9**: 131-146. Corrections in *Eugenics Quarterly*, **10**: 84-85.

Post, R. 1962. Population Differences in Visual Acuity: A Review with Speculative Notes on Selection Relaxation. *Eugenics Quarterly*, **9**: 189-212.

Post, R. 1963. Colorblindness "Distribution in Britain, France and Japan: A Review with Notes on Selection Relaxation. *Eugenics Quarterly*, **10**: 110-118.

Post, R. 1964. Hearing Acuity Variation Among Negroes and Whites. *Eugenics Quarterly*, 11: 65-81.

Rainer, J., and Firschein, I. 1959. Mating and Fertility Patterns in Families with Early Total Deafness. *Eugenics Quarterly*, 6: 117-127.

Ray, A. 1969. Color Blindness, Culture and Selection. *Social Biology*, 16: 203-208.

Reed, S., and Reed, E. 1965. *Mental Retardation: A Family Study*. Saunders, Philadelphia.

Reed, S. 1966. The Evolution of Human Intelligence: Some Reasons Why It Should be a Continuing Process. *American Scientist*, 53: 317-326.

Reed, S., Reed, E., and Higgins, J. 1962. The Relationships of Human Welfare to Marriage Selection. *Journal of Heredity*, 53: 153-156.

Reynolds, V. 1966. Open Groups in Hominid Evolution. *Man*, 2: 130-132, 1967.

Rosenthal, D. 1970. *Genetic Theory and Abnormal Behavior*. McGraw-Hill, New York, 318 pp.

Salzano, F. 1963. Letter on "Population Differences in Red and Green Color Vision Deficiency: A Review and A Query on Selection Relaxation." *Eugenics Quarterly*, 10: 81-83.

Shaw, R., and Smith, A. 1969. Is Tay-Sachs Disease Increasing? *Nature*, 224: 1214-1215.

Spuhler, J. 1967. Behavior and Mating Patterns in Human Populations. Pp. 241-268 in Spuhler, J. (ed.) 1967. *Genetic Diversity and Human Behavior*. Aldine, Chicago. 291 pp.

Stevens, B. 1969. Probability of Marriage and Fertility of Psychiatric Patients in England. *Populations Studies*, 23: 435-454.

Tajfel, H. 1966. Cooperation Between Human Groups. *Eugenics Review*, 58: 77-84.

Tinbergen, N. 1968. On War and Peace in Animals and Man. *Science*, 160: 1411-1418.

PART V

Natural Selection and the Future Genetic Composition of Human Populations

Should man practice eugenics? Should he control his own evolution? Most biologists, when facted with these questions, take a position somewhat similar to that of Mirsky (1964), Rockefeller University biologist, who states that "before man can embark on a program of eugenics he will need far more knowledge than he possesses today." That man needs more knowledge is incontrovertible. However, Mirsky's statement implies that man is not currently engaged in such a program.

Man is already influencing his evolution. He's affecting the genetic composition of future generations by his medical practices (vaccinations, genetic counseling, treatment of hereditary diseases, etc.), his governmental taxing policies, his governmental welfare programs, and his living arrangements. The question now is not *whether* man should direct his own evolution, but rather *how* man should direct his own evolution. Hardin, (1962), in discussing this question, points out that man's only options are these:

"(1) to do so consciously or unconsciously;
(2) to minimize or maximize the role of chance;
(3) to control individual actions on a directive basis, or merely by statistical biasing mechanisms; and
(4) to set up one system (of government and selection), or many."

Bentley Glass (1967) has argued that the problem we face is not so much a prob- of how to change our genetic nature as it is the problem of deciding what kind of human being we wish to have populate the earth. Do we want populations of human beings who are uniform with respect to physique, health, intellect, personality or do we want to maintain genetic diversity within and/or between human populations with respect to these characteristics? If we want to maintain genetic diversity we need to decide how much genetic diversity we want and how the cost of maintaining such diversity will be paid. Society certainly needs a clearer idea as to the goals it wishes to attain and the values it wishes to maintain if society as a whole is to embark on a conscious eugenics program. Yet to argue that society must be in complete agreement before instituting an eugenics program is a patent rationalization for inactivity embraced for other reasons. We do not agree completely on anything but that has not prevented society from acting in other matters (Davis, 1966). We have passed laws governing birth control, marriage, and divorce, tax exemptions for religious institutions, restrictions on the sale of alcohol, and rules governing adoption of children, although we were and still are far from being in complete agreement in these matters.

The population—environment crisis characterized by growing human populations in deteriorating environments that are becoming increasingly unstable is forcing societies to question their goals and to institute programs aimed at controlling numbers. Many of the questions which need to be answered before a democratic society can consciously and deliberately institute an effective eugenics program are questions which are in the process of being asked and answered as societies try to determine what an optimum population size is and the means by which the optimum number of human beings can be maintained in a society over time.

A number of eugenics programs have been proposed to improve the genetic heritage of future generations. Virtually all of the eugenics proposals fall into one or more of the following categories:

1. Genetic surgery: genetic reprogramming of existing cells (directed mutation)
2. Cloning: producing exact replicas (identical twins) of individuals.
3. Artifical insemination using donor sperm (AID): selection via the male.
4. Authoritation control by society over reproduction: the granting of licenses to reproduce by an eugenics board of control.
5. The Eugenic Hypothesis: the organization of democratic society in such a way that the most successful individuals will have the most children.

Genetic surgery—the idea that scientists will be able to correct faulty genes within the cells of the individual via directed mutation—has been discussed by Hotchkiss (1965), Sinsheimer (1969), and Tatum (1966). Cloning—the idea that scientists will be able to take individual cells from the body of a human being and, by treatment with the appropriate chemicals, induce regeneration not of just a limb but of the whole individual—has been discussed by Lederberg (1966). While theoretically possible, both of these techniques are science fiction at the present time because society lacks the necessary technology. Some of the immense technological problems that must be overcome before genetic surgery would become practical enough to base an eugenics program on are discussed by Muller (1965). It may very well be that the costs in terms of individual human misery (because of errors of genetic surgery or errors in development associated with cloning) and economics (both methods require very sophisticated technology and large amounts of professional time per individual case) will preclude the widespread use of these techniques even if all the technological barriers are overcome.

The late Herman J. Muller, who won the Nobel prize in 1946 for his discovery that X-rays cause gene mutations, has been the chief advocate of artificial insemination as a way of improving the genetic make-up of human populations. Artificial insemination as proposed by Muller (1959, 1964, and his 1967 paper in this section) involves the establishment of a sperm bank, which would store sperm from males who have exceptional qualities (for example, high intelligence, creativity, cooperative personality, good physique, and physical health). After sufficient time had elapsed to insure that the donor was objectively evaluated with respect to the phenotypic expression of his genes, the sperm would be made available for artificial insemination. Couples who are motivated to provide their children with the best possible genetic heritage would be able to request the stored sperm of any specific donor and would make their selection on the basis of the information on record concerning each donor. Muller argued that the quality of offspring resulting from artificial insemination would provide the stimulus for an ever increasing proportion of the population to utilize artificial insemination when having children.

Artificial insemination as a means of achieving eugenic goals has some advantages. First, there are no technical difficulties to be overcome (unless long-term storage of sperm beyond a couple of years is required). Artificial insemination is being performed every day in some hospital or clinic in the United States. Second, no action on the part of society is required, only the voluntary initiative of the couple involved. Third, there is already a continual demand for donor sperm by couples where the male is either sterile or is carrying a defective gene that he does not want to pass on to his children. Estimates as to the annual number of children born in the United States as the result of artificial inseminations run as high as 10,000 per year. Although the degree of selection currently being employed by physicians when they choose a donor is not as great as that advocated by Muller, the donors are usually professionals in the field of medical practice or research and, taken as a group, are therefore above average with respect to mental ability and physical health. Because of questions concerning the legality of the procedure of artificial insemination in some states and the timidity of the medical profession, an objective analysis of the genetic consequences of artificial insemination has not been carried out. Because of the lack of information as to the genetic consequences, it is doubtful that artificial insemination will be adopted by a significant enough proportion of the population in the near future to bring about a detectable genetic improvement of the population. Yet the selection of one of the two parents on the basis of eugenic criteria has considerable potential for the genetic improvement of the human species.

Many of the eugenics programs proposed in the past have been authoritarian in nature. The genocide policies of Nazi Germany carried out under the guise of eugenics led many people to adopt a distorted view of eugenics and to believe that all eugenics programs are by their very nature, authoritarian. While some eugenics programs require mutual coercion (see the paper by Bajema in this section) there are other programs (artificial insemination as advocated by H. S. Muller, the Eugenic Hypothesis as advocated by Osborn) that are designed to work within the framework of democratic societies that allow freedom of parenthood.

The Eugenic Hypothesis—the establishment of a modified welfare state as a means of bringing about eugenic selection by the environment—was first advocated by Frederick Osborn in the late 1930's. According to Osborn natural selection has shifted from operating mainly through deaths to operating mainly through births in such societies as the United States and Sweden. If societies were organized in such a way that the most successful individuals had the most children, then the genes underlying the traits held desirable by society would increase each generation. Osborn (1940) outlined three steps that he considered necessary for the development of such a society:

"1. General improvement of the environment. A voluntary system of eugenics cannot operate under conditions of extreme poverty, ill health, ignorance, and isolation. The first eugenic requirement is to raise the poorest environments in order that . . . selection may take place at a relatively high environmental level.

2. Completion of the change now taking place to freedom of parenthood [availability of contraceptives, abortion, etc.] throughout the whole population.

3. Finally, the introduction of eugenic measures of a psychological and cultural sort which will tend to encourage births among parents most responsive to the possibilities of their environment, and to diminish births among those least responsive, thus bringing about a process of eugenic selection through variations in size of family."

The Eugenic Hypothesis, a eugenics program designed to work within the framework of a democracy, involves no coercive action and does not require the establishment of an eugenic board of control which would decide who can reproduce and who cannot. The environment makes this selection and the individual may not even be conscious of the fact that he or she is under "pressure" to reproduce or not to reproduce. The Eugenic Hypothesis has been criticized by some who argue that freedom of parenthood cannot exist in an overpopulated world and/or that freedom of parenthood will lead to genetic deterioration (Darwin, 1956, 1959; Hardin, 1968).

In addition to containing papers discussing the probable genetic effects of various proposed eugenics programs this section also contains papers discussing the probable effects of current trends in fertility and mortality (trends brought about by programs adopted for other than genetic considerations) on the future genetic composition of human populations. The genetic impact of family planning practices is discussed by Matsunaga and Medawar discusses the relaxation of natural selection due to modern medical practice.

In spite of the research conducted by Higgins, Reed, and Reed (1962) and the editor (Bajema, 1963) many geneticists believe that natural selection is currently operating against human intelligence (see Muller's paper in this section). Data concerning the operation of natural selection in relation to human personality is essentially nonexistent. The most informed speculation seems to take the position that natural selection is favoring those personalities that most people consider least desirable (see Hardin's paper in this section). Rene Dubos (1965), a Rockefeller University biologist, in discussing the effect of man's future environment on the direction and intensity of natural selection in relation to human personality patterns, states:

> "Most disturbing perhaps are the behavioral consequences likely to ensue from overpopulation. The ever-increasing complexity of the social structure will make some form of regimentation unavoidable; freedom and privacy may come to constitute antisocial luxuries and their attainment to involve real hardships. In consequence, there may emerge by selection a stock of human beings suited genetically to accept as a matter of course a regimented and sheltered way of life in a teeming and polluted world, from which all wilderness and fantasy of nature will have disappeared. The domesticated farm animal and the laboratory rodent in a controlled environment will become true models for the study of man."

Since a number of societies seem to be on the verge of instituting policies aimed at controlling population size the editor has included two papers which discuss the genetic implications of population control (see the papers by Hardin and by Bajema).

The papers in this section, although speculative in nature, give us a better understanding of how rapidly changes in the direction and intensity of natural selection are taking place in modern societies and of how difficult it is to ascertain just what the genetic consequences of a particular program might be.

18. Do Advances in Medicine Lead to Genetic Deterioration?

PETER BRIAN MEDAWAR

I have chosen to put my title and subject in the form of a question: Do advances in medicine lead to genetic deterioration? This question is in itself a special and particularly insistent form of a more general question, which I think it is very appropriate that we should ask at a centennial celebration honoring a great center of medical research. The more general question is this: What is medicine tending to; what are the ultimate goals of medicine in terms of human welfare generally, not just narrowly in terms of health; and what will be the ultimate effects on society of medical practice and medically inspired legislation? I suppose that two extreme views are held in answer to these questions, and though they are mutually contradictory, they are, in fact, sometimes held by the same people, though admittedly not at the same time.

First there is the view that medicine is abolishing—indeed has already abolished—the main causes of ill health; that the goal of medical research is to put its own practitioners out of business; that this goal will be achieved and the major task of medicine will one day be accomplished so that eventually medicine will wither away, as Lenin said the state would wither away. But some people who say this can also, in different moods, be heard to say (without any apparent awareness of inconsistency) that medicine by its very success in preserving the weak and the defective and the medically dependent is imposing a huge and growing burden on society and, in particular, on the medical services of society; that medicine, far from withering away, will, as George Orwell said the state would, become ever more pervasive and intrusive and demanding until one day all the world will become a hospital and even the best of us will only be ambulatory patients in it.

Both these views are grossly exaggerated. But there is enough truth in the first to create a sense of satisfaction, which is instantly dispelled by reflecting that there is also a good deal of truth in the second. The second—the problem that arises because of what we may generally and rather crudely call preservation of the unfit—is particularly disturbing where there is a strong hereditary, genetic, inborn, element

SOURCE. Peter B. Medawar, "Do Advances in Medicine Lead to Genetic Deterioration?" *Mayo Clinic Proceedings*, 40: 23-33 (Minnesota, 1965). P. B. Medaware is Director, National Institute for Medical Research, Great Britain.

in any particular human ailment, because to preserve those who are ill or who are more or less incapable for genetic reasons is also to preserve genetic factors which led to their being in the condition in which they are. This is why the question I put in my title—Do advances in medicine lead to genetic deterioration?—is a particularly cogent one. There is no mystery about what is meant by "deterioration." By deterioration I mean any restriction of liberty or of competence which a particular genetic constitution may bring in its train—being obliged to eat some foods, for example, or being forbidden to eat others. Thus, all human beings suffer from a grave and sometimes fatal inborn defect, namely that of being unable to synthesize ascorbic acid (vitamin C). A very small minority of human beings have another, in some ways comparable, defect—being unable to synthesize an enzyme which allows them to make use of milk sugar: they are victims of galactosemia. I couple these two inborn afflictions with deliberate irony to show only that the difference between them is a difference of degree. We do not normally think of the inability to synthesize ascorbic acid as a deficiency, because we all suffer from it and because the remedy—eating enough vitamin C—is pretty well universally available. We do think of galactosemia as a grievous affliction, because it affects only a very small minority, and the circumstance which gives rise to it, namely eating food that contains milk sugar, is pretty well ubiquitous.

For a long time I thought this question, Do advances in medicine lead to genetic deterioration? was *one* question having, if only one had the wit to find it, *one* answer. But I now know on further reflection that there are *two* questions here and that they have *two* different answers. I think the failure to analyze the problem sufficiently to see that there are two questions here is responsible for some of the bitter differences of opinion that are held about what the answers should be. The first question, or first form of the problem, is as follows. The physical environment was at one time very much more perilous than it now is, through shortage of food and through the hazards of infection and of predators, and so forth. As the environment improves, as civilization advances, we are increasingly protected from this hostile and dangerous environment in which the human race was reared. Therefore, we are losing those genetic endowments which have conferred fitness, the ability to cope, with these hard, difficult, tough environments.

That is one form of deterioration. At the same time—and this is the second point—as the environment improves, as it becomes easier for us, we are preserving and increasing the representation of those genetic endownments which would have been highly deleterious in that old, hard, difficult environment that we have grown out of. These two propositions are not the same. They raise different questions which have, as I shall hope to convince you, two quite different answers. The first of them refers to the loss of genetic makeups or endowments which conferred fitness in the old environment of mankind. The second refers to the gain of genetic makeups which confer or bring in their train unfitness. That is the distinction which must be borne in mind—the loss of makeups conferring fitness, the gain of makeups conferring or bringing about unfitness.

I want to deal now with the first—the loss of the qualities which have conferred fitness. It seems perfectly obvious that if we eliminate or ameliorate undernourishment, specific dietary deficiencies, the attacks of pathogenic microorganisms, the attacks of predators and all enemies of the species, and if we are no longer exposed to extremes of heat and cold or wet and dry in an air-conditioned world—if we improve the environment in these ways, we are nullifying or discounting the inborn advantages, the alleged genetic virtues, enjoyed by those whose genetic makeups equip them especially well to cope with these various hazards. Therefore, there must be a genetic softening up, a genetic lowering of the defenses. This seems perfectly obvious, and at first sight one feels it must be so; but plausible though the argument seems, I believe it to be completely fallacious. The apparent rightness of the argument depends on the assumption that there is something intrinsically meritorious in possessing the genetic equipment to combat particular hazards even though those hazards are no longer with us. This is not the case. It so happens that the genetic, or inborn, qualities that do confer resistance to particular infections or affections or stresses of one kind or another confer resistance to them alone. They are often achieved by some kind of genetic trick or device which it is a positive drawback to possess in environments in which these particular hazards and perils and challenges no longer exist. The classic demonstration of this argument, the truth of which is conceded by even the gloomiest of our genetic Cassandras, concerns inborn resistance to malaria. Malaria is still the major killing disease in the world. The World Health Organization estimates that some two and a half million people die of malaria every year and perhaps a hundred times that number are afflicted. Now, a whole variety of cheap genetic tricks are known, or are thought with various degrees of conviction, to owe their existence to their ability to confer some measure of inborn resistance to malaria: the genetic factor which causes the substitution of an abnormal hemoglobin, hemoglobin S, for normal hemoglobin and thus gives rise to sickle-cell trait; the analogous gene which substitutes hemoglobin C for normal hemoglobin; the molecular lesion which is responsible for Cooley's anemia, or thalassemia major, a blood disease widely prevalent, as the name thalassemia indicates, in the Mediterranean area; glucose-6-phosphate dehydrogenase deficiency, an enzymic deficiency which is not in itself particularly harmful, though it does lay its possessors open to certain disabilities—hypersensitivity, for example, to primaquine and to sulfonamide drugs and also an inability to indulge in the rather bizarre addictive habit of eating naphthalene mothballs.

By far the best worked-out of these various genetic devices which confer inborn resistance to malaria is sickle-cell trait. The person who has inherited the gene responsible for this condition from one parent only does enjoy a very considerable extra measure of so-called natural resistance to subtertian malaria. I call this, very deliberately, a cheap genetic trick because when these carriers marry each other, then on the average one quarter of their children are afflicted with sickle-cell anemia, which is a grave and in most cases fatal disease.

These various genetic devices all serve a purpose, and a very important purpose, where malaria is a hazard, and they persist where malaria is still, or has in the not too remote past been a hazard. They are useless or actively harmful when the environment improves in the sense of the eradication of malaria. This relationship has the very important corollary, that the improvement of the environment, in this case the eradication of malaria, leads to genetic *improvement;* that is to say, it increases our capabilities, it enlarges freedom, even if it is only one of the humbler freedoms, such as freedom to eat mothballs.

Our genetic Cassandras argue that the particular case which I have just outlined to you is a special and unrepresentative case. I believe it is not. There is no doubt at all, from all the records of mortality we possess, that since the domestication of man five or ten thousand years ago, infectious disease has been overwhelmingly the most important cause of mortality and the major selective force acting on the human population. There is not the slightest reason to think that the genetic devices which confer some degree of natural or inborn immunity to infectious diseases—poliomyelitis, tuberculosis, smallpox, or great pox—are signs of any general genetic virtue or confer any advantage whatsoever when those hazards are not present. Inborn resistance is very much more likely to be purchased by some kind of little metabolic trick which will be quite harmful in a normal environment. For example, it may be a deficiency of some growth factor which the microorganism needs in order to proliferate. The reason why newborn rats cannot be infected with the malarial organism Plasmodium berghei is not that they are innocently virtuous animals but that they do not manufacture enough para-aminobenzoic acid to support the growth of the malarial organism.

Of course it could be argued that the proposition I am putting forward—that the improvement of the environment leads to genetic improvement—is true only of infectious disease, and that though infectious diseases are important they are not the only afflictions that affect mankind and they are certainly not all-importnat. But I suspect that the proposition I have just outlined may be true of many disorders of metabolism—in other words, that the devices which procure some kind of resistance or natural immunity to noninfectious diseases may be a nuisance or may even be harmful when the danger of contracting these diseases no longer exists. There is certainly no general virtue or merit in belonging to one blood group rather than another. It just so happens that those people who secrete the AB substances into their body fluids enjoy a naturally greater resistance to duodenal ulceration. Again, it has been suggested that the genetic difference which shows up as the ability or inability to taste the compound phenylthiourea is, in itself, a very trivial outward manifestation of a metabolic difference which has some quite important effect upon one's liability to contract ordinary nodular goiter. Still more speculative, and not yet beyond the level of after-dinner conversation, is the idea that under conditions of chronic and prolonged caloric deficiency (quantitative undernutrition) a certain degree of inborn resistance, or ability to cope, can be purchased by metabolic devices which in an environment of plenty such as our own environment manifest themselves as diabetes mellitus. Certainly there is a hereditary element in diabetes, and it is quite a problem to know why the genetic factors responsible for it should be so very common.

Thus, in general, there is a good case for saying that improvement of the environment improves mankind genetically and that it is better not to have an inborn resistance to many particular hazards and stress in the environment if those hazards and stresses are no longer present. This view arouses tremendous tempermental opposition. It seems only reasonable, people say, that if the environment softens, we too must soften; and there is an extremely strong element of pure Puritan bigotry in this attitude (central heating is not merely debilitating, it is also sinful). But another element which has contributed very greatly to the opposition, in England anyway, can be traced to certain low-grade literary and philosophical propaganda which grew up at about the time of the romatic revival and which has been embodied in what the *Lancet* calls the Myth of Merry England. The myth is that primitive man or pre-industrial man lived in a state of happy innocence-vigorous, healthy, robust, free, and having a sort of inner tranquility and radiance which cynics say comes nowadays only to those possessing a substantial private income from trustee investments. Let me make it clear, therefore, that the primates generally, the zoological group to which we belong, have not been a great success as animals in spite of their great intelligence; nor until quite recently have human beings been anything that you would really call a success as animals. Until a few hundred years ago the foothold of human beings on the world was not a completely secure one, and the doubling time of the human population, which is our crude biological measure of prowess, was of the order of thousands or tens of thousands of years. Man, in fact, owes his success not so much to genetic devices as to exogenetic devices, to devices contributing to the exosomatic evolution or psychosocial evolution of which Dr. Eiseley speaks. The people who first called our attention to this were, as I now realize having listened to Dr. Eiseley, Alfred Russel Wallace and also of course Herbert Spencer, a considerable philosophic figure whom we think of nowadays only really to laugh at, though he did in fact say some extremely sensible things.

I want to turn now to the other half of the question. I have been discussing genetic devices which confer fitness in bad environments and I have been saying that their loss will lead, not to genetic deterioration but, on the contrary, to improvement. Now the other half of the story is the deterioration which may result from the increase in frequency of genetic makeups which would have been disadvantageous and perhaps fatally disadvantageous in the old bad environment. Today the sheer virtuosity and exuberance of medical skill and care in keeping alive people who fifty, twenty-five, or even five years ago would certainly have died, is bound, so it is argued, to place a genetic burden on society, because in spite of treatment such people have a greater or lesser restriction of freedom; they have a lowered general capability. Insofar as medical science preserves these weaker people and insofar as their weaknesses have a genetic origin, there is a real danger—and this is the central problem of eugenics—of genetic deterioration. Of course, such people will grow up and will very naturally have children, and therefore they will perpetuate the genetic factors which were responsible for such shortcomings as they have.

There is no doubt that this is happening. The question is what is to be done about it. Some kinds of genetic affliction are not yet open to eugenic "treatment." I shall not consider, when I speak of these inborn disorders, those of chromosomal origin, important though they are (and mongolism is one of them), because people so afflicted are not kept alive primarily by medical intercession, and medicine is not yet to blame for any increase in frequency which may be coming about (in point of fact, their frequency, I think, is very slightly declining). Nor shall I consider so-called dominant disorders of very early onset which cause death very early in life or before birth, because their spread, genetically speaking, is obviously self-limited. But certain grave problems arise from dominant genetically determined disorders of late onset like Huntington's chorea or like familial intestinal polyposis, which leads to cancerous growth. I will say something about this in just a moment. But from our point of view, from the eugenic point of view, collectively the most important genetic defects are those of so-called recessive determination, those in which the causative genes must be inherited from both parents, or must be present in both gametes which unite to form the individual, if the disease is to become manifest. Among these—individually all are rare, collectively they are numerous—are, for example, phenylketonuria, due to an inability to metabolize phenylalanine; galactosemia, which I have already mentioned, due to an inability to handle milk sugar; fibrocystic disease of the pancreas; and congenital hypogammaglobulinemia, a disease which interferes with the synthesis of antibody proteins. All these afflictions (and many of the people who suffer from them would until very recent years have died) are the product of specific molecular or metabolic lesions; and if their victims are now kept alive so that they can eventually reproduce, the genetic factors responsible for them will come to acquire a frequency higher than the equilibrium level reached when the forces of natural selection, which eliminate the gene, and of mutation, which constantly cause the gene to reappear in the population, just cancel out. If the number of harmful genes goes up by the propagation of their bearers, so in the normal course of events must the number of the afflicted go up, and so, in turn, the number of genes. Thus one has a self-exacerbating problem.

I want very briefly to consider these recessive diseases and what one can do about them. Obviously it is humane and proper to work toward the prevention of the diseases they are responsible for and to try to cure them. But in the genetic context, prevention and cure have quite special and rather different meanings from those they hold in an ordinary medical context, and one must be aware of the consequences of those differences. With these recessive diseases it is theoretically possible to prevent almost all cases of the overt disease from ever happening, by forestalling and preventing the union of gametes containing the *same* harmful recessive gene—in other words, by preventing the genetic conjunction which causes the disease to become manifest. I am not, of course, suggesting that one should prevent the conjunction of gametes containing any harmful recessive genes, because it is universally conceded that all gametes carry a certain number—possibly five or six— of harmful recessive genes, so that this would be equivalent to recommending the suicide of the human race. I am suggesting that there should somehow be prevented the conjunction of gametes known to contain or suspected to contain the *same* harmful recessive gene, and the reasoning underlying this approach is extremely simple. Consider a recessive disease of rare occurrence. Let the overt occurrence, the frequency in the population, be, say, one in 10,000 (one in 10,000 is decidedly on the high side for some of these diseases). If, in one person in 10,000, *two* recessive genes of the same kind have come together, then the number of carriers of *one* gene—that is, the number who have inherited the gene from one parent only and do not show outward signs of the disease—will be very much higher, or approximately one in 50. Now, most cases of the disease arise in children of a marriage between two of these carriers. When two carriers happen to marry each other, one quarter of their children, on the average, will contract the disease and one half of their children will, like themselves, be carriers. Thus, if one can identify the carriers, and that is a very big *if* and if one can discourage them—in the nicest possible way, of course— from marrying each other; or, if they insist upon marrying each other, if one can discourage them against or warn them of the consequences of having children, then the overt incidence of these diseases would crash downward. In a sort of liberal frenzy one is at first inclined to say that this is a cruel and an absolutely unwarranted intrusion, a wicked infringement of the private liberty and rights of the individual. I do not think it is. Firstly, for quantitative reasons, if the overt incidence of the disease we are talking about is one in 10,000, only one in 2500 marriages will be contraindicated because of the danger that the children will contract the disease, and the marriage is contraindicated only in the sense that each of the parties to the intended marriage would have to try to make do with one in 49 other people instead. I hope you will not think I am being flippant; I think that this kind of marriage counselling and guidance is a perfectly logical extension of the veto on closely consanguineous marriages, a veto which is at least partly based on genetic reasons. Thus, in general, if the overt incidence in the population of a certain recessive condition is X, say one in 10,000, the proportion of marriages contraindicated is 4X, one in 2500. A more realistic figure would be an incidence of the actual disease of one in 40,000, in which case only one in 10,000 marriages would have to be discouraged. Of course, this depends on being able to identify the carrier. In theory, if the condition is a fully recessive one, it should be impossible to discern who the carriers, the people who inherited the recessive gene from one parent only, are; but in practice, dominance and recessiveness are matters of degree, and in many cases, notably in the case of phenylketonuria, it is in fact possible to identify the carriers. Time will not allow me to mention how an analogous kind of reasoning might be applied to people who suffer from dominant afflictions of late onset. In the case of Huntington's chorea or genetic intestinal polyposis, the important thing is to try to identify beforehand and early enough in life those in whom the affliction is likely to appear, and then it is only just and proper to warn them of the consequences of having children; and the consequences, genetically speaking, are that half of their children would be afflicted with a disability similar to their own, and, what is perhaps more important, all of them would grow up in the dread of contracting the disease. Many people would not consent to be the cause of such unhappiness.

To end up, I must emphasize the difference between prevention of a disease in the genetic sense, the sense that I have just been describing, and prevention in the sense in which one would use it of the spread not of a gene but of a microorganism of some kind. If one discourages marriages between carriers of some particular recessive gene, one is not preventing the spread of the offending gene; on the contrary, one is, of course, *promoting* the spread of the offending gene. Although we are preventing the outward manifestation of the disease, which depends on the conjunction of two such genes in one individual, we are promoting the spread of the gene because we are withdrawing the one obstacle which has hitherto resisted the spread of that gene in the population, namely, of course, the death or infertility (which is genetic death) of the people who carried it. Thus, until the new regimen of marriage counselling is adopted, the frequency of the gene would be kept down in the population by natural selection, which would counteract the tendency of mutation to bring the gene into the population. But if one adopted the kind of eugenic solution which I am suggesting, the frequency of the gene, no longer opposed, would increase steadily, and the problems to which it gives rise would become very slowly but progressively more urgent. It is also important to remember that when one uses the term "cure" in connection with inborn afflictions, that word too is used in a rather special sense, because one is not curing the genetic lesion itself but one is only curing its more or less remote somatic consequences; the cure of an inborn metabolic disease, genetically speaking, involves supplying from the outside something which the body cannot build up, if that substance is necessary, or withholding from the environment something which the body cannot break down, if that should be harmful. In general, cure consists in providing an environment in which a harmful genetic constitution cannot manifest itself. We cope perfectly well with our grave inborn inability to synthesize vitamine C by taking pains to eat the stuff; but with other, rarer, afflictions the problem is grave and exacting. But it can be solved. Phenylketonuria can be cured, if it is identified early enough, by withholding phenylalanine-containing food from the diet (which is a great deal more difficult than it sounds). Galactosemia can be prevented by withholding milk. A person suffering from phenylketonuria who has been cured in this way still suffers a disability and a restriction of freedom, but *it is not phenylketonuria:* it is being unable to eat what he likes. Whichever solution one adopts, whether it is prevention in the special sense of trying to prevent that conjunction of genes which leads to the overt manifestation of the disease, or whether it is cure, the frequency of the offending gene is going to rise in the population because one is withholding the action of natural selection, the force that until then kept the frequency of the gene down against the constant, nagging pressure of mutation to reintroduce it into the population.

Thus there arises a well-founded fear of building up in the human population a huge and increasing genetic liability like the national debt. I myself am not dismayed by this prospect, and I see no very good reason why anybody else should be dismayed. The point is that the rate of genetic deterioration brought about by the methods I have just described is extremely slow. The unit of the time-scale of evolution is a generation, and the order of the length of time one is thinking of when one speaks of genetic deterioration is not tens of years but hundreds of years. During these future hundreds of years, solutions will certainly be found to cope with these difficulties. If you will forgive my comparing two obviously incommensurable quantities, the rate of accession of knowledge and so the rate of increase of our power to cope with difficulties of this kind is enormously greater than the rate of evolutionary change. Solutions will be found. They might take the form, as many people have suggested, of direct genetic intervention, a sort of genetic repair process of a kind known to be able to cure certain genetic shortcomings of bacteria. At any rate, the point I want to make is that we must not at this time arrogate to ourselves the task of trying to provide solutions for all the problems that may afflict mankind in the future. I think the time will certainly come when our present-day medicine seems to future generations as inept as Galen's seems to us, and we are not yet qualified to prescribe for the medical welfare of our great grandchildren. In a slogan, I should say that present skills are sufficient for present ills.

But only one word of warning, I think, must be said. It is extremely likely that the subject I have been discussing here will feature prominently in the second Centennial Symposium and celebration of the Mayo Clinic and Foundation, and I should be very, very interested to know how big the problem will have become by then, and just how our descendants will be trying to solve it.

19. The Quality of People: Human Evolutionary Changes

JAMES F. CROW

By far the most important recent human evolutionary change is the radical alteration of birth and death rates throughout the world. We face the virtual certainty of a world too full of people unless there is the catastrophe of wide scale nuclear, or perhaps biological, war—the twin problems, as they have been called, of overpopulation and no population at all.

The increasing population and its relation to natural resources and the quality of life is discussed by others at this session. My assignment is to consider changes in the proportion of different kinds of people, so I shall ignore the total number. Yet I do not want to begin my topic without noting that it is of considerably less urgency. We can afford a longer consideration of the problems of genetic change because these changes are measured in generations. Nevertheless, there are trends that can be discerned and there are real possibilities of scientific discoveries that may radically affect human evolution. The questions of current human evolution—what the trends are and what we could or should do about them—merit our careful attention.

Any circumstances that might require immediate social action are most likely to arise from new technology. For example, some widely used and presumably harmless chemical may turn out to be extremely mutagenic, necessitating immediate control measures. Alternatively, someone may discover a simple means for controlling the sex ratio. Individual preferences may be too capricious to entrust with a matter as important as the next generation's sex ratio. On the other hand, we also need to be more pedestrian and look at what is now happening in the human population.

Rates of Evolutionary Change

J. B. S. Haldane (1949) discussed this subject in a typically interesting and original way, and what I am about to say comes mainly from this source. By ordinary

SOURCE. James Crow, "The Quality of People: Human Evolutionary Changes," BioScience, 16:863-867 (Washington, D.C.: 1966).

standards, evolution is an exceedingly slow process. The horse tooth, during some 35 million years, changed in paracone height at a rate of about 4% per million years. This is a measure of size; changes in shape were similarly slow. The ratio of paracone height to ectoloph length evolved at about the same rate as the teeth became higher crowned. The body size of dinosaurs changed at about the same rate. Changes in human head measurements were somewhat faster. On the other hand, the size of a number of invertebrates changed at much slower rates.

Haldane suggested, I don't know how seriously, that we define a unit of evolutionary rate, the "Darwin." A Darwin corresponds to a change by a factor e in a million years, or a doubling or halving in 693,000 years. This corresponds to a change of one millionth per year or 0.1% per thousand years. The horse tooth changed at a rate of about 4 centidarwins or 40 millidarwins. Rates as high as a Darwin must be quite unusual, except where there is human intervention. Changes in the size of dogs, milk production of cattle, and energy yield in maize are several orders faster. The evolution of industrial melanism in British moths and of insecticide resistance throughout the world are also preceeding at high rates.

Changes that are happening now and which do not seem too striking, if prolonged, would amount to enormous evolutionary rates. Changes in height and weight in the United States in the past century (and I do not mean to imply that the cause is mainly genetic) would have to be measured in hundreds of Darwins.

More basically, evolutionary rates might be measured not as morphological, physiological, or behavior changes but as gene substitutions. We can approach this (measurement of gene substituttions) by measuring rates of change in terms of the variability of the population at a given time. The data for estimation of population variances are less reliable, but Haldane noted that in horse teeth the mean changed by about a standard deviation in 1.5 million years. For a highly heritable trait, and with some rather tenuous assumptions, this would be the equivalent of a gene substitution every few million years at a locus involved in this trait. I say equivalent, for two changes in gene frequency of 0.5 are equivalent to one full substitution (from frequency 0 to 1); it may well be that evolution of size characters is accomplished by slight frequency changes at many loci rather than large frequency changes of fewer loci.

Haldane's rates suggested in 1949 agree remarkably well with modern molecular findings. In recent years it has been possible to determine amino acid sequences of three proteins in various mammals. Knowing the genetic code and the amino acid changes, one can easily compute the minimum number of mutations required to produce the observed differences. One can apply a simple Poisson correction to get an estimate of the most probable number, but for small numbers of changes, such as between man and horse, the actual number is probably close to the minimum.

Hemoglobin, cytochrome C, and ribonuclease, although they differ somewhat, average roughly one gene substitution per million years (for a review see Epstein and Motulsky, 1965). Thus, we arrive at about the same rate for structural genes determining mammalian proteins as Haldane suggested for genes determining morphological measurements in the horse. I do not want to emphasize the actual values, since these are very crude for the morphological traits and, for the biochemical, based only on three—perhaps not representative—proteins. I do want to note the order of magnitude agreement and to emphasize the extremely slow rate of gene turnover. Stated in terms of individual nucleotides, the rate is several hundred times less.

These slow rates show how little effective selective pressure is applied to individual gene substitution or to changing quantitative traits. If we ignore the last century, human height changed very little in a million years. An average selection differential of 0.001 inch (that is, if those who survived and reproduced averaged 0.001 inch taller than the average) would have changed the height by several feet, even with low heritability.

This suggests that for many, and probably most, traits there is little selection toward systematic change in a fixed direction. Most natural selection is not changing things. Rater it is acting to remove deviants in both directions from the mean, or adjusting to fluctuations in the environment, eliminating recurrent harmful mutations, or maintaining polymorphisms. Considerable selection is needed to maintain the genetic *status quo*, even without any progressive evolutionary changes.

Has Technological Advance Removed the Opportunity for Natural Selection?

The demographic transition from high rates of births and deaths to lower rates for both clearly alters the situation. One way to inquire into the effect of this change is to ask how it affects the *differential* contribution of individuals to the next generation. A high fertility and high death rate alone are not enough; in order to have selection, some individuals must leave more descendants than others.

I have proposed an "Index of Opportunity for Selection" that tells how much potential genetic selection is inherent in the pattern of births and deaths (Crow, 1958). Of course these differences may not be heritable—hence I speak of *opportunity* for selection. If a trait is completely heritable and perfectly correlated with fitness, the index tells its rate of increase. Otherwise, and this is the situation in practice, it provides only an upper limit.

The Index, I, is defined as V/x^2, where V is the variance and x the mean number of progeny per parent. This can be separated into components due to mortality and to fertility differences (Crow, 1958). It can also be adjusted to some extent for differences in age of reproduction, but I have not done this.

The changes in the two components with a changed environment are strikingly shown in some data from Chile (see Table 1). The data and calculations are from Dr. Cruz-Coke, who has kindly permitted me to quote them.

Table 1. The Effect of Demographic Differences in Three
Contemporary Chilean Populations on the
Index of Opportunity for Selection

	Town	Village	Nomads
Mean number of children	4.3	5.9	6.1
Variance in number of children	8.5	7.5	6.4
Proportion surviving to adulthood	.87	.75	.42
Index (mortality), I_m	.15	.33	1.38
Index (fertility), I_f	.45	.22	.17
Index (total), I	.67	.62	1.78

The three populations were: (1) A seacoast town population, mostly industrial; (2) An upland group of villages, mostly farmers; and (3) Nomadic shepherd tribes living in the highlands. These three groups, all living in a relatively small area, show at one time the kinds of changes that often occur secularly in the same population.

It is no surprise that the Index of Opportunity for Selection due to deaths decreases as the death rate lowers, and this is shown strikingly in these data. At the same time the Index due to fertility differences changed in the other direction. Somehow, although the family size is less in the villages and still less in the towns, the variability in family size increases so that the fertility Index is largest in the towns. The total Index has decreased.

Table 2 shows the trends in the fertility component of the Index in the United States. The pattern has changed from uniform high fertility (x small, I_f small) through low fertility with considerable variability from family to family (x small, I_f large) to the most recent situation where the family size is more uniform and I_f is again small.

Table 2. Changes in Mean Number of Children, $\bar{x}$, and the
Fertility Component of the Index of Opportunity for Selection,
I_f, in the United States

Date of Mother's Birth	$\bar{x}$	I_f
1839	5.5	.23
1861-65	3.9	.78
1901-05	2.3	1.14
1911-15	2.2	.97
1921-25	2.5	.63

Note: This includes both white and nonwhite and both married and unmarried. Families in the 1921-25 cohort may not be completed. 1839 data from Baber and Ross (1924), others from U.S. Census.

Meanwhile, the index due to mortality has dropped from about 1 for those born a century ago to less than 0.1 for current death rates.

These values quantify what has already been pointed out, that the opportunity for selection from differential postnatal mortality is greatly reduced. Prenatal mortality has changed much less, and probably contributes at least 0.3 to the total index. The opportunity for selection from fertility differences first increased and then decreased.

If all the differential viability and fertility were genetic, the total index would still be large enough so that a considerable amount of selection could be occurring. But the more important question of how much genetically effective selection is taking place cannot be answered. One can only get answers to specific aspects of special instances.

Changing Criteria of Selection

The demographic changes that I have been talking about not only change the opportunity for selection but also have had qualitative effects by changing the criteria for selection. A trait that was advantageous in the past may no longer be so.

Many of the effects of these changes are unknown. Even in a particular country the changing effect of selection on intelligence, for example, is usually not clear. When all the world is considered, hardly anything meaningful can be said.

Prediction of future evolutionary trends is difficult because man himself plays such a decisive role. The largest influence in man's future is man himself—the things that the individual and society do, intentionally or unwittingly.

Yet, some trends are clear. Bacterial and protozoon diseases have been drastically reduced in many parts of the world. A few decades ago a gene producing a decreased susceptibility to smallpox would have had a great selective advantage. Now, in much of the world such a gene is of little value. We can expect that throughout the world selection for resistance to infection will become less and less important.

The greater mobility of contemporary populations will also have genetic consequences. There is certain to be less inbreeding as persons tend to find mates away from their home environs. This should decrease the incidence of rare recessive diseases and cause some increase in general health and vigor—although the latter may not be measurable directly.

A second consequence of mobility may be enhanced degree of assortative marriage. The greater participation in higher education, the stratification of students by aptitude, the growth of communities with similar interests and attainments all can lead to increased correlations between husband and wife. Added to this is the greater range of choice created by affluence and mobility so that any inherent preferences for assortative marriage are more easily realized.

The effect of assortative marriage is to increase the population variability. There is already a high correlation in IQ between husband and wife, and this may well increase. To the extent that this trait is heritable there will be greater variability next generation than would otherwise be the case. This means more geniuses as well as more at the other end of the scale.

That life is getting softer for much of the world is another clear trend. Genes that used to cause death, debility, and lowered fertility may no longer do so. Sometimes this is unimportant; genes that enabled a man to eat by hunting with a bow and arrow, to kill or at least not be killed by neighboring tribesmens' spears, to survive unclothed in a rigorous climate are largely irrelevant—provided we never have to return to that kind of life. Genes for resistance to diseases that have been eradicated are likewise unimportant.

The disease-resistance mechanism may leave a residue of troublesome genes, e.g., those causing sickle cell or other abnormal hemoglobins. A further anthropologist studying man after *Plasmodium falciparum* has been exterminated may find the high incidence of hemoglobin S a complete puzzle unless he is also a student of history. It has frequently been suggested that some of our present polymorphisms are relics of earlier disease-resistant mechanisms.

The population always carries a number of deleterious genes. Many of these are recessive, or nearly so, and owe their incidence to mutation or to opposing selective forces. They may or may not be at equilibrium; probably many are not, for the conditions determining the equilibrium may change faster than the time necessary for the equilibrium to be reached. In any case, an environmental change that makes a gene less likely to cause death or sterility will cause that gene to be more frequent in later generations than it would otherwise be. A certain consequence of such relaxed selection against harmful genes will be their increase. How great is this effect?

The Effect of Relaxed Selection in Simple Cases

I shall consider first a rare recessive gene causing a disease that can now be treated. Let p be the relative frequency of this gene. Then, with random mating, the proportion of persons homozygous for this gene will be p^2. Let the probability of death or infertility in untreated cases (relative to the normal population) be s and in treated cases t. Then, as a result of the treatment, a fraction $(s\text{-}t)p^2$ recessive genes will be transmitted to the next generation that, in the absence of the treatment, would be eliminated by death or failure to reproduce. If p' is the proportion of recessive genes that would otherwise be present in the next generation, the proportion of harmful genes in the next generation is $p' + (s\text{-}t)p^2$. The incidence of the disease is the square of this.

Ordinarily p and p' are very similar, for the gene frequency is likely to be somewhat near equilibrium. Letting $p' = p$, the incidence in the next generation is

$p^2(1 + ip)^2$, where $i = s\text{-}t$ is the improvement produced by the treatment, measured in terms of survival and reproduction.

The proportion by which the incidence is increased is

$$\frac{p^2 (1 + ip)^2 - p^2}{p^2} = 2ip$$

approximately, where p is small.

A familiar example is phenylketonuria, which can now be treated with considerable success by a low phenylalanine diet. If untreated persons hardly ever reproduce, s is nearly 1. If the treatment is fully successful, $t = 0$; so i = 1. The gene frequency, p, in this case is about .01, thus the proportion of increase is $2ip = .02$.

An increased incidence of 2% per generation would mean about 40 generations for the incidence to double. This is more than a thousand years. The genetic consequences of the successful treatment of diseases caused by rare recessive genes are slight.

The situation is quite different with a rare dominant gene. In this case, if p is the frequency of the gene in this generation, and p' the frequency in the next generation in the absence of treatment, the fraction of deleterious genes added as a consequence of the treatment is ip. Thus, the proportion in the next generation is $p' + ip$, or if p and p' are approximately equal, the proportion by which the incidence is increased is simply i.

The reason for this is that in contrast to a recessive any noneliminated dominant gene is expressed in the immediately following generation. Then, if the gene is so rare that homozygous mutants can be ignored, the increase in the next generation as a consequence of the treatment, expressed as a fraction of the frequency this generation, is exactly i. This is easily understood for a lethal or sterilizing mutant. In this case all individuals showing the abnormal phenotype are new mutants; thus a completely effective treatment would double the incidence, adding last generation's new mutants to those currently arising.

If the mutant has a constant but reduced penetrance, P, the proportion of increase is approximately iP. In summary, the proportion increase in incidence in the next generation for simple inheritance is approximately

Recessive $2pi$
Dominant i
Dominat with penetrance P Pi

The difference in the immediate consequences of successful treatment of dominant and recessive diseases is lessened if there is genetic counseling or if the person knows the mode of inheritance of his diseases. If the disease is caused by a recessive gene, the increase is slight, especially if the gene is rare and consanguineous marriages continue to decrease in frequency. Thus there is little incentive for a person who has been treated not to have children. The risk to the child is very small, being equal to the gene frequency in the population. On the other hand, if the disease is dominant, the risk to the child is 0.5. If the disease is severe, disabling, or painful and the treatment unsatisfactory, a person who has the disease will not want to inflict it on his children. On the other hand, if the disease is mild or the treatment fully effective, a person may not hesitate to expose his child to a 50% risk. For these reasons, if genetic information is available, the consequences of the successful treatment of many dominant diseases may not be severe for the next few generations.

However, I must introduce two cautions in this perhaps over-optimistic discussion of simple examples. One is that the increase is geometric, not arithmetic, and over a long period, this will become important. The other is that with such treatments (and more generally with other environmental improvements that reduce the selection against harmful mutants) there is no turning back. The improvement must be permanent. A return to the original conditions leads to the immediate full impact of all the mutants that have accumulated during the period of treatment or improved environment. As new methods for repairing genetic weaknesses are discovered and used, we shall have to devote an increasing share of our energies to the repair of one another's genetic deficiencies.

I have talked about simple inheritance. What about more complex cases involving several genes and complicated interrelationships with the environment? I can only state rather vague generalities. To the extent that the genes involved are dominat, the impact of the improvement on the future incidence will be more immediate. If the condition depends on interactions of several genes, especially if some are recessive, the increase in early generations will be greater if the abnormality or disease is common. The most important point, I think, is to realize that, unless there is some conscious genetic intervention, an environmental improvement must be permanent.

The human population has a store of genetic variability, as does any Mendelian population, leading to various undesirable phenotypes such as I have discussed. This store is maintained by various selective factors, many of which may have changed radically in recent years. The store is added to by new mutations. When all these are in balance, there is a complex equilibrium among all the selective forces and mutation.

Whether or not conditions have been so stable in the past that most genes were near equilibrium frequencies, it is clear that environmental factors relevant to the selective value of many genes are changing rapidly—far more rapidly than the gene frequencies can change. The improved environment and the consequent relaxed intensity of selection against many deleterious genes will cause these genes to increase. In nature mutant genes are ordinarily eliminated at the same rate at which they occur. Probably, as Muller has often emphasized, they are now occuring faster than they are being eliminated. The opportunity for postnatal death selection is much reduced, and it is hardly possible that birth selection has completely replaced this as a means of eliminating unwanted genes.

The situation with monogenic traits is reasonably clear, as I have already emphasized. But there is good reason to think that most mutants have small effects. To the extent that selection against them is relaxed, these mutants will slowly increase.

The total mutation rate is such that 10% or more zygotes carry a new mutation. In a balanced state, this number of mutants must be eliminated in each generation by selection. This means 10% of the zygotes who fail to leave descendants, qualified by whatever fraction of mutants are not harmful in all gene combinations or are rendered less harmful by environmental improvements.

The rate of accumulation of mutants with relaxed selection is slow relative to the rate of environmental change. Can environmental change continue to stay ahead for the indefinite future, or must we soon begin genetic steps if the human phenotype is not to deteriorate? And should we be content merely to keep ourselves from getting worse?

Eugenics

This brings us to the touchy subject of eugenics. To what extent should man try to influence his genetic future?

An immediate difficulty is to avoid the bias of our own society. What constitutes a good phenotype is not likely to be thought to be the same in Africa, China, and Greenland.

As in almost any serious issue there are questions of ends and mean: means—some culture may condone infanticide, many permit abortion, many more permit contraception; and ends—Do we value abstract intelligence more than such traits as generosity or cooperativeness? Do we want uniformity or variety? (I vote for variety.) How do we plan for a genotype suited to the crowded cities and wholly different technology of the future?

I see little basis for an immediate agreement on positive goals or for means, certainly not on a global basis. On the other hand, there is probably good agreement on negative goals; hardly anyone would defend the present incidence of hemophilia, provided that the means for its reduction were acceptable.

Genetic counseling is beginning to be a precise and useful subject. Greater knowledge of hereditary diseases, more certain diagnoses, heterozygote detection, and cytogenetic advances have added greatly to the proportion of cases where the counselor can provide precise information. Even in this relatively noncontroversial area there are questions; for example, to what extent should the counselor give advice as opposed to information alone?

Recently, it has been possible to obtain and cultivate cells from the amniotic fluid of the fetus. As the techniques improve, as they surely will, it will become increasingly feasible to detect genetic abnormalities before the child is born, and at earlier stages of pregnancy.

An immediate possibility is the diagnosis of chromosome abnormalities. A prospective parent who knows that he is a carrier of a translocation realizes that if he has a child, it has a high risk of multiple anomalies. Under these circumstances he either decides not to have any children or has to gamble against other high odds that the child will be abnormal. If abortion were sanctioned, the pregancy would be terminated when the fetal cells were found to be chromosomally abnormal.

An analysis of embryonic cells could also reveal the sex of the child. If the mother were known to be a carrier of a severe sex-linked recessive disease, the sons would each have a 50% chance of developing the disease, but the daughters would be free of it. Here the ethical issue is somewhat different, for if all males were aborted, only half would have had the disease.

With future advances, it is very likely that chemical tests for various diseases will be developed that could be applied during pregnancy. I am sure that in societies where abortion is sanctioned considerable human suffering can be prevented in this way.

Ways in which man can change his phenotype, either as a result of a genetic change or through developmental engineering (euphenics), are certain to come. Some are here already. Others will probably become practical before they have been fully discussed. The use or nonuse may depend on accidents of place and time rather than on a consensus thoughtfully arrived at.

The spectacular success of molecular biology and the genetics of microorganisms immediately suggest for man such possibilities as removal, addition, or replacement of genes, DNA transformations, directed mutation, and suitably designed episomes. All these have one thing in common: they are not yet feasible, therefore their proponents are relieved of any responsibility for their misuse.

In contrast, H. J. Muller strongly believes that we should make use of what we now know He fears that discussions of future sophisticated genetic techniques foster escapism and postponement. "It would be intellectually dishonest and morally reprehensible of us to exploit mankind's eventual success in this enterprise as an excuse for not giving our support to the great re-educational process that could make possible, by means now available, a most significant advance in the genetic constitution of our species" (Muller, 1965). He is referring to artificial insemination.

Perhaps—as Lederberg (1964) has emphasized—the real impact of molecular biology in the foreseeable future will be on the control of human development rather than the genetic makeup. To me, these are not antithetical but complementary. The great difficulties are agreement on ends, whether the means are appropriate, and the extent to which we dare trust mankind to plan even short time changes.

The great impediment to short range genetic prediction is exactly that which has made long range evolution practical—Mendelian heredity. The randomization that occurs in every generation through the process of meiosis and fertilization means that only probabilistic predictions are possible. Plant breeders have long circumvented this difficulty in many species by either vegetative propagation or, what amounts to the same thing, by inbreeding and crossing.

Until recently, such possibilities for animals have been visionary. Now they seem probable enough to discuss as future possibilities. Transplants of nuclei from one cell to another, in particular into an egg cell, are now possible in some vertebrates. If we have doubts about the wisdom of encouraging the development of such technology in mammals, these will surely not be shared by dairy cattle breeders, for the minimum practical consequence of such a procedure would be control of the sex ratio.

I do not pretend that a widely practiced eugenics based on clonal propagation raises easier issues than one based on artificial insemination. Catering to public whims might be even more dangerous—for carbon copies of popular heros might be less salutary than a random 50% of their genes. To cite just one more problem—that of the sex ratio—how is this to be regulated? It could hardly be left to chance. Even if regulated, should society strive for equality at the age of marriage and risk protests from widows of age 50; or should there be an excess of males at younger ages, which would also create problems? Would society prefer social change to biological regulation; e.g., polyandry in the 20's and polygyny in the 50's?

Haldane, Muller, and more recently Lederberg (1966) have discussed the possibilities of clonal reproduction in man. My reason for mentioning it here is the guess that clonal reproduction may come sooner than many of the molecular possibilities that are more often discussed. Despite the above mentioned cautions, the possibilities for decreasing human misery and increasing human well being are enormous.

It is clear that biological and chemical possibilities for influencing human evolution and development are certain to come, probably before we have thought them through. Eugenics could be a far more potent force in the future than previously. In the past it has been tolerated partly becuase it was not likely to make an appreciable genetic change. The early eugenics was genetically naive and was connected with various dubious and even tragic political movements. I think the time is here when the subject should be reopened and discussed by everyone—not just biologists—with a serious consideration of the consequences of misjudgments as well as the possibilities for good.

References

Crow, J. F. 1958. Some possibilities for measuring selection intensities in man. *Human Biol.*, 30: 1-13.

Epstein, C. J., and A. G. Motulsky. 1965. Evolutionary origins of human proteins. *Progr. Med. Genet.*, 4: 85-127.

Haldane, J. B. S. 1949. Suggestions as to the quantitative measurement of rates of evolution. *Evolution*, 3: 51-56.

Lederberg, J. 1964. Biological future of man. In *Man and His Future*, G. Wolstenholme, Ed. Little Brown, & Co., Boston.

Lederberg, J. 1966. Experimental genetics and human evolution. *Bull. At. Scientist*, **22** (8): 4-11.

Muller, H. J. 1965. Means and aims in human genetic betterment. In *The Control of Human Heredity and Evolution*, T. M. Sonneborn, Ed. The Macmillan Co., N.Y.

20. The Population Explosion, Conservative Eugenics, and Human Evolution

LEONARD ORNSTEIN

Crow's essay (1966) on "The Quality of People: Human Evolutionary Changes" posed a number of important problems concerning the evolutionary future of the human species. Particularly important are the problems resulting from the slow increase in frequency of mutant genes in human populations which must inexorably follow in the most obvious democratic and humane termination of the population explosion.

Among knowledgeable biologists, much confidence is placed in our present understanding of the broad outlines of the evolutionary process in populations of sexual organisms. Evolution is seen to result from the interplay of the genetic endownment of the members of an interbreeding population with that population's environment. The genetic endowment is continuously subject to recombinations as a result of processes associated with sexual reproduction. Occasional random physical rearrangements, duplications, etc., provide opportunities for linking together of especially adaptive recombinations. Rare random point mutations of the coded genetic message provide the seeds for adaptive innovation. Those mutuations and rearrangements that provide their carriers with some adaptive advantage will, on the average, gradually replace the parental stock. Those mutations which, on the average, confer a selective disadvantage to their carriers are generally eliminated—or held at low incidence—in competition with the parental stock. Thus, evolution results from a balanced interaction of random mutation and selection (Dobzhansky, 1951).

In the absence of mutation, a nonhuman species would ultimately be expected to become extinct if its fixed genetic resources were inadequate to permit it to adapt to the kinds and magnitudes of environmental changes to which it might be

SOURCE. Leonard Ornstein, "The Population Explosion, Conservative Eugenics and Human Evolution," *BioScience,* 17: 461-464 (Washington, D. C., 1967).

This article appeared originally in a somewhat different form in the *Bulletin of the Atomic Scientists* (Ornstein, 1967).

exposed over periods of millions of years. In contrast, man, with his unique ability to revamp his environment, now depends upon cultural and technological mutations to provide even greater plasticity for coping with natural environmental change than that provided by adaptive innovations which are sparked by random genetic mutation. Therefore, the absence or elimination of mutation in humans need not constitute a biological threat.

In the absence of selection, however, *any* species (including *Homo sapiens*) would be expected to degenerate gradually through a process closely related to genetic drift (Dobzhansky, 1951). The resulting increasing accumulation of mutations would produce wider and wider departures of the individual phenotypes from one another and from the parental type. The mechanism is quite simple. The overwhelming majority of mutations which occur in a selective environment are deleterious and are therefore eliminated, tending to keep the population relatively monotypic. Selection plugs the multiple leaks that are forever occurring in the genetic dike and channels the flow of life thermodynamically uphill along adaptive paths. In the absence of selection, the dike would slowly crumble and the flow would dissipate down a multitude of exentropic gulleys, producing a vastly refashioned species. The only directions or styles that would be apparent in this kind of evolutionary process would be those which reflect changes in those genetic code words which, for whatever reasons, mutate at the highest rates—and these mutations will almost always represent phenotypic departures from the parental type. And later generations of mutations would represent still further departures from this more heterotypic base.

Now what can reasonably be meant by "in the absence of selection"? If death were eliminated and fertility maintained indefinitely, there would be, at least initially, no selection—however, in an extremely short period of time, a species would exhaust any "real" environmental resources—and this hypothetical kind of "elimination of selection" thus would be too short-term to be relevant to our present discussion. But there are two other ways to eliminate selection: (a) a mechanism which maintains a stable population size, in the presence of random *fluctuations* of fertility, by random elimination of offspring, independent of the genetic endowments of the individuals; and (b) a mechanism which maintains a stable population size by uniformly or randomly limiting fertility of individuals, independent of their genetic endowment.

If the elimination of offspring occurs before birth (e.g., abortion and some forms of birth control), it is essentially equivalent to a limitation on fertility, or what is commonly called birth control.

With this frame of reference, we will now consider two facts of life which, although in and of themselves are quite encouraging from a humanistic point of view, nonetheless conspire to generate the next great threat to humanity.

There is growing agreement that some form of birth control will provide the only reasonable solution to the problems of the population explosion and the limited resources of our planet. All efforts to increase food production to relieve, at least temporarily, population pressures should be encouraged. But no one supposes that exponential population growth can be matched by food technology. Therefore, in the long run, enthusiasm for such efforts is not likely to be permitted to reduce attempts at population control. Those who divert attention from the real and pressing problem by appeal to science fiction—excepting "to reap the resources of the universe"—will hopefully soon begin to appreciate the sobering energy cost estimates of even a trip to a nearby star (von Hoerner, 1962).

The ranks of those who at least pay lip service to the principles of equalitarianism and classless or open-class societies are swelling rapidly both within bona fide democracies and in major totalitarian states. It seems not only undesirable from a humanistic point of view that this trend should be reversed, but reversal is also unlikely.

The expected solution, if any, to the population explosion therefore will probably involve the almost universal application of birth control with *voluntary* (although socially and/or economically rewarded or coerced) individual commitment to the maintenance of a reproductive rate of two offspring per pair of adults, independent of the genetic (or other) endowment of the parents. If such a program is successful in maintaining a stable population and in avoiding racial, class, and individual biases in the rates of reproduction (and therefore in the composition of future generations), *the human species may eliminate selection and thus be on the road to ultimate biological degradation and probable extinction!*

Is there a democratic way out? Lederberg (1966) has stated that "It would be a tour de force to demonstrate any change (increase) in the frequency of a specific harmful gene in a human population that could be unambiguously traced to relaxation of natural selection against it. In comparison to the pace of medical progress, these exigencies are trivial." Crow also has pointed out that "An increased incidence [of homozygotes for a rare recessive harmful gene] of 2% per generation would mean about 40 generations for the incidence to double. This is more than a thousand years. The genetic consequences of the successful treatment of diseases caused by rare recessive genes are slight." The tone of such remarks is calculated to lull the reader into a state of evolutionary complacency. Yet, on the following page, Crow's tone turns. "However, I must introduce two cautions in this perhaps overoptimistic discussion of simple examples. One is that the increase is geometric, not arithmetic, *and over a long period will become important*" (italic supplied).

Effective tools for recognizing the human carriers of recessive genes (the great majority of new mutations are recessive and harmful) have only been discovered within the past few years. Changes in the frequency of such genes due to mutation in large breeding populations (the human population now is effectively a very large breeding population) occur very slowly. Therefore an extremely large random sample of each of two successive generations would probably be required to demonstrate a change unambiguously. Lederberg's first statement is therefore correct.

His second statement, however, requires more careful examination. "Trivial" by what standards? Examine the case of a disease such as diabetes. Assume that diabetes is due to the presence of a recessive gene in the homozygous state. Prior to the discovery of insulin, a large fraction of diabetics died before reaching sexual maturity (or soon enough thereafter to lower the probability of survival of their offspring). The human efforts, in terms of research and medical care, and the economic and other social costs of the production of drugs which control diabetes clearly are trivial when compared to the suffering of millions of diabetics and of their families that was endured before the development and use of such drugs. Medicine has, in this case, effectively begun to neutralize the harmfulness of diabetes genes. Their frequencies will therefore automatically increase among future generations due to unopposed mutation pressure (Dobzhansky, 1961). In a similar way, eye glasses, artificial kidneys, and all the devices, transplants, and drugs of the coming euphenic revolution will reduce or eliminate the harmfulness of many genes. But the genetic base from which harmful mutations arise has until now been kept relatively homogeneous by selection. Therefore, the *numbers of kinds of* mutations that *at present* can occur are constrained by the homogeneity of that base. As this special variant of genetic drift slowly takes over, the base will become more and more heterogeneous, and euphenic correction of each new mutation will become more and more a problem of the custom engineering of individual medical or biochemical crutches or prostheses. Insulin solves the problem of a very large number of diabetic individuals and the social cost per individual is very small.[1] This is likely to gradually become less and less the case for the correction of newly arising mutations. It may not be possible to predict, with any accuracy, the relative rates of progress of medical and euphenic research as compared to the rates of increase of problems with which medicine will have to deal as a result of the elimination of selection. In the short run, the benefits from the development of crutches will clearly outweight the cost, but in the long run (and how long is problematical) the costs are likely to become prohibitive. It takes little effort to conjure up glimpses of the bizarre brave new world—a world of enormous individual variability, each individual (human?)—uniquely wired-up and supported by his own special set of transplants and external biochemical plant. A glimpse into a relatively modern hospital will convince one of the rapidity with which this vision is being realized at present, although admittedly for a relatively tiny fraction of the world population. But later, a major portion of technology would be medical technology and virtually all society's resources would be consumed by that technology. An individual that would be recognizable as a member of *Homo sapiens* would be rare indeed. Are such exigencies "trivial"?

[1]The production of insulin has been coupled with, rather than competitive with, food production. It is, however, perhaps instructive to note that until very recently the maintenance of an average diabetic over a 30-year period required the production and destruction of about 1000 head of cattle.

The cultural relativist might argue that provided such a culture does not exhaust its resources in trying to keep itself alive, its values and way of life may be just as good for its members as ours are for us. I would counter that if we can now, by judicious planning, provide greater adaptive flexibility and fewer biological and economic burdens for our descendants, *as judged by our standards of value*, then we cannot entertain the relativist's rationalizations with a clear conscience.

What alternatives exist? Lederberg (1966) states that "Eugenics is relatively inefficacious since its *reasonable* aims are a necessarily slow shift in the population frequencies of *favorable* genes" (italic supplied). He and others (e.g., see Dobzhansky, 1962) have rightly emphasized the problems of defining "favorable" genes in our present state of genetic ignorance. The problems of defining "unfavorable" genes may often be equally difficult. If the genes for schizophrenia were responsible, in the heterozygous state, the attributes of the kind of intelligence which we believe we value, eugenic attempts to reduce the frequency of schizophrenics from their present levels of 1% to 2% might reduce average intelligence of the population as a whole. This might produce an undesired and unexpected by-product which would outweigh the desired reduction in human misery and in the social burden that elimination of schizophrenia should represent. Would schizophrenia genes average out as favorable or unfavorable? And would we want to increase or decrease their frequency? Because of the difficulty in defining "favorable" genes as well as "unfavorable" genes, I question Lederberg's implication of the absence of other reasonable (short-term, i.e., within the next 10,000 years?) aims of eugenics.

In discussing the evolutionary process, Crow reminds us ". . . that for many, and probably most, traits there is little selection toward systematic change in a fixed direction. Most natural selection is not changing things. Rather it is acting to remove deviants in both directions from the mean, or adjusting to fluctuations in the environment, eliminating recurrent harmful mutations, or maintaining polymorphisms. Considerable selection is needed to maintain the genetic status quo, even without any progressive evolutionary changes." And later on he asks, ". . . must we soon begin genetic steps if the human phenotype is not to deteriorate? And should we be content merely to keep ourselves from getting worse?"

I believe we must begin by being "content merely to keep ourselves from getting worse," and that perhaps *the only reasonable short-term and conservative aims* of eugenics, taking into account the impending reduction in natural selection and our present state of ignorance of human genetics, are: (1) the approximate maintenace of the present distribution of gene frequencies and frequencies of "linked" combinations of genes, and (2) the reduction of the frequency of those rare mutations which clearly confer severe phenotypic disabilities that are not easily compensated by present medical technology. Such conservative aims should be vigorously pursued, provided that the individual and social costs are not excessive.

If we had methods for decoding and reading the complete set of genetic messages of each and every individual and for recording this data in a central computer file, it would be possible, in principle, to examine the message sets of any two prospective mates to compute recommendations as to the number of offspring they should have in order to help contribute to the maintenance of the genetic status quo. For most couples, the recommended number of children would be two; for many, one or three; and in rare cases, none or more than three. As Crow (1958) previously demonstrated, variances in reproductive rate of this sort can provide very "considerable selection." The computer would be programmed to take past frequencies of both intentional and accidental departures from the recommended values (continuously up-dated from birth records) into account in formulating recommendations. Some such program of conservative eugenics is probably the only kind of eugenic program that would have a chance to start to function successfully in democratic societies.

Some reasonable eugenic measures have begun to be put into effect to hold down the frequencies of the rare genes that produce severe disabilities. Those with the highest natural mutation rates pose the greatest threat, and it is just those which tend to be among the first to be singled out for attention. This is the kind of genetic counselling program to which informed and humane physicians are often privately committed.

It is clear that the evolutionary process itself has selected, in some cases, for reduction of effective mutation rate to compensate for increase in generation time and decrease in number of offspring per mating. We are now beginning to understand something of the workings of some mutation-rate control mechanisms such as excision of nucleotide codons which constitute coding errors and replacement with correct codons (using an unmutuated complementary strand as a model?). Increased redundancy in the genetic code (e.g., polyteny and polyploidy and gene duplications) may have provided *natural* means for reduction of effective mutation rates through the action of such genetic reading and editing mechanisms. And in so far as we can discover artificial means to reduce natural mutation rates, the rate of genetic drift can be slowed.

New high-resolution electrophoretic techniques for separation of proteins (which are the direct translation of genetic messages) and techniques for the fingerprinting of the peptide digests of pure proteins begin to permit us to collect significant amounts of data on gene frequencies. These techniques more often than not permit the identification of heterozygous carriers of otherwise phenotypically recessive genes. Routine cataloguing of the accessible proteins of blood cells and serum and other body fluids of each individual (to be followed up later by routine analyses of the proteins of samples of tissue biopsies or of cultures from such biopsies) will begin to lay a foundation for the kind of genetic analyses of human populations that is required to guide conservative eugenics.

Although the pace of genetic drift in large populations is initially very slow, the development of the kind of biomedical information-retrieval system and genetic decoding techniques required to stem the tide of drift may also be very slow, and attempts to discover practical means for reducing mutation rates may be even slower in reaching fruition. Therefore, the sooner a very much more substantial social commitment is made to the pursuit of such ends, *including making adequate genetic education a required part of all high school curricula*, the more secure will be the future of humanity.

Removing the spectre of suicide by nuclear, chemical, or biological warfare and putting a damper on the population explosion (which includes worldwide democratic application of birth control and the elimination of poverty) come first and second on my personal list of social priorities. Attending to our evolutionary future comes a very close third. Learning to live with leisure and computers follows. A 1000 BEV Alternating Gradient Synchnotron, trips to the moon and planets, listening for messages from outer space (Project OZMA), etc., all seem trivial by comparison. As for large-scale applications of "algeny" and positive eugenics to the improvement of mankind, I believe, with Lederberg (1966), Dobzhansky (1962), and Hotchkiss (1965) that these must wait *at least* until we are both technically more profcient and genetically vastly more knowledgeable.

References

Crow, J. F. 1958. Some possibilities for measuring selection intensities in man. *Human Biology*, 30: 1

Crow, J. F. 1966. The quality of people: human evolutionary changes. *BioScience*, 16: 863-867.

Dobzhansky, T. 1951. *Genetics and the Origin of Species.* Columbia University Press, New York.

Dobzhansky, T. 1962. *Mankind Evolving.* Yale University Press, New Haven, Conn.

Hotchkiss, R. D. 1965. Portents for a genetic engineering. *J. Heredity*, 56: 197.

Lederberg, J. 1966. Experimental genetics and human evolution. *Bull. Atom. Sci.*, 22: 4.

Ornstein, L. 1965. Subnuclear particles: a question of social priorities. *Science*, 149: 584.

Ornstein, L. 1967. The population explosion, conservative eugenics and human evolution. *Bull. Atom. Sci.*, 23 (6).

von Hoerner, S. 1962. The general limits of space travel. *Science*, 137: 18.

21. Possible Genetic Consequences of Family Planning

EI MATSUNAGA

Family planning means to have children in a desired number, each child at a desired time. Although the parental desire in this regard may be different in individual families, a more or less uniform pattern of reproduction would emerge from practicing family planning, since this need is conditioned by life circumstances that are rather common to all members of the society. Thus, if an appreciable fraction of a population practices it, a change in demographic trends, namely, decline in variances would be expected not only in the number of children per family but also in birth order and parental ages in live-birth data. As family planning is usually used to limit family size, a second sign would be decline in the mean number of children and the mean birth order, while the mean of the parental ages may scarcely be affected.

The purpose of this paper is to review the various aspects of demographic data and vital records now available in Japan with special attention to their potential effects upon future generations.

Recent Demographic Transition in Japan

Japan has achieved an unprecedented drop in births during a short period after World War II. It is true that this has been done mostly by induced abortions, legalized by the Eugenic Protection Law in 1948 for economic reasons as well as from physical considerations for maternal health. But less is known about the fact that the trend toward lowered fertility has been, as pointed out by demographers in this country, emerging since around 1920. Table 1 shows the mean and the variance of maternal age for selected years from 1925 to 1960, together with the standardized

SOURCE. Ei Matsunaga, "Possible Genetic Consequences of Family Planning," *The Journal of the American Medical Association,* **198**: 533-540 (Chicago, Illinois, 1966).

From the Department of Human Genetics, National Institute of Genetics, Mishima, Yata, Japan.

Read before the Regional Seminar of the International Planned Parenthood Federation, Western Pacific Region, Hoken Kaikan, Tokyo, May 26, 1966.

Table 1. Standardized Birth Rates* and Means and Variances
of Maternal Age in Live-Births Data in Japan, 1925-1960

Year	Standardized Birth Rate†	Maternal Age, Yr.	
		Mean	Variance
1925	35.3	28.4	45.48
1930	32.4	28.6	42.50
1937	29.8	28.9	39.34
1938	26.0	29.0	39.71
1939	25.4	29.3	38.62
1940	27.7	29.3	37.29
1947	30.7	29.1	36.36
1948	30.0	28.7	37.75
1949	29.7	28.5	35.22
1950	25.3	28.2	33.45
1952	20.8	28.1	30.04
1954	17.4	27.8	27.04
1956	15.8	27.6	24.67
1958	15.2	27.3	21.85
1960	14.6	27.1	19.93

*From Institute of Population Problems.[1]
†Rates are given as number of births per 1,000 population.

birth rates, which are from data in the report[1] by the Institute of Population Problems. The birth rate has been steadily declining, from 35.3 in 1925 to 14.6 in 1960, with the exception of 1940, during which the government encouraged population increase, and of the following period of the postwar baby boom. In accord with this trend the variance of maternal age does show a remarkable decline throughout the period; it decreased from 45.5 in 1925 to 19.9 in 1960. It is of some interest to note that the decline in the variance of maternal age was hardly affected by the government's policy in 1940 or by the postwar baby boom. These results clearly indicate that the practice of family planning has been diffusing in Japan since some 40 years ago.

Although the direct genetic effect of the above change is not immediately obvious, it may come into light if we review what kinds of change have taken place in the distribution of live births by maternal age and birth order. Table 2 illustrates the relevant data for 1947, 1953, and 1960. The number of live births to rank 4 or higher decreased from 36% in 1947 to only 10% in 1960. During the same period mothers aged 19 or less decreased from 2.3% to 1.2%, and mothers aged 35 or more decreased from 19.8 to 5.8%, while the mean age of women at first marriage was raised from 22.9 to 24.2 years. Briefly, a rapid transition to a family pattern with two to three children, born when the mother's age was 20 to 34 years, has taken place in Japan.

Table 2. Population Trends in Japan, 1947 to 1960*

Population Trend	1947	1953	1960
Total No. of live births	2,679,000	1,868,000	1,606,000
Total No. of reported cases of induced abortions	. . .	1,068,000	1,063,000
Mean age of women at first marriage, yr	22.9	23.4	24.2
Percentage of live births of birth order 1 to 3	64.1	75.1	90.3
Percentage of live births by age of mother			
Under 19 yr	2.3	1.7	1.2
Over 35 yr	19.8	12.1	5.8
Infant death rate, per 1,000 live births	76.7	48.9	30.7
Infant death rate from congenital malformations, per 10,000 live births	14.7†	21.1	19.0

*Data are based on the Annual Vital Statistics from the Ministry of Health and Welfare.

†This figure seems to be an underestimate; the corresponding rate was 23.7 in 1950, and thereafter has been declining.

The demographic transition outlined above must have resulted in important changes in the frequencies at birth of those congenital defects which are correlated with parental age and birth order. Since we are concerned here only with the genetic aspects, reference may be made to two categories of the defects; one is of those due to new mutations, on both chromosomal and genic levels, and the other is determined by genetic factors in combination with environmental influence.

Among the first category of defects, the best known is mongolism (Down's syndrome), whose occurrence depends upon maternal age. The increase with maternal age is rather slow until the mother reaches the age of 35 years, and then it rises almost exponentially as the mother approaches the menopause. Similar but less pronounced dependency upon maternal age has been noted for Klinefelter's syndrome (XXY), triple-X female (XXX), and trisomy 18 syndrome.[2,3] As to the occurrence of gene mutation, there is some evidence of positive correlation with paternal age for certain rare dominant anomalies, notably achondroplasia and acrocephalosyndactyla.[3] However, the data are still insufficient for evaluation of the increased rate of the risk according to paternal age.

The second category includes a variety of congenital defects in addition to Rh-erythroblastosis, for which the risk increases with birth order. Recent studies in Canada[4,5] have shown increased risks to children of older mothers, apart from birthorder effect, for cerebral palsy and congenital malformations of the circulatory system, and increased risks with advancing birth order, independent of maternal-age effect, for strabismus and other congenital malformations of the nervous system and sense organs. On the other hand, there were apparently no special risks, except for injuries at birth, to firstborns when maternal-age effect was eliminated or to children of very young mothers provided they were firstborns. The risks of diseases of the nervous system and sense organs were increased with higher birth rank to the children of very young mothers, however. The average relative-risk figures for all these defects, in Canadian data, were noted to range from 1.6 to 2.4, as compared with the standard rates either for younger mothers or for children of lower birth rank.

Consequently, the variety of congenital defects mentioned above must have been reduced in Japan as a result of the decreasing frequencies of births of higher rank as well as those of both older and very young mothers. Considering the changes from 1947 to 1960, the reduction value may have been about one third for mongolism, not so much for XXY and XXX types of chromosomal aberrations, more than one half for Rh-erythroblastosis, and perhaps of the order of one tenth for the rest of the defects. It should be noted that the reduction in the defects attributable to new mutations is beyond doubt eugenic for the population, whereas the reduction in other diseases may be dysgenic, since selection intensity against the genes is relaxed. However, so long as the pattern of small family size continues, the manifestation of the diseases would be prevented.

Secular Change in Sibship Size and Frequency of Consanguineous Marriages

While there are a variety of factors which affect the frequencies of consanguineous marriages, one of the most important is the mean size of sibships in a population. It is obvious that with a smaller sibship mean a person has fewer relatives—for example, fewer cousins than was the case when the mean was larger—but how frequencies of consanguineous marriages will be affected by the change in the variance of sibship size is not immediately recognized. Taking account of both mean and variance of sibship size, Nei and Imaizumi[6] were able to formulate the frequencies of various types of consanguineous marriages. We may cite here with a slight modification (in the denominator, they used $N-1$ instead of N), their formula for first-cousin marriages:

$$f = \frac{\bar{x}_2^2 \, (V_1 + \bar{x}_1^2) - \bar{x}_1 \, (V_2 + \bar{x}_2^2)}{\bar{x}_1 \, \bar{x}_2} \cdot \frac{2}{N},$$

where $\bar{x}_1$, V_1 and $\bar{x}_2$, V_2 stand for the mean and variance of sibship sizes of parental and children's generations, respectively, and N for the so-called isolate size in the sense of Dahlberg[7] or the size of the population within which marriages are contracted at random: the formula (f) represents the expected frequency of first-cousin marriages by chance. It is to be noted that the variability of sibship size in the parental generation increases the frequency of first-cousin marriages, while the variability in the children's generation reduces it. The same argument can be applied to the frequencies of other types of consanguineous marriages. Although the formula is based on certain assumptions (ie, 1 : 1 sex ratio at maturity, disregarding those who will not marry, no polygamy and no remarriage, and no correlation in sibship size between parents and children), it may be used to give us an idea about the expected change we are concerned with. In a recent survey by the Population Problems Research Council of the Mainichi Newspapers, Tokyo,[8] however, a significantly negative correlation has been found between sibship size of married women and number of their live children.

At present, we have no means to estimate the isolate size, N, without resorting to the observed frequencies of consanguineous marriages in the population studies. It is evident, however, that industrialization and urbanization have been increasingly breaking down the geographical barriers among populations that had been isolated, so that the scope of finding one's mate must have become considerably wider than before. As to the secular change in sibship sizes, two sets of data may be used to deduce the trend; the one is the 1960 census report[9] on the number of children born to all married women with at least one child and the other is the vital statistics of live births by live-birth order from 1950 to 1963 (*Annual Vital Statistics of Japan*, 1950-1963). Table 3 shows that the mean and variance of the number of children had been almost stable for Japanese mothers now older than 55 years, the mean being slightly larger than five and the variance about seven, while for younger mothers the mean has been decreasing gradually and the variance has been decreasing relatively faster. Table 4 shows the secular change in the live-birth order of all newborns during the past 15 years; the mean and variance have decreased from about three and four in 1950 to less than two and one, respectively, in 1963.

Table 3. Mean and Variance of No. of Children Ever Born to
All Married Women With at Least One Child, by Age*

Age of Mother, yr	No. of Children	
	Mean	Variance
80-	5.159	6.798
75-79	5.132	6.842
70-74	5.261	7.152
65-69	5.209	7.000
60-64	5.191	7.093
55-59	5.130	6.966
50-54	4.846	6.054
45-49	4.249	4.514
40-44	3.563	3.021

*Data are based on the 1960 census report in Japan.

Table 4. Secular Changes in Mean and Variance of
Live-Birth Orders of All Live-Born Children, 1950 to 1963*

Year	Mean	Variance	Variance Square of Mean
1950	2.878	3.847	0.465
1954	2.578	2.646	.398
1955	2.492	2.512	.405
1956	2.381	2.338	.412
1957	2.272	2.124	.412
1958	2.148	1.874	.406
1959	2.039	1.664	.400
1960	1.957	1.461	.382
1961	1.890	1.306	.366
1962	1.819	1.106	.334
1963	1.784	0.999	0.314

*Data are based on Annual Vital Statistics from the Ministry of Health and Welfare.

Considering the high mortality for children in the earlier times, especially in large
families, it may roughly be estimated that the mean and variance of sibship size had
been approximately four and six, respectively, until about 1930, and then both
have been steadily declining, at first rather slowly but very rapidly after World War
II. In accord with such secular changes, various values of means and variances of
sibship sizes have been substituted in the following formula:

$$e = \frac{\bar{x}_2^2 \left(V_1 + \bar{x}_1^2\right) - \bar{x}_1 \left(V_2 + \bar{x}_2^2\right)}{\bar{x}_1 \, \bar{x}_2}$$

were e is the expected number of potentially marriageable first cousins for each individual. Table 5 shows that the expected number of such first cousins decreases
from 16.5 for an expanding population with the mean of four and the variance of 6
to 2.5 only for a stationary population with the mean of two and the variance of 1;
the reduction in the mean noticeably affects the reduction in the number of the
first cousins, while the reduction in the variance will have its effect one generation
later. It can be shown that the expected rates of reduction are more pronounced for
first cousins once removed and for second cousins.

Japan is well-known for the high frequency of consanguineous marriages. Although it varies according to localities investigated, it is usually about 2% to 5% for
urban areas and of the order of 10% for rural areas. Table 6 shows as an example
the results of our recent survey in the city of Ohdate, Akita Prefecture, in northern
Japan.[10] The data cover all couples which were registered in the ward offices and in
which the wife's age was in the range from 30 to 40 years. The frequency of the

Table 5. Changes in Expected No. of Potentially Marriageable
First Cousins for Each Person in Hypothetical Population
With Trend Toward Smaller Sibship

	Parental Generation*		Children's Generation†		Expected No. of Potentially Marriageable First-Cousins
	$\bar{x}_1$	V_1	$\bar{x}_2$	V_2	
Population with large family	4	6	4	6	16.5
	4	6	3.5	5	14.3
	3.5	5	3	4	10.5
Population in transition to smaller family size	3	4	3	3	9.0
	3	3	2.5	2.5	6.5
	2.5	2.5	2	2	4.0
	2	2	2	1	3.5
Population with small family size	2	1	2	1	2.5

*$\bar{x}_1$ and V_1 = mean and variance of sibship size for parental generation.
†$\bar{x}_2$ and V_2 = mean and variance of sibship size for children's generation.

major types of consanguineous marriages was about 2% for couples living in the
central part of the city, while it was as high as about 8% for those in the peripheral
part. When the data are broken down by age of wives, no secular change is apparent
in the frequencies of consanguineous marriages. This is not surprising, because these
wives were born during the period from 1922 to 1932, so that their sibship sizes as
well as those of their parents must have been still large. There have certainly been
other socioeconomic factors, particularly in rural areas, which favored consanguinity. Nevertheless, the rapid reduction in both the mean and the variance of sibship
size, on the one hand, and the modern trend toward the breakdown of isolates, on
the other hand, should in the near future result in a significant reduction in the frequency of consanguineous marriages. Unless other factors dominate, the reduction
rate may presumably be of the order of 4/5 in proportion with the reduction expected in the number of potentially marriageable relatives for each individual, say
from 15 to 3.

The genetic consequence would be of advantage for the society, since consanguineous marriages lead to increased risks of illness, premature death, and congenital abnormality among the offspring. The reduction rate in mortality and morbidity
for the population as a whole depends upon the present frequencies of consanguineous marriages and their reduction rates, as well as the increased rates in the relative
risks to the children from such marriages. The results of recent informative
studies[11] show that the increased rates in the relative risks are not very large. In

Table 6. Frequency of Consanguineous Marriages, by Ages of Wives
30 to 40 Years Old, Registered in City of Ohdate (1962)*

Relationship	Age of Wives, Yr.					Total
	30-31	32-33	34-35	36-37	38-40	
Center of City (Old City)						
First cousins	3	1	2	0	4	10
First cousins once removed	3 } 2.3%	0 } 0.9%	2 } 2%	1 } 1%	1 } 2.6%	7 } 1.7%
Second cousins	1	2	3	2	3	11
Other	0	0	0	2	2	4
Not related	299	347	350	297	299	1,592
Total	306	350	357	302	309	1,624
Periphery of City (New City)						
First cousins	6	12	13	13	11	55
First cousins once removed	6 } 5.8%	10 } 9.9%	7 } 8.1%	7 } 8.3%	10 } 7%	40 } 7.8%
Second cousins	9	13	11	9	4	46
Other	9	5	4	5	6	29
Not related	334	313	347	314	324	1,632
Total	364	353	382	348	355	1,802

*In Table 2 presented at the World Population Conference held in Belgrade in 1965,[11] the arrangement of the data was not proper, this has been corrected in this Table.

certain Japanese cities child mortality, including stillbriths, was found to be 10% to 11% among the offspring of first cousins, against 8% to 9% among the controls, whereas the risk of death caused by congenital abnormalities was relatively high but less than double. Assuming double average risk for the children from first-cousin marriages and a decrease in their frequency from 5% to 1%, the reduction rate in death rate for the population may be estimated to be about 4%. This may seem of relatively small magnitude but may not be insignificant from the point of view of the community as a whole. On the other hand, the decrease in consanguineous marriages would be dysgenic, as it should result in increase in rare recessive genes carried by heterozygous persons, that had otherwise been eliminated in homozygous form.

Possible Dysgenic Effect of Differential Fertility

The practice of family planning usually spreads more rapidly in some social strata than in others, and more rapidly among better-educated than among less-educated portions of the society. This would in all probability result in fertility differences, and if these were correlated with some genetically determined traits, the genetic composition of the offspring population would be altered.

The first question to be answered is to what extent the family-planning practice in Japan has been unevenly distributed with respect to certain characteristics of the population. In Table 7 some relevant data are reproduced from the reports[7,12,13] of seven consecutive surveys conducted by the Population Problems Research Council of the Mainichi Newspapers, showing secular changes in the frequencies of

Table 7. Secular Changes in Spread of Contraceptive Practice,
Not Including Induced Abortions, Among Different Strata in Japan, 1950 to 1965*

	Percentage of "Current Users"						
	1950	1952	1955	1957	1959	1961	1965
Occupation of the husband							
Farmers and fishermen	11.5	17.0	25.4	30.5	34.9	37.7	47.0
Manual workers		23.9	35.8	37.6	40.1	36.7	50.0
Medium and small enterprisers		24.7	37.4	39.0	40.4	39.1	51.0
Salaried workers	25.9	36.9	39.7	49.1	50.7	50.3	54.4
Others		35.2	41.0	47.0	...	...	...
Duration of husband's education, yr							
Less than 9	14.2	18.2	28.2	33.4	37.6	37.1	...
10-12	25.4	37.0	37.7	46.5	43.9	45.4	...
More than 13	37.3	47.0	48.8	52.5	54.0	56.2	...
Duration of wife's education, yr							
Less than 9	13.0	20.1	28.2	33.3	35.0	35.6	46.9
10-12	32.4	38.7	46.1	48.4	51.6	49.6	58.1
More than 13	36.0	59.1	47.8	53.2	51.9	60.4	65.2
Total	19.5	26.3	33.6	39.2	42.5	42.5	51.9

*Data are based on surveys by the Population Problems Research Council, Mainichi Newspapers, Tokyo.[8,12,13]

use of various kinds of contraceptive methods, not including induced abortions, among married women aged 16 to 49. For each survey some 3,000 couples were sampled from the whole country by appropriate methods of stratification, so that the results for a specific year are comparable. The term "current users" does not necessarily include those couples who had practiced contraception in the past. From the Table it is clear that the practice of contraception has been rapidly spreading among all social strata; the average frequency of current users has increased from 20% in 1950 to 52% in 1965. There have been, in fact, variations in the frequencies according to husband's occupation and couple's educational background. But perhaps the most remarkable feature of the data is that the increased rate has been particularly high among those portions of the society in which the initial frequencies were the lowest, so that the variation due to social stratification appears to have been considerably lessened. This impression is verified by computing the coefficient of variation for each year of the surveys among categories of husband's occupation and of couple's education (Table 8). There is no doubt that the high literacy of the people, active mass communication of the knowledge about contraception, and mass campaigns by the regional centers for public health services have greatly contributed to the spread of family-planning practice in Japan.

Table 8. Secular Changes in Coefficient of Variation
With Respect to Differential Spread of Contraception

Cause of Variation	Coefficient of Variation, %						
	1950	1952	1955	1957	1959	1961	1965
Husband's occupation	. . .	30.3	17.2	18.5	15.9	15.4	6.0
Husband's education	45.1	42.9	27.0	22.1	18.3	20.7	. . .
Wife's education	45.6	49.6	26.7	23.1	20.9	25.6	16.3

The second question is concerned with the extent of variation in fertility according to the social strata. Since fertility should be measured for women who had completed reproduction, the outcome of the current trends is still to be seen. Here we may refer to the data obtained from the fertility surveys made in 1952 and 1962 by the Institute of Population Problems.[14] Table 9 shows the average number of live births per wife aged over 45 in the two surveys by husband's education and occupation, together with the corresponding coefficients of variation. It is seen that the coefficients of variation in fertility due to the two characteristics of the husbands have both decreased from 16% in 1952 to about 10% in 1962. In this connection, it is to be noted that in Japan the age at marriage tends to be higher with the longer span of education, which in turn correlates positively with the rank in the occupational status. Therefore, it would hardly be possible to evaluate the net effect of family planning upon the differential fertility.

The third question is referred to the possible genetic consequence of the observed differences in fertility. Among a variety of mental traits the best defined and the most important is intelligence. The extent of possible dysgenic effect with respect to intelligence depends, not only upon its heritability and correlation with the length of education or the occupational status, but also upon the extent of variation in fertility differences, both among social strata and within a social stratum.

Table 9. Secular Changes in Differential Fertility in Japanese
Wives Aged 45 Years, by Husbands' Education and Occupation*

	Average No. of Live Births	
	1952 Survey	1962 Survey
Total	4.47	3.91
Duration of husband's education, yr		
Less than 9	4.62	4.05
10-12	3.62	3.60
More than 13	3.47	3.21
Coefficient of variation, %	16.0	11.6
Husbands' occupation		
Farmers and fishermen.....................	5.06	4.19
Workers on own account		
in nonprimary industries..................	4.08	4.02
Manual workers	3.79	3.82
Nonmanual workers	3.57	3.37
Coefficient of variation, %	15.9	9.2

*Data are based on consecutive surveys by The Institute of Population Problems.

Although present methods of measuring intelligence are imperfect, there seems to
be a high correlation between the intelligence quotient and school performance for
Japanese children, and those with higher scores are more prone to enter into the
higher school courses. We do not know, however, to what extent intelligence is correlated with the length of education or the occupational status. The extent of heritability within the normal range of intelligence still remains to be recognized. Although we do know the extent to which variation in fertility difference among social strata is rapidly decreasing in Japan, we have no information about the differential fertility by intelligence within each stratum. The loss of variation among
strata does not necessarily mean the loss of variation within a stratum. In the absence of our knowledge on many important points, it is difficult to evaluate the extent, if any, of dysgenic effect of differential fertility upon the intelligence of the
future population.

Possible Relaxation of Selection Intensity

As is well-known, the modern demographic trend toward smaller family size in
western Europe as well as in Japan has been accompanied by a reduction in childhood mortality, so that a greater proportion of the children born have survived and

reached maturity than had been formerly the case. This must have resulted in a relaxation of natural selection against most, if not all, of the mutant genes affecting viability.

While the above result is mainly attributed to factors other than family planning itself, such a consequence would be brought about if genetic losses of children were compensated as a result of family planning. If every family had exactly the same number of children that survivied and reached maturity, natural selection due to genetic difference between families would be removed, the component due only to within-sibship difference remaining; in this model the selection rates against harmful genes affecting fitness are reduced to about one half to two thirds of the normal rate.[15] Though such compensation is likely to occur in some families, the effect would be counterbalanced if contraception were employed in other families where the genetic risk is known to be high to the subsequent child. This aspect will be treated in the next section.

Crow[16] has proposed a method for measuring the total selection intensity by an index, I, representing the ratio of the variance in progeny number to the square of its mean, and has shown that I can be separated into two components, I_m and I_f, due to differential mortality and fertility, respectively. Following his expression,

$$I = I_m + \frac{1}{p_s} I_f, \text{ and}$$

$$I_m = \frac{p_d}{p_s}, I_f = \frac{V_f}{\bar{x}_s^2},$$

where p_d (counted at birth) is the proportion of premature deaths, p_s is the proportion of those survived and having varying number of progeny, and $\bar{x}_s$ and V_f are the mean and variance of the number of births per surviving parents. Later, using vital statistics data for different populations Spuhler[17] found that the total selection intensity remains relatively high in industrialized populations; differential fertility seems to keep it high despite the relatively low preadult mortality. However, it is to be noted that the above formulation is based on the assumption that all variation in mortality and fertility has a genetic basis and fitness is completely heritable; the index provides, therefore, not the net intensity of total selection, but only its upper limit. Further, the genetic evaluation for varying values for I_f may be quite different, depending upon whether the population studied uses birth-control measures or is still under natural conditions.[18] Therefore, caution should be taken in the conclusion of the results.

We have seen that the family-planning practice in Japan has resulted in a rapid reduction in the variance of progeny number relative to its mean. In this situation, it would be pertinent to separate the index I_f further into two components, I_i and I_f', the former being the selection intensity due to infertility and the latter due to the variation in the number of children for fertile parents. Thus, the formula may be written as follows:

$$I_f = I_i + \frac{1}{1-p_0} I_f',$$

$$I_i = \frac{p_0}{1-p_0}, \quad I_f' = \frac{V_f'}{\bar{x}_s'^2},$$

where p_0 is the proportion of parents having no children, x_s' and V_f' are respectively the mean and variance of the number of births per fertile couple. If family planning were practiced only by potentially fertile parents, this would result in a reduction in the value for I_f' but not for I_i, while if otherwise infertile couples could become fertile by medical-care service, the value for I_i would be affected.

In order to show the secular changes in the respective values for I_i and I_f', Table 10 represents some relevant data that were based on the 1960 census report. It should be mentioned that the census report is limited because of the failure to include those married women who died during the reproductive period. The proportion of married women having no children has been slowly decreasing, from about 10% for the women now aged over 60 to less than 8% for those aged 40 to 49, resulting in a gradual decline in the value for I_i from about 0.12 to 0.08, while the value for I_f' appears to have been almost constant. These data are concerned with married women who had completed reproduction. On the other hand, it is clear from Table 4, where the expression of variance divided by square of the mean is

Table 10. Secular Changes in Selection Intensity due to Fertility Differences*

Age of Married Women	$\bar{x}_s$†	V_f	p_0	I_i	I_f'	I_f
80-	4.670	8.462	0.096	0.106	0.255	0.388
75-79	4.574	8.661	.109	.122	.260	.414
70-74	4.683	8.992	.108	.121	.258	.410
65-69	4.680	8.717	.100	.111	.258	.398
60-64	4.688	8.791	.098	.108	.263	.400
55-59	4.675	8.458	.088	.096	.265	.387
50-54	4.466	7.320	.080	.087	.258	.367
45-49	3.919	5.455	.076	.082	.250	.353
40-44	3.286	3.682	0.077	0.083	0.238	0.341

*Data are based on the 1960 census report in Japan.

†$\bar{x}_s$ = mean No. of children born to married women; V_f = variance in No. of children born to married women; p_0 = proportion of married women without children; I_i = index of selection intensity due to infertility; I_f' = index of selection intensity due to variation in No. of children for women having at least one child; I_f = index of selection intensity due to differential fertility.

equivalent to I_f', that the value for I_f' has been steadily declining for the currently reproducing women from 0.47 in 1950 to 0.31 in 1963.

There is no doubt that the above tendency will become more evident in the near future. It is not clear, however, to what extent the observed reductions in these values are a reflection of the reduction in the net intensity of selection. We do know that some cases of infertility are genetically determined; they are mostly resistant to medical cure, so that the net intensity of selection due to infertility seems unlikely to be subjected to relaxation. Since the loss of variation in the number of children among *fertile* women represents social conformity more than genetic homogeneity, the population may scarcely lose its variability as to fecundity and could resume high fertility if social needs required this.

Further Eugenic Application of Birth-Control Measures

The Eugenic Protection Law primarily aims at two quite different objects, the prevention of hereditary diseases and the protection of maternal health. In Table 11 are summarized all reported cases of induced abortions in Japan during five years

Table 11. No. of Induced Abortions Reported, 1960 to 1964, by Stated Reason*

Year	Stated Reason					Total
	Hereditary Disease	Leprosy	Maternal Health	Violence	Unknown	
1960	1,109	191	1,059,801	310	1,845	1,063,256
1961	995	225	1,031,910	284	1,915	1,035,329
1962	698	85	982,296	226	2,046	985,351
1963	556	93	952,142	166	2,135	955,092
1964	646	99	875,808	243	1,952	878,748
Total	4,004	693	4,901,957	1,229	9,893	4,917,776
%	0.08	0.01	99.68	0.03	0.20	100

*Data are based on statistics from the Ministry of Health and Welfare.

from 1960 to 1964, classified by the reasons stated in the table (*Annual Reports for the Eugenic Protection Statistics*, 1960-1964). The total number of cases amounts to about 4.9 million, of which only 4,004 were reported to be "for prevention of hereditary diseases." Taking account of the negligible frequency, it may appear that the law had scarcely contributed to the initial purpose of preventing hereditary diseases. However, there are some reasons to believe that there may be many more cases of induced abortions for some kind of eugenic reasons than stated in the official report. Because of the traditional family system, a hereditary disease in the family is particularly shameful for the Japanese. Further, the distinction in the reasons for "prevention of hereditary diseases" from "protection of maternal health" seems to play for the physician practically no role. These considerations make it very hard to evaluate an official report of this nature.

In countries like Japan, where family planning is practiced on a large scale, it may be hoped that various measures for birth control could readily be used for eugenic purpose. This should be conditioned by popular education on genetic matters of public-health importance and by genetic counselling as an integral part of medical-care services. If the birth of a child with some hereditary disorder could deter the parents from further reproduction, this would reduce not only the absolute number but also the relative frequency of the affected individuals in the population.[19] The opportunity for such selective limitation would be large enough for a number of defects that are recognizable early in infancy. For many other diseases with relatively late onset, the affected individuals may survive and marry, but they would still have the opportunity for using birth control. A preliminary result of our recent attempt to find out whether the parents having had a child with mental defect would limit further reproduction failed to provide evidence that such selection is occurring to an appreciable extent. But we may hope that the increased opportunities for eugenic application of family-planning practice, if guided by proper genetic counselling, could have an important effect upon future generations.

Summary and Conclusions

The recent demographic transition in Japan must have resulted in some reduction in the frequencies of a variety of congenital defects that are correlated with parental age and birth order. These defects may be classified into two categories, one comprising new mutations and the other determined by both genetic and environmental factors. The decrease in the diseases of the latter category may be regarded as dysgenic in the sense that selection against the genes is relaxed, but so long as the pattern of small family size continues the manifestation of those diseases would be prevented. The same transition should in the near future result in a significant reduction in consanguineous marriages and hence to some extent in a reduction in mortality and morbidity. Again, this effect would be dysgenic, because of the increase in the carriers of rare detrimental genes.

The variations in the distribution of family-planning practice, as well as variations in fertility among some social strata, are rapidly decreasing, but because of the lack of our knowledge on many important points, it is difficult to evaluate the extent of the assumed dysgenic effect upon some mental traits that are correlated with social stratification. The reduction in childhood mortality correlated with smaller family size must have resulted in relaxation of selection against most, if not all, of the detrimental genes affecting viability. The rate of infertility has been slowly decreasing, but the net intensity of selection due to infertility seems unlikely to undergo relaxation. The loss of variation in the number of children among potentially fertile women is obvious, but this reflects social conformity more than genetic homogeneity, so that the fecundity of future populations may scarcely be affected.

In the overall assessment of a benefit or harm for future generations, the possible effects upon the immediate or near future must be distinguished from those taking place in the long run. Within the limit of our present knowledge, the balance for the former appears to be far more in the direction of benefit, while for the latter it may be rather in the reverse. We may hope that the increased opportunities of using birth-control means for eugenic purpose, if guided by proper genetic counselling, could counterbalance the presumed dysgenic effects.

This work, contribution 587 from the National Institute of Genetics, was supported by grant RF 61113 from the Rockefeller Foundation and a grant from the Toyo Rayon Foundation for the Promotion of Science and Technology.

References

1. *Standardized Vital Rates for All Japan: 1920-1960.* Research Series 155, Tokyo: Ministry of Health and Welfare, Institute of Population Problems, 1963, p. 17.

2. Penrose, L. S.: Review of Court Brown, W. M., et al: *Abnormalities of the Sex Chromosome Complement in Man, Ann. Hum. Genet.* 28:199-200 (Nov) 1964.

3. Lenz, W.: Epidemiologie von Missbildungen, *Pädiatrie Pädologie,* 1:38-50, 1965.

4. Newcombe, H. B., and Tavendale, O. G.: Maternal Age and Birth Order Correlations: Problems of Distinguishing Mutational From Environmental Components, *Mutat. Res.* 1:446-467, 1964.

5. Newcombe, H. B.: Screening for Effects of Maternal Age and Birth Order in a Register of Handicapped Children, *Ann. Hum. Genet.* 27:367-382, 1964.

6. Nei, M., and Imaizumi, Y.: *Random Mating and Frequency of Consanguineous Marriages,* Annual Report of the National Institute of Radiological Sciences of Japan (1962), 1963, p. 48.

7. Dahlberg, G.: Inbreeding in Man, *Genetics,* 14:421-454, 1929.

8. *Summary of Eighth National Survey of Family Planning,* Population Problems Series 19, Tokyo: Mainichi Newspapers, Population Problems Research Council, 1965, pp. 32, 57-58.

9. *1960 Population Census of Japan,* Tokyo: Office of the Prime Minister, Bureau of Statistics, 1962, vol. 2 pp. 372-373.

10. Matsunaga, E.: Measures Affecting Population Trends and Possible Genetic Consequences, Paper B, 12/I/E/22, read before the United Nations World Population Conference, Belgrade, Yugloslavia, Aug. 31, 1965.

11. Schull, W. J., and Neel, J. V.: *The Effect of Inbreeding on Japanese Children,* New York: Harper & Row, Publishers, Inc., 1965, pp. 92-96.

12. *Fifth Public Opinion Survey on Birth Control in Japan.* Population Problems Series 16, Tokyo: Mainichi Newspapers, Population Problems Research Council, 1959, p. 21.

13. *Sixth Opinion Survey on Family Planning and Birth Control: A Preliminary Report,* Population Problems Series 18, Tokyo: Mainichi Newspapers, Population Problems Research Council, 1962, pp. 25-26.

14. Aoki, H.: Report of the Fourth Fertility Survey in 1962, *J Population Problems* 90: 1-54 (March) 1964.

15. King, J. L.: The Effect of Litter Culling—or Family Planning—on the Rate of Natural Selection, *Genetics,* 51:425-429 (March) 1965.

16. Crow, J. F.: Some Possibilities for Measuring Selection Intensities in Man, *Hum. Biol.* 30: 1-13 (Feb) 1958.

17. Spuhler, J. N.: Empirical Studies on Quantitative Human Genetics," in *Proceedings of the UN/WHO Seminar on the Use of Vital and Health Statistics for Genetic and Radiation Studies,* New York: United Nations, 1962, pp. 241-252.

18. Burgeois-Pichat, J.: in discussion Spuhler, J. N.[17]

19. Goodman, L. A.: Some Possible Effects of Birth Control in the Incidence of Disorders and on the Influence of Birth Order, *Ann. Hum. Genet.* 27:41-52, 1963.

22. The Tragedy of the Commons

GARRETT HARDIN

At the end of a thoughtful article on the future of nuclear war, Wiesner and York (*1*) concluded that: "Both sides in the arms race are . . . confronted by the dilemma of steadily increasing military power and steadily decreasing national security. *It is our considered professional judgment that this dilemma has no technical solution.* If the great powers continue to look for solutions in the area of science and technology only, the result will be to worsen the situation."

I would like to focus your attention not on the subject of the article (national security in a nuclear world) but on the kind of conclusion they reached, namely that there is no technical solution to the problem. An implicit and almost universal assumption of discussions published in professional and semipopular scientific journals is that the problem under discussion has a technical solution. A technical solution may be defined as one that requires a change only in the techniques of the natural sciences, demanding little or nothing in the way of change in human values or ideas of morality.

In our day (though not in earlier times) technical solutions are always welcome. Because of previous failures in prophecy, it takes courage to assert that a desired technical solution is not possible. Wiesner and York exhibited this courage; publishing in a science journal, they insisted that the solution to the problem was not to be found in the natural sciences. They cautiously qualified their statement with the phrase, "It is our considered professional judgment. . . ." Whether they were right or not is not the concern of the present article. Rather, the concern here is with the important concept of a class of human problems which can be called "no technical solution problems," and, more specifically, with the identification and discussion of one of these.

It is easy to show that the class is not a null class. Recall the game of tick-tack-toe. Consider the problem, "How can I win the game of tick-tack-toe?" It is well

SOURCE. Garrett Hardin, "The Tragedy of the Commons," Science, 162 (Dec. 13, 1968): 1243-1248 (Washington, D. C.)

known that I cannot, if I assume (in keeping with the conventions of game theory) that my opponent understands the game perfectly. Put another way, there is no "technical solution" to the problem. I can win only by giving a radical meaning to the word "win." I can hit my opponent over the head; or I can drug him; or I can falsify the records. Every way in which I "win" involves, in some sense, an abandonment of the game, as we intuitively understand it. (I can also, of course, openly abandon the game—refuse to play it. This is what most adults do.)

The class of "No technical solution problems" has members. My thesis is that the "population problem," as conventionally conceived, is a member of this class. How it is conventionally conceived needs some comment. It is fair to say that most people who anguish over the population problem are trying to find a way to avoid the evils of over-population without relinquishing any of the privileges they now enjoy. They think that farming the seas or developing new strains of wheat will solve the problem—technologically. I try to show here that the solution they seek cannot be found. The population problem cannot be solved in a technical way, any more than can the problem of winning the game of tick-tack-toe.

What Shall We Maximize?

Population, as Malthus said, naturally tends to grow "geometrically," or, as we would now say, exponentially. In a finite world this means that the per capita share of the world's goods must steadily decrease. Is ours a finite world?

A fair defense can be put forward for the view that the world is infinite; or that we do not know that it is not. But, in terms of the practical problems that we must face in the next few generations with the foreseeable technology, it is clear that we will greatly increase human misery if we do not, during the immediate future, assume that the world available to the terrestrial human population is finite. "Space" is no escape (2).

A finite world can support only a finite population; therefore, population growth must eventually equal zero. (The case of perpetual wide fluctuations above and below zero is a trivial variant that need not be discussed.) When this condition is met, what will be the situation of mankind? Specifically, can Bentham's goal of "the greatest good for the greatest number" be realized?

No—for two reasons, each sufficient by itself. The first is a theoretical one. It is not mathematically possible to maximize for two (or more) variables at the same time. This was clearly stated by von Neumann and Morgenstern (3), but the principle is implicit in the theory of partial differential equations, dating back at least to D'Alembert (1717-1783).

The second reason springs directly from biological facts. To live, any organism must have a source of energy (for example, food). This energy is utilized for two purposes: mere maintenance and work. For man, maintenance of life requires about

1600 kilocalories a day ("maintenance calories"). Anything that he does over and above merely staying alive will be defined as work, and is supported by "work calories" which he takes in. Work calories are used not only for what we call work in common speech; they are also required for all forms of enjoyment, from swimming and automobile racing to playing music and writing poetry. If our goal is to maximize population it is obvious what we must do: We must make the work calories per person approach as close to zero as possible. No gourmet meals, no vacations, no sports, no music, no literature, no art. . . . I think that everyone will grant, without argument or proof, that maximizing population does not maximize goods. Bentham's goal is impossible.

In reaching this conclusion I have made the usual assumption that it is the acquisition of energy that is the problem. The appearance of atomic energy has led some to question this assumption. However, given an infinite source of energy, population growth still produces an inescapable problem. The problem of the acquisition of energy is replaced by the problem of its dissipation, as J. H. Fremlin has so wittily shown (4). The arithmetic signs in the analysis are, as it were, reversed; but Bentham's goal is still unobtainable.

The optimum population is, then, less than the maximum. The difficulty of defining the optimum is enormous; so far as I know, no one has seriously tackled this problem. Reaching an acceptable and stable solution will surely require more than one generation of hard analytical work—and much persuasion.

We want the maximum good per person; but what is good? To one person it is wilderness, to another it is ski lodges for thousands. To one it is estuaries to nourish ducks for hunters to shoot; to another it is factory land. Comparing one good with another is, we usually say, impossible because goods are incommensurable. Incommensurables cannot be compared.

Theoretically this may be true; but in real life incommensurables *are* commensurable. Only a criterion of judgment and a system of weighting are needed. In nature the criterion is survival. Is it better for a species to be small and hideable, or large and powerful? Natural selection commensurates the incommensurables. The compromise achieved depends on a natural weighting of the values of the variables.

Man must imitate this process. There is no doubt that in fact he already does, but unconsciously. It is when the hidden decisions are made explicit that the arguments begin. The problem for the years ahead is to work out an acceptable theory of weighting. Synergistic effects, nonlinear variation, and difficulties in discounting the future made the intellectual problem difficult, but not (in principle) insoluble.

Has any cultural group solved this practical problem at the present time, even on a intuitive level? One simple fact proves that none has: there is no prosperous population in the world today that has, and has had for some time, a growth rate of zero. Any people that has intuitively identified its optimum point will soon reach it, after which its growth rate becomes and remains zero.

Of course, a positive growth rate might be taken as evidence that a population is

below its optimum. However, by any reasonable standards, the most rapidly growing populations on earth today are (in general) the most miserable. This association
(which need not be invariable) casts doubt on the optimistic assumption that the
positive growth rate of a population is evidence that it has yet to reach its optimum.

We can make little progress is working toward optimum population size until we
explicitly exorcize the spirit of Adam Smith in the field of practical demography.
In economic affairs, *The Wealth of Nations* (1776) popularized the "invisible
hand," the idea that an individual who "intends only his own gain," as it were, "led
by an invisible hand to promote . . . the public interest" (5). Adam Smith did not
assert that this was invariably true, and perhaps neither did any of his followers. But
he contributed to a dominant tendency of thought that has ever since interfered
with positive action based on rational analysis, namely, the tendency to assume that
decisions reached individually will, in fact, be the best decisions for an entire society. If this assumption is correct it justifies the continuance of our present policy of
laissez-faire in reproduction. If it is correct we can assume that men will control
their individual fecundity so as to produce the optimum population. If the assumption is not correct, we need to reexamine our individual freedoms to see which ones
are defensible.

Tragedy of Freedom in a Commons

The rebuttal to the invisible hand in population control is to be found in a scena-
rio first sketched in a little-known pamphlet (6) in 1833 by a mathematical amateur
named William Forster Lloyd (1794-1852). We may well call it "the tragedy of the
commons," using the word "tragedy" as the philospher Whitehead used it (7): "The
essence of dramatic tragedy is not unhappiness. It resides in the solemnity of the
remorseless working of things." He then goes on to say, "This inevitableness of destiny can only be illustrated in terms of human life by incidents which in fact involve unhappiness. For it is only by them that the futility of escape can be made
evident in the drama."

The tragedy of the commons develops in this way. Picture a pasture open to all.
It is to be expected that each herdsman will try to keep as many cattle as possible
on the commons. Such an arrangement may work reasonably satisfactorily for centuries because tribal wars, poaching, and disease keep the numbers of both man and
beast well below the carrying capacity of the land. Finally, however, comes the day
of reckoning, that is, the day when the long-desired goal of social stability becomes
a reality. At this point, the inherent logic of the commons remorselessly generates
tragedy.

As a rational being, each herdsman seeks to maximize his gain. Explicitly or im-
plicitly, more or less consciously, he asks, "What is the utility *to me* of adding

one more animal to my herd?" This utility has one negative and one positive component.

1. The positive component is a function of the increment of one animal. Since the herdsman receives all the proceeds from the sale of the additional animal, the positive utility is nearly +1.

2. The negative component is a function of the additional overgrazing created by one more animal. Since, however, the effects of overgrazing are shared by all the herdsmen, the negative utility for any particular decision-making herdsman is only a fraction of −1.

Adding together the component partial utilities, the rational herdsman concludes that the only sensible course for him to pursue is to add another animal to his herd. And another; and another. . . . But this is the conclusion reached by each and every rational herdsman sharing a commons. Therein is the tragedy. Each man is locked into a system that compels him to increase his herd without limit—in a world that is limited. Ruin is the destination toward which all men rush, each pursuing his own best interest in a society that believes in the freedom of the commons. Freedom in a commons brings ruin to all.

Some would say that this is a platitude. Would that it were! In a sense, it was learned thousands of years ago, but natural selection favors the forces of psychological denial (8). The individual benefits as an individual from his ability to deny the truth even though society as a whole, of which he is a part, suffers. Education can counteract the natural tendency to do the wrong thing, but the inexorable succession of generations requires that the basis for this knowledge be constantly refreshed.

A simple incident that occurred a few years ago in Leominster, Massachusetts, shows how perishable the knowledge is. During the Christmas shopping season the parking meters downtown were covered with plastic bags that bore tags reading: "Do not open until after Christmas. Free parking courtesy of the major and city council." In other words, facing the prospect of an increased demand for already scarce space, the city fathers reinstituted the system of the commons. (Cynically, we suspect that they gained more votes than they lost by this retrogressive act.)

In an approximate way, the logic of the commons has been understood for a long time, perhaps since the discovery of agriculture or the invention of private property in real estate. But it is understood mostly only in special cases which are not sufficiently generalized. Even at this late date, cattlemen leasing national land on the western ranges demonstrate no more than an ambivalent understanding, in constantly pressuring federal authorities to increase the head count to the point where overgrazing produces   erosion and weed-dominance. Likewise, the oceans of the world continue to suffer from the survival of the philosophy of the commons. Maritime nations still respond automatically to the shibboleth of the "freedom of the seas." Professing to believe in the "inexhaustible resources of the oceans," they bring species after species of fish and whales closer to extinction (9).

The National Parks present another instance of the working out of the tragedy of the commons. At present, they are open to all, without limit. The parks themselves are limited in extent—there is only one Yosemite Valley—whereas population seems to grow without limit. The values that visitors seek in the parks are steadily eroded. Plainly, we must soon cease to treat the parks as commons or they will be of no value to anyone.

What shall we do? We have several options. We might sell them off as private property. We might keep them as public property, but allocate the right to enter them. The allocation might be on the basis of wealth, by the use of an auction system. It might be on the basis of merit, as defined by some agreed-upon standards. It might be by lottery. Or it might be on a first-come, first-served basis, administered to long queues. These, I think, are all the reasonable possibilities. They are all objectionable. But we must choose—or acquiesce in the destruction of the commons that we call our National Parks.

Pollution

In a reverse way, the tragedy of the commons reappears in problems of pollution. Here it is not a question of taking something out of the commons, but of putting something in—sewage, or chemical, radioactive, and heat wastes into water; noxious and dangerous fumes into the air; and distracting and unpleasant advertising signs into the line of sight. The calculations of utility are much the same as before. The rational man finds that his share of the cost of the wastes he discharges into the commons is less than the cost of purifying his wastes before releasing them. Since this is true for everyone, we are locked into a system of "fouling our own nest," so long as we behave only as independent, rational, free-enterprisers.

The tragedy of the commons as a food basket is averted by private property, or something formally like it. But the air and waters surrounding us cannot readily be fenced, and so the tragedy of the commons as a cesspool must be prevented by different means, by coercive laws or taxing devices that make it cheaper for the polluter to treat his pollutants than to discharge them untreated. We have not progressed as far with the solution of this problem as we have with the first. Indeed, our particular concept of private property, which deters us from exhausting the positive resources of the earth, favors pollution. The owner of a factory on the bank of a stream—whose property extends to the middle of the stream—often has difficulty seeing why it is not his natural right to muddy the waters flowing past his door. The law, always behind the times, requires elaborate stitching and fitting to adapt it to this newly perceived aspect of the commons.

The pollution problem is a consequence of population. It did not much matter how a lonely American frontiersman disposed of his waste. "Flowing water purifies itself every 10 miles," my grandfather used to say, and the myth was near enough to the truth when he was a boy, for there were not too many people. But as population became denser, the natural chemical and biological recycling processes became overloaded, calling for a redefinition of property rights.

How To Legislate Temperance?

Analysis of the pollution problem as a function of population density uncovers a not generally recognized principle of morality, namely: *the morality of an act is a function of the state of the system at the time it is performed (10)*. Using the commons as a cesspool does not harm the general public under frontier conditions, because there is no public; the same behavior in a metropolis is unbearable. A hundred and fifty years ago a plainsman could kill an American bison, cut out only the tongue for his dinner, and discard the rest of the animal. He was not in any important sense being wasteful. Today, with only a few thousand bison left, we would be appalled at such behavior.

In passing, it is worth noting that the morality of an act cannot be determined from a photograph. One does not know whether a man killing an elephant or setting fire to the grassland is harming others until one knows the total system in which his act appears. "One picture is worth a thousand words," said an ancient Chinese; but it may take 10,000 words to validate it. It is as tempting to ecologists as it is to reformers in general to try to persuade others by way of the photographic shortcut. But the essence of an argument cannot be photographed: it must be presented rationally—in words.

That morality is system-sensitive escaped the attention of most codifiers of ethics in the past. "Thou shalt not . . ." is the form of traditional ethical directives which make no allowance for particular circumstances. The laws of our society follow the pattern of ancient ethics, and therefore are poorly suited to governing a complex, crowded, changeable world. Our epicyclic solution is to augment statutory law with administrative law. Since it is practically impossible to spell out all the conditions under which it is safe to burn trash in the back yard or to run an automobile without smog-control, by law we delegate the details to bureaus. The result is administrative law, which is rightly feared for an ancient reason—*Quis custodiet ipsos custodes?*—"Who shall watch the watchers themselves?" John Adams said that we must have "a government of laws and not men." Bureau administrators, trying to evaluate the morality of acts in the total system, are singularly liable to corruption, producing a government by men, not laws.

Prohibition is easy to legislate (though not necessarily to enforce); but how do we legislate temperance? Experience indicates that it can be accomplished best through the mediation of administrative law. We limit possibilities unnecessarily if we suppose that the sentiment of *Quis custodiet* denies us the use of administrative law. We should rather retain the phrase as a perpetual reminder of fearful dangers we cannot avoid. The great challenge facing us now is to invent the corrective feed-backs that are needed to keep custodians honest. We must find ways to legitimate the needed authority of both the custodians and the corrective feedbacks.

Freedom to Breed Is Intolerable

The tragedy of the commons is involved in population problems in another way. In a world governed solely by the principle of "dog eat dog"—if indeed there ever was such a world—how many children a family had would not be a matter of public concern. Parents who bred too exuberantly would leave fewer descendants, not more, because they would be unable to care adequately for their children. David Lack and others have found that such a negative feedback demonstrably controls the fecundity of birds (11). But men are not birds, and have not acted like them for millenniums, at least.

If each human family were dependent only on its own resources; *if* the children of improvident parents starved to death; *if*, thus, overbreeding brought its own "punishment" to the germ line—*then* there would be no public interest in control-ling the breeding of families. But our society is deeply committed to the welfare state (12), and hence is confronted with another aspect of the tragedy of the com-mons.

In a welfare state, how shall we deal with the family, the religion, the race, or the class (or indeed any distinguishable and cohesive group) that adopts overbreed-ing as a policy to secure its own aggrandizement (13)? To couple the concept of free-dom to breed with the belief that everyone born has an equal right to the commons is to lock the world into a tragic course of action.

Unfortunately this is just the course of action that is being pursued by the United Nations. In late 1967, some 30 nations agreed to the following (14):

> "The Universal Declaration of Human Rights describes the family as the natural and fundamental unit of society. It follows that any choice and decision with regard to the size of the family must irrevocably rest with the family itself, and cannot be made by anyone else."

It is painful to have to deny categorically the validity of this right; denying it, one feels as uncomfortable as a resident of Salem, Massachusetts, who denied the reality of witches in the 17th century. At the present time, in liberal quarters, something like a taboo acts to inhibit criticism of the United Nations. There is a feeling that the United Nations is "our last and best hope," that we shouldn't find fault with it; we shouldn't play into the hands of the archconservatives. However, let us not forget what Robert Louis Stevenson said: "The truth that is suppressed by friends in the readiest weapon of the enemy." If we love the truth we must openly deny the validity of the Universal Declaration of Human Rights, even though it is promoted by the United Nations. We should also join with Kingsley Davis (*15*) in attempting to get Planned Parenthood-World Population to see the error of its ways in embracing the same tragic ideal.

Conscience Is Self-Eliminating

It is a mistake to think that we can control the breeding of mankind in the long run by an appeal to conscience. Charles Galton Darwin made this point when he spoke on the centennial of the publication of his grandfather's great book. The argument is straightforward and Darwinian.

People vary. Confronted with appeals to limit breeding, some people will undoubtedly respond to the plea more than others. Those who have more children will produce a larger fraction of the next generation than those with more susceptible consciences. The difference will be accentuated, generation by generation.

In C. G. Darwin's words: "It may well be that it would take hundreds of generations for the progenitive instinct to develop in this way, but if it should do so, nature would have taken her revenge, and the variety *Homo contracipiens* would become extinct and would be replaced by the variety *Homo progenitivus*" (*16*).

The argument assumes that conscience or the desire for children (no matter which) is hereditary—but hereditary only in the most general formal sense. The result will be the same whether the attitude is transmitted through germ cells, or exosomatically, to use A. J. Lotka's term. (If one denies the latter possibility as well as the former, then what's the point of education?) The argument has here been stated in the context of the population problem, but it applies equally well to any instance in which society appeals to an individual exploiting a commons to restrain himself for the general good—by means of his conscience. To make such an appeal is to set up a selective system that works toward the elimination of conscience from the race.

Pathogenic Effects of Conscience

The long-term disadvantage of an appeal to conscience should be enough to condemn it; but it has serious short-term disadvantages as well. If we ask a man who is exploiting a commons to desist "in the name of conscience," what are we saying to him? What does he hear?—not only at the moment but also in the wee small hours of the night when, half asleep, he remembers not merely the words we used but also the nonverbal communication cues we gave him unawares? Sooner or later, consciously or subconsciously, he senses that he has received two communications, and that they are contradictory: (i) (intended communication) "If you don't do as we ask, we will openly condemn you for not acting like a responsible citizen"; (ii) (the unintended communication) "If you *do* behave as we ask, we will secretly condemn you for a simpleton who can be shamed into standing aside while the rest of us exploit the commons."

Everyman then is caught in what Bateson has called a "double bind." Bateson and his co-workers have made a plausible case for viewing the double bind as an important causative factor in the genesis of schizophrenia (*17*). The double bind may not always be so damaging, but it always endangers the mental health of anyone to whom it is applied. "A bad conscience," said Nietzsche, "is a kind of illness."

To conjure up a conscience in others is tempting to anyone who wishes to extend his control beyond the legal limits. Leaders at the highest level succumb to this temptation. Has any President during the past generation failed to call on labor unions to moderate voluntarily their demands for higher wages, or to steel companies to honor voluntary guidelines on prices? I can recall none. The rhetoric used on such occasions is designed to produce feelings of guilt in noncooperators.

For centuries it was assumed without proof that guilt was a valuable, perhaps even an indispensable, ingredient of the civilized life. Now, in this post-Freudian world, we doubt it.

Paul Goodman speaks from the modern point of view when he says: "No good has ever come from feeling guilty, neither intelligence, policy, nor compassion. The guilty do not pay attention to the object but only to themselves, and not even to their own interests, which might make sense, but to their anxieties" (*18*).

One does not have to be a professional psychiatrist to see the consequences of anxiety. We in the Western world are just emerging from a dreadful two-centuries-long Dark Ages of Eros that was sustained partly by prohibition laws, but perhaps more effectively by the anxiety-generating mechanisms of education. Alex Comfort has told the story well in *The Anxiety Makers* (*19*); it is not a pretty one.

Since proof is difficult, we may even concede that the results of anxiety may sometimes, from certain points of view, be desirable. The larger question we should ask is whether, as a matter of policy, we should ever encourage the use of a technique the tendency (if not the intention) of which is psychologically pathogenic. We hear much talk these days of responsible parenthood; the coupled words are incorporated into the titles of some organizations devoted to birth control. Some people have proposed massive propaganda campaigns to instill responsibility into the nation's (or the world's) breeders. But what is the meaning of the word responsibility in this context? Is it not merely a synonym for the word conscience? When we use the word responsibility in the absence of substantial sanctions are we not trying to browbeat a free man in a commons into acting against his own interest? Responsibility is a verbal counterfeit for a substantial *quid pro quo*. It is an attempt to get something for nothing.

If the word responsibility is to be used at all, I suggest that it be in the sense Charles Frankel uses it (*20*). "Responsibility," says this philosopher, "is the product of definite social arrangements." Notice that Frankel calls for social arrangements—not propaganda.

Mutual Coercion Mutually Agreed Upon

The social arrangements that produce responsibility are arrangements that create coercion, of some sort. Consider bank-robbing. The man who takes money from a bank acts as if the bank were a commons. How do we prevent such action? Certainly not by trying to control his behavior solely by a verbal appeal to his sense of responsibility. Rather than rely on propaganda we follow Frankel's lead and insist that a bank is not a commons; we seek the definite social arrangements that will keep it from becoming a commons. That we thereby infringe on the freedom of would-be robbers we neither deny nor regret.

The morality of bank-robbing is particularly easy to understand because we accept complete prohibition of this activity. We are willing to say "Thou shalt not rob banks," without providing for exceptions. But temperance also can be created by coercion. Taxing is a good coercive device. To keep downtown shoppers temperate in their use of parking space we introduce parking meters for short periods, and traffic fines for longer ones. We need not actually forbid a citizen to park as long as he wants to; we need merely make it increasingly expensive for him to do so. Not prohibition, but carefully biased options are what we offer him. A Madison Avenue man might call this persuasion; I prefer the greater candor of the word coercion.

Coercion is a dirty word to most liberals now, but it need not forever be so. As with the four-letter words, its dirtiness can be cleansed away by exposure to the light, by saying it over and over without apology or embarrassment. To many, the word coercion implies arbitrary decisions of distant and irresponsible bureaucrats; but this is not a necessary part of its meaning. The only kind of coercion I recommend is mutual coercion, mutually agreed upon by the majority of the people affected.

To say that we mutually agree to coercion is not to say that we are required to enjoy it, or even to pretend we enjoy it. Who enjoys taxes? We all grumble about them. But we accept compulsory taxes because we recognize that voluntary taxes would favor the conscienceless. We institute and (grumblingly) support taxes and other coercive devices to escape the horror of the commons.

An alternative to the commons need not be perfectly just to be preferable. With real estate and other material goods, the alternative we have chosen is the institution of private property coupled with legal inheritance. Is this system perfectly just? As a genetically trained biologist I deny that it is. It seems to me that, if there are to be differences in individual inheritance, legal possession should be perfectly correlated with biological inheritance—that those who are biologically more fit to be the custodians of property and power should legally inherit more. But genetic recombination continually makes a mockery of the doctrine of "like father, like son" implicit in our laws of legal inheritance. An idiot can inherit millions, and a trust fund can keep his estate intact. We must admit that our legal system of private property plus inheritance is unjust—but we put up with it because we are not convinced, at the moment, that anyone has invented a better system. The alternative of the commons is too horrifying to contemplate. Injustice is preferable to total ruin.

It is one of the peculiarities of the warfare between reform and the status quo that it is thoughtlessly governed by a double standard. Whenever a reform measure is proposed it is often defeated when its opponents triumphantly discover a flaw in it. As Kingsley Davis has pointed out (21), worshippers of the status quo sometimes imply that no reform is possible without unanimous agreement, an implication contrary to historical fact. As nearly as I can make out, automatic rejection of proposed reforms is based on one of two unconscious assumptions: (i) that the status quo is perfect; or (ii) that the choice we face is between reform and no action; if the proposed reform is imperfect, we presumably should take no action at all, while we wait for a perfect proposal.

But we can never do nothing. That which we have done for thousands of years is also action. It also produces evils. Once we are aware that the status quo is action, we can then compare its discoverable advantages and disadvantages with the predicted advantages and disadvantages of the proposed reform, discounting as best we can for our lack of experience. On the basis of such a comparison, we can make a rational decision which will not involve the unworkable assumption that only perfect systems are tolerable.

Recognition of Necessity

Perhaps the simplest summary of this analysis of man's population problems is this: the commons, if justifiable at all, is justifiable only under conditions of low-population density. As the human population has increased, the commons has had to be abandoned in one aspect after another.

First we abandoned the commons in food gathering, enclosing farm land and restricting pastures and hunting and fishing areas. These restrictions are still not complete throughout the world.

Somewhat later we saw that the commons as a place for waste disposal would also have to be abandoned. Restrictions on the disposal of domestic sewage are widely accepted in the Western world; we are still struggling to close the commons to pollution by automobiles, factories, insecticide sprayers, fertilizing operations, and atomic energy installations.

In a still more embryonic state in our recognition of the evils of the commons in matters of pleasure. There is almost no restriction on the propagation of sound waves in the public medium. The shopping public is assaulted with mindless music, without its consent. Our government is paying out billions of dollars to create supersonic transport which will disturb 50,000 people for every one person who is whisked from coast to coast 3 hours faster. Advertisers muddy the airwaves of radio and television and pollute the view of travelers. We are a long way from outlawing the commons in matters of pleasure. Is this because our Puritan inheritance makes us view pleasure as something of a sin, and pain (that is, the pollution of advertising) as the sign of virtue?

Every new enclosure of the commons involves the infringement of somebody's personal liberty. Infringements made in the distant past are accepted because no contemporary complains of a loss. It is the newly proposed infringements that we vigorously oppose; cries of "rights" and "freedom" fill the air. But what does "freedom" mean? When men mutally agreed to pass laws against robbing, mankind became more free, not less so. Individuals locked into the logic of the commons are free only to bring on universal ruin; once they see the necessity of mutual coercion, they become free to pursue other goals. I believe it was Hegel who said, "Freedom is the recognition of necessity."

The most important aspect of necessity that we must now recognize, is the necessity of abandoning the commons is breeding. No technical solution can rescue us from the misery of overpopulation. Freedom to breed will bring ruin to all. At the moment, to avoid hard decisions many of us are tempted to propagandize for conscience and responsible parenthood. The temptation must be resisted, because an appeal to independently acting consciences selects for the disappearance of all conscience in the long run, and an increase in anxiety in the short.

The only way we can preserve and nurture other and more precious freedoms is by relinquishing the freedom to breed, and that very soon. "Freedom is the recognition of necessity"—and it is the role of education to reveal to all the necessity of abandoning the freedom to breed. Only so, can we put an end to this aspect of the tragedy of the commons.

References

1. J. B. Wiesner and H. F. York, *Sci. Amer.*, 211 (No. 4), 27 (1964).

2. G. Hardin, *J. Hered.* 50, 68 (1959); S. von Hoernor, *Science* 137, 18 (1962).

3. J. von Neumann and O. Morgenstern, *Theory of Games and Economic Behavior* (Princeton Univ. Press, Princeton, N.J., 1947), p. 11.

4. J. H. Fremlin, *New Sci.*, No. 415 (1964), p. 285.

5. A. Smith, *The Wealth of Nations* (Modern Library, New York, 1937), p. 423.

6. W. F. Lloyd, *Two Lectures on the Checks to Population* (Oxford Univ. Press, Oxford, England, 1833) reprinted (in part) in *Population, Evolution, and Birth Control*, G. Hardin, Ed. (Freeman, San Francisco, 1964), p. 37.

7. A. N. Whitehead, *Science and the Modern World* (Mentor, New York, 1948), p. 17.

8. G. Hardin, Ed. *Population, Evolution, and Birth Control* (Freeman, San Francisco, 1964), p. 56.

9. S. McVay, *Sci. Amer.* **216** (No. 8), 13 (1966).

10. J. Fletcher, *Situation Ethics* (Westminster, Philadelphia, 1966).

11. D. Lack, *The Natural Regulation of Animal Numbers* (Clarendon Press, Oxford, 1954).

12. H. Girvetz, *From Wealth to Welfare* (Stanford Univ. Press, Stanford, Calif., 1950).

13. G. Hardin, *Perspec. Biol. Med.* **6**, 366 (1963).

14. U. Thant, *Int. Planned Parenthood News*, No. 168 (February 1968), p. 3.

15. K. Davis, *Science* **158**, 730 (1967).

16. S. Tax, Ed., *Evolution after Darwin* (Univ. of Chicago Press, Chicago, 1960), vol. 2, p. 479.

17. G. Bateson, D. D. Jackson, J. Haley, J. Weakland, *Behav. Sci.* **1**, 251 (1956).

18. P. Goodman, *New York Rev. Books* **10**(8), 22 (23 May 1968).

19. A. Comfort, *The Anxiety Makers* (Nelson, London, 1967).

20. C. Frankel, *The Case for Modern Man* (Harper, New York, 1955), p. 203.

21. J. D. Roslansky, *Genetics and the Future of Man* (Appleton-Century-Crofts, New York, 1966), p. 177.

23. The Genetic Implications of American Life Styles in Reproduction and Population Control

CARL J. BAJEMA

Introduction

Each generation of mankind is faced with the awesome responsibility of having to make decisions concerning the quantity and quality (both genetic and cultural) of future generations. Because of its concern for its increasing population in relation to natural resources and quality of life the United States appears to be on the verge of discarding its policy favoring continued population growth and adopting a policy aimed at achieving a zero rate of population growth by voluntary means. The policy that a society adopts with respect to population size will have genetic as well as environmental consequences.

Human populations adapt to their environments genetically as well as culturally. These environments have been and are changing very rapidly with most of the changes being brought about by man himself. Mankind, has, by creating a highly technological society, produced a society in which a large proportion of its citizens cannot contribute to its growth or maintenance because of the limitations (both genetic and environmental) of their intellect. The modern technological societies of democratic nations offer their citizens a wide variety of opportunities for self-fulfillment but find that many of their citizens are incapable of taking advantage of the opportunities open to them.

American society, if it takes its responsibility to future generations seriously, will have to do more than control the size of its population in relation to the environment. It will have to take steps to insure that individuals yet unborn will have the best genetic and environmental heritage possible to enable them to meet the challenges of the environment and to take advantage of the opportunities for self-fulfillment made available by society.

SOURCE. Carl J. Bajema, "The Genetic Implications of Population Control With Special Reference to American Life Styles in Reproduction," Enlarged version of paper presented at the First National Congress on Optimum Population and Environment (Chicago, Illinois: June 7-11, 1970).

American society cannot ignore the question of genetic quality for very long because it, like all other human societies, must cope with two perpetual problems as it attempts to adapt to its environment. First, it must cope with a continual input of harmful genes into its population via mutation (it has been estimated that approximately one out of every five newly fertilized human eggs is carrying a newly mutated gene that was not present in either of the two parents). The genetic status quo can be maintained in a human population only if the number of new mutant genes added to the population is counterbalanced by an equal number of the mutant genes not being passed on due to the nonreproduction or decreased reproduction of individuals carrying these mutant genes. Otherwise the proportion of harmful genes in the population will increase. Second, American society has to adapt to a rapidly changing environment. For instance, the technologically based computer age sociocultural environment being created by American society has placed a premium on the possession of high intelligence and creativity. Our society requires individuals with high intelligence and creativity to help it make the appropriate social and technological adjustments in order to culturally adapt to its rapidly changing environment. On the other hand individuals require high intelligence and creativity in order that they, as individuals, can cope with the challenges of the environment and take advantage of the opportunities for self-fulfillment present in our society.

The proportion of the American population that already is genetically handicapped—that suffers a restriction of liberty or competence because of the genes they are carrying—is not small. Therefore the genetic component of the human population-environment equation must be taken into account as we attempt to establish an environment that has a high degree of ecological stability and that maximizes the number of opportunities for self-fulfillment available to each individual human being.

The Genetic Consequences of American Life Styles in Sex and Reproduction

American life styles with respect to sex and reproduction are currently in a tremendous state of flux and are changing rapidly. This makes it difficult to accurately predict the genetic consequences of these life styles. Yet, because these life styles determine the genetic make-up of future generations of Americans it is necessary that we evaluate the genetic consequences of past and present trends and speculate concerning the probable genetic consequences of projecting these trends into the future. Only then will we be able to determine the severity of the problem and to determine what steps, if any, need to be taken to maintain and improve the genetic heritage of future generations.

American society has developed modern medical techniques which enable many individuals with severe genetic defects to survive to adulthood. Many of these individuals can and do reproduce, thereby passing their harmful genes on to the next generation increasing the frequency of these genes in the population. At present there is no indication that heredity counseling decreases the probability that these individuals will have children. The life styles of these individuals with respect to reproduction is creating a larger genetic burden for future generations of Americans to bear.

The effect of American life styles in sex and reproduction on such behavioral patterns as intelligence and personality is much less clear. For instance, during most of man's evolution natural selection has favored the genes for intelligence. The genes for higher mental ability conferred an advantage to their carriers in the competition for survival and reproduction both within and between populations. Thus the more intelligent members of the human species passed more genes on to the next generation than did the less intelligent members, with the result that the genes for higher intelligence increased in frequency. As Western societies shifted from high birth and death rates toward low birth and death rates, however, a breakdown in the relation of natural selection to achievement or "success" took place.

The practice of family planning spread more rapidly among the better educated strata of society resulting in negative fertility differentials. At the period of extreme differences, which in the United States came during the great depression, the couples who were poorly educated were having about twice as many children as the more educated couples. The continued observation of a negative relationship between fertility and such characteristics as education, occupation and income during the first part of this century led many scientists to believe that this pattern of births was a concomitant of the industrial welfare state society and must make for the genetic deterioration of the human race. This situation was, in part, temporary. The fertility differentials have declined dramatically since World War II so that by the 1960's some cohorts of college graduate women were having 90% as many children as the U.S. average. A number of recent studies of American life styles in reproduction, when taken collectively, seem to indicate that, as the proportion of the urban population raised in a farm environment decreases, as the educational attainment of the population increases, and as women gain complete control over childbearing (via contraception and induced abortion) the relationship between fertility and such characteristics as income, occupation, educational attainment, and intelligence (I.Q.) will become less negative and may approach zero.

The overall net effect of current American life styles in reproduction appears to be slightly dysgenic—to be favoring an increase in harmful genes which will genetically handicap a larger proportion of the next generation of Americans. American life styles in reproduction are, in part, a function of the population policy of the United States. What will be the long-range genetic implications of controlling or not controlling population size in an industrialized welfare state democracy such as America?

The Genetic Implications of Policies Favoring Continued Population Growth

Most contemporary human societies are organized in such a way that they encourage population growth. How is the genetic make-up of future generations affected by the size of the population? What will be the ultimate genetic consequences given a society that is growing in numbers in relation to its environment? One possible consequence is military aggression coupled with genocide to attain additional living space. This would result in genetic change insofar as the population eradicated or displaced differs genetically from the population that is aggressively expanding the size of its environment. The displacement of the American indians by west Europeans is an example of this approach to the problem of population size in relation to the environment. Throughout man's evolution such competition between different populations of human beings has led to an increase in the cultural and genetic supports for aggressive behavior in the human species. Violence as a form of aggressive behavior to solve disagreements among populations appears to have become maladaptive in the nuclear age. It will probably take a nuclear war to prove this contention.

If one assumes that military aggression plus genocide to attain additional living space is not an option open to a society with a population policy encouraging growth in numbers then a different type of genetic change will probably take place. Most of the scientists who have attempted to ascertain the probable effect that overcrowding in a welfare state will have on man's genetic make-up have concluded that natural selection would favor those behavior patterns that most people consider least desirable. For instance, Rene Dubos (1965), in discussing the effect of man's future environment on the direction and intensity of natural selection in relation to human personality patterns states that:

"Most disturbing perhaps are the behavioral consequences likely to ensure from overpopulation. The ever-increasing complexity of the social structure will make some form of regimentation unavoidable; freedom and privacy may come to constitute antisocial luxuries and their attainment to involve real hardships. In consequence, there may emerge by selection a stock of human beings suited to accept as a matter of course a regimented and sheltered way of life in a teeming and polluted world, from which all widerness and fantasy of nature will have disappeared. The domesticated farm animal and the laboratory rodent in a controlled environment will then become true models for the study of man."

The genetic and cultural undesirability of either of these two alternative outcomes for mankind makes it imperative that societies move quickly to adopt policies aimed at achieving and maintaining an optimum population size that maximizes the dignity and individual worth of a human being rather than maximizing the number of human beings in relation to the environment.

Genetic Implications of Policies Favoring Control of Population Size by Voluntary Means

There is strong evidence that contemporary societies can achieve control of their population size by voluntary means, at least in the short run.

What will be the distribution of births in societies that have achieved a zero population growth rate? In a society where population size is constant—where each generation produces only enough offspring to replace itself—there will still be variation among couples with respect to the number of children they will have. Some individuals will be childless or have only one child for a variety of reasons—biological (genetically or environmentally caused sterility), psychological (inability to attract a mate, desire to remain childless), etc. Some individuals will have to have at least three children to compensate for those individuals who have less than two children. The resulting differential fertility—variation in the number of children couples have—provides an opportunity for natural selection to operate and would bring about genetic change if the differences in fertility among individuals are correlated with differences (physical, physiological, or behavioral) among individuals.

The United States is developing into a social welfare state democracy. This should result in an environment that will evoke the optimal response from the variety of genotypes (specific combinations of genes that individuals carry) present in the population. It is questionable, however, as to whether a social welfare state democracy creates the type of environment that will automatically bring about a eugenic distribution of births resulting in the maintenance or enhancement of man's genetic heritage. It is also questionable as to whether a social welfare state democracy (or any society for that matter) will be able to achieve and maintain a zero population growth rate—a constant population size—by voluntary means.

Both Charles Darwin and Garrett Hardin have argued that universal compulsion will be necessary to achieve and maintain zero population growth. They argue that appeals to individual conscience as the means of restraining couples from having more than two children won't work because those individuals or groups who refused to restrain themselves would increase their numbers in relation to the rest with the result that these individuals or groups with their cultural and/or biological supports for high fertility would constitute a larger and larger proportion of the population of future generations and *Homo contracipiens* would be replaced by *Homo progenetivis*.

Hardin (1968) raises this problem in his classic paper, The Tragedy of the Commons, when he states:

"If each human family were dependent only on its own resources; if the children of improvident parents starved to death; if, thus, overbreeding brought its own 'punishment' to the germ line—then there would be no public interest in controlling the breeding of families. But our soiciety is deeply committed to the welfare state, and hence confronted with another aspect of the tragedy of the commons.

"In a welfare state, how shall we deal with the family, the religion, the race, or the class (or indeed any distinquishable and cohesive group) that adopts overbreeding as a policy to secure its own aggrandizement? To couple the concept of freedom to breed with the belief that everyone born has an equal right to the commons is to lock the world into a tragic course of action."

The only way out of this dilemma according to Hardin is for society to create reproductive responsibility via social arrangements that produce coercion of some sort. The kind of coercion Hardin talks about is mutual coercion, mutually agreed upon by the majority of the people affected. Compulsory taxes are an example of mutual coercion. Democratic societies frequently have to resort to mutual coercion to escape destruction of the society by the irresponsible. Mutual coercion appears to be the only solution to the problem of pollution. If Hardin is right it may also be the only solution for any society that is attempting to control the size and/or the genetic make-up of its population.

Hardin's thesis has been questioned on the basis that children are no longer the economic assets they once were in agarian societies. Rufous Miles has argued that given today's post industrial economy children are expensive pleasures; they are economic liabilities rather than assets. Miles (1970) points out that:

"There is no conflict, therefore, between the economic self-interest of married couples to have small families and the collective need of society to preserve "the commons." It is in both their interests to limit procreation to not more than a replacement level. Unfortunately, couples do not seek their self-interest in economic terms alone, but in terms of total satisfactions. They are "buying" children and paying dearly for them. The problem, therefore, is compounded of how to persuade couples to act more in their own economic self-interest and that of their children; how to assist them in obtaining more psychological satisfactions from sources other than large families; and how to replace the outworn and now inimical tradition of the large family with a new "instant tradition" of smaller families."

As pointed out earlier in this paper there is some evidence to support the contention that as American society becomes more urbanized, achieves higher levels of educational attainment, and allows its citizens to exercise complete control over their fertility, reproductive patterns will develop which will lead to a zero or negative population growth rate and an eugenic distribution of births. If this prediction is correct then there will be no need for the adoption of mutual coercion—compulsory methods of population control—by American society in order to control the size and/or genetic quality of its population. If, on the other hand, these reproductive patterns do not develop or are transitory it may very well be that reproduction will have to become a privilege rather than a right in social welfare state democracies in order to insure that these societies and their citizens do not have to suffer the environmental and genetic consequences of irresponsible reproduction.

What might the genetic consequences be if a society had to resort to mutual coercion—had to employ compulsory methods of population control—to control its numbers?

Genetic Implications of Compulsory Population Control

There are a number of methods by which compulsory population control can be achieved (Berelson, 1969). Mutual coercion could be institutionalized by a democratic society to ensure that couples who would otherwise be reproductively irresponsible are restricted to having only 2 children. Compulsory abortion and/or sterilization could be employed to guarantee that no woman bears more children than she has a right to under the rules set up by society.

A democratic society forced to employ mutual coercion to achieve zero population growth will probably assign everyone the right to have exactly two children. Because of the fact that some individuals will have only one child or will not reproduce at all it will be necessary to assign the births needed to achieve replacement level to other individuals in that population. The assignment of these births could be made at random via a national lottery system. The result would probably be genetic deterioration. While those individuals who have less than two children would constitute a sample of the population with above average frequencies of various genetic defects the selective removal of their genes would probably not be sufficient to counterbalance the continual input of mutations. Thus the result would probably be genetic deterioration even if the environment remained constant. If the environment were changing (this is about the one thing we can count on—a constantly changing environment) the population would become even more genetically ill-adapted because those individuals in the society that are best adapted to changing environments and to the new environments would not be passing more genes on to the next generation on a per person basis than those individuals less well adapted.

What kinds of eugenics programs could be designed for a democratic society where mutual coercion is institutionalized to ensure that couples who would otherwise be irresponsible are restricted to having 2 children?

One compulsory population control program designed to operate in a democratic society that has eugenic implications is the granting of marketable licenses to have children to women in whatever number necessary to ensure replacement of the population (say 2.2 children per couple). The certificate unit might be the deci child, or one-tenth of a child, and the accumulation of ten of these units by purchase, inheritance, or gift would permit a woman in maturity to have one child. If equality of opportunity were the norm in such a society those individuals with genetic make-ups that enable them to succeed (high intelligence, personality, etc.) would be successful in reaching the upper echelons of society and would be in the position of being able to purchase certificates from the individuals who were less successful because of their genetic limitations. The marketable baby license approach to compulsory population control, first discussed by Kenneth Boulding (1964) in his book *The Meaning of the Twentieth Century*, relies on the environment, especially the sociocultural environment, to do the selecting automatically based on economics. The marketable baby license approach would probably bring about a better genetic adaptation between a population and its environment. Remember, the direction and rate of genetic change is, to a great extent, a function of the social structure of the human population. The marketable baby license approach ensures that those people selected for in society are those people who are most successful economically. To ensure genetic improvement society would have to make sure that achievement and financial reward are much more highly correlated than they are at the present.

Another compulsory population program that a democracy might adopt would be to grant each individual the right to have two children and to assign the childbearing rights of those individuals unable or unwilling to have two children to other individuals based on their performance in one or more contests (competition involving mental ability, personality, sports, music, arts, literature, business, etc.). The number of births assigned to the winners of various contests would be equal to the deficit of births created by individuals having less than two children. Society would then determine to a great extent the direction of its future genetic (and cultural) evolution by determining the types of contests that would be employed and what proportion of the winners (the top 1% or 5%) would be rewarded with the right to have an additional child above the two children granted to all members of society.

A society might even go further and employ a simple eugenic test—the examination of the first two children in order to assure that neither one was physically or mentally below average—which a couple must pass before being eligible to have additional children (Glass, 1967). The assignment of additional births to those individuals who passed the eugenic tests then could be on the basis of a lottery, marketable baby licenses, or contests, with the number of licenses equaling the deficit of births created by individuals who, at the end of their reproductive years (or at time of death if they died before reaching the end of their reproductive years) did not have any children or who only had one child.

The programs designed to bring about an eugenic distribution of births that have been discussed so far may prove to be incapable of doing much more than counteracting the input of harmful mutations. In order to significantly reduce the proportion of the human population that is genetically handicapped, a society may have to require that each couple pass certain eugenic tests before being allowed to become the genetic parents of *any* children. If one or both of the prospective genetic parents fail the eugenic tests the couple could still be allowed to have children via artificial insemination and/or artificial inovulation utilizing human sperm and eggs selected on the basis of genetic quality. Such an approach would enable society to maintain the right of couples to have at least two children while improving the genetic birthright of future generations at the same time.

Successful control of the size and/or genetic quality of human populations by society may require restrictions on the right of individual human beings to reproduce. The right of individuals to have as many children as they desire must be considered in relation to the right of individuals yet unborn to be free from genetic handicaps and to be able to live in a high-quality environment. The short-term gain in individual freedom attained in a society that grants everyone the right to reproduce and to have as many children as they want can be more than offset by the long-term loss in individual freedom by individuals who as a consequence, will be born genetically handicapped and/or forced to live in an environment that has deteriorated due to the pressure of human numbers.

Conclusion

Each generation of mankind faces anew the awesome responsibility of making decisions which will affect the quantity and genetic quality of the next generation. A society, if it takes is responsibility to future generations seriously, will take steps to ensure that individuals yet unborn will have the best possible genetic and cultural heritage to enable them to meet the challenges of the environment and to take advantage of the opportunities for self-fulfillment present in that society.

The way in which a society is organized will determine, to a great extent, the direction and intensity of natural selection especially with respect to behavioral patterns. The genetic make-up of future generations is also a function of the size of the population and how population size is regulated by society. The genetic implications of the following three basic types of population policies were explored in this paper:

1. Policies favoring continued population growth;
2. Policies aimed at achieving zero population growth by voluntary means; and
3. Policies aimed at achieving zero population growth by compulsory measures (mutual coercion mutually agreed upon in a democratic society).

If societies adopt compulsory population control measures it will be for the control of population size and not for the control of the genetic make-up of the population. However, it is but a short step to compulsory control of genetic quality once compulsory programs aimed at controlling population size have been adopted. I personally hope that mankind will be able to solve both the quantitative and qualitative problems of population by voluntary means. Yet one must be realistic and consider the alternatives. This is what I have attempted to do in this paper by reviewing the genetic implications of various population control programs.

Literature Cited

Bajema, C. 1963. Estimation of the Direction and Intensity of Natural Selection in Relation to Intelligence by Means of the Intrinsic Rate of Natural Increase. *Eugenics Quarterly*, 10: 175-187.

Bajema, C. 1966. Relation of Fertility to Educational Attainment in a Kalamazoo Public School Population: A Follow-up Study. *Eugenics Quarterly*, 13: 306-315.

Bajema, C. 1968. Relation of Fertility to Occupational Status, IQ, Educational Attainment and Size of Family of Origin: A Follow-up Study of a Male Kalamazoo Public School Population. *Eugenics Quarterly*, 15: 198-203.

Berelson, B. 1969. Beyond Family Planning. *Science*, 163:533-543.

Boulding, K. 1964. *The Meaning of The Twentieth Century: The Great Transition*. Harper & Row, N.Y.

Darwin, C. 1958. *The Problems of World Population*. Cambridge Univ. Press, Cambridge. 42pp.

Dubos, R. 1965. *Man Adapting*. Yale Univ. Press, New Haven.

Duncan, O. 1965. Farm Background and Differential Fertility. *Demography*, 2:240-249.

Freedman, R., and Slesinger, D. 1961. Fertility Differentials for the Indigenous Non-Farm Population of the United States. *Population Studies*, 15:161-173.

Glass, B. 1967. *What Man Can Be*. Paper presented at the American Association of School Administrators Convention, Atlantic City, N.J. 23pp.

Goldberg, D. 1959. The Fertility of Two-Generation Urbanites. *Population Studies*, 12:214-222.

Goldberg, D. 1960. Another Look at the Indianapolis Fertility Data. *Milbank Fund Quarterly*, 38:23-36.

Goldberg, D. 1965. Fertility and Fertility Differentials: Some Observations on Recent Changes in the United States. Pp. 119-142. Sheps, M. and Ridley, J. (eds.) 1965. *Public Health and Population Change*. Univ. of Pittsburgh Press, Pitsburgh. 557pp.

Hardin, G. 1968. The Tragedy of the Commons. *Science*, 162:1243-1248.

Hulse, F. 1961. Warfare, Demography and Genetics. *Eugenics Quarterly*, 8:185-197.

Kirk, D. 1969. The Genetic Implications of Family Planning. *Journal of Medical Education*, 44: Suppl 2:80-83.

Miles, R. 1970 . Whose Baby is the Population Problem? *Population Bulletin*, 16: 3-36.

Waller, J. 1969. The Relationship of Fertility, Generation Length, and Social Mobility to Intelligence Test Scores, Socioeconomic Status and Educational Attainment. Doctoral Thesis, University of Minnesota, Minneapolis. 100pp.

24. A Return to the Principles of Natural Selection

FREDERICK OSBORN

Editor's Note: It is generally agreed that much further research is needed if eugenics is to make progress. But it is difficult to agree on what types of research are needed so long as there is no theoretical framework covering the operation of eugenic selection and so long as we are without objective criteria for measuring the types of people being selected. The following article is of immediate interest because it proposes answers to these questions. The Quarterly would like comments from its readers on the practical soundness of these proposals compared with other suggestions.

The aim of eugenics should be to restore the principles—but not all the methods—of natural selection as it operated for 500,000 years of prehuman and human evolution.

Natural Selection—Selection for an Environment

During the long period of prehistory the measure of selection was ability to cope successfully with the environment. Selection was related directly to achievement, namely, to survival. The area of selection included both the physical and the social environments. When prehuman primates first began to use tools, the area of physical selection was progressively reduced and that of social selection increased. The first tools may have been only small stones in their natural shapes, but they gave their users a new weapon in their struggle for survival. It took intelligence to use tools, even stones. When natural selection began to favor those who among other things used tools most effectively, intelligence became the critical factor in the selective process. Intelligence could invent new tools, develop new uses for them, and so increase its selection advantage. At the same time, prehuman man began to develop responsible social relationships based on kinship. Success in the social group,

SOURCE. Frederick Osborn, "A Return to the Principles of Natural Selection," *Eugenics Quarterly*, 7: 204-211 (Chicago, Illinois: The University of Chicago Press, 1960).

the care of others, became a factor in survival. "The structure of modern man must be the result of the change in the terms of natural selection that came with the tool using way of life.—The reason that the human brain makes the human way of life possible is that it is the result of that way of life."[1]

The Decline of Natural Selection

The conditions in which natural selection operated were altered when man gave up hunting and gathering in favor of agriculture and the domestication of animals. Since food was distributed among large groups of people, individual abilities did not count so much toward survival in times of drought and famine. At the same time, during long periods when food was relatively abundant, ties of kinship made people care for the weak and inefficient. Selection was becoming less related to achievement. With the coming of civilization the relation between selection and achievement declined and probably continued to decline during the 6,000 years of recorded history.

The real breakdown in the relation of natural selection to achievement, or "success," in an environment has come in the last hundred years. Until the middle of the nineteenth century, almost half of all infants born died before they reached the age of reproduction. Selection by deaths still had a chance to operate. At the same time married women were having children about up to the biologic maximum, and there could have been little selection by births except for biological fitness. Then two things happened almost simultaneously in the more technically advanced countries. The death rate began going down, and the introduction of birth control began slowly to reduce the birth rate. The reduction in deaths caused an immediate increase in the growth of populations. The first result of the reduction in births was greatly to increase differential fertility between social, economic and educational groups. The use of birth control began among the more educated people and was rapidly extended among them. It spread much more slowly among the less educated. At the period of extreme differences, which in the United States came during the great depression, large groups of the more educated couples were having only half the birth rate of the less educated. If we assume, as we have some ground for doing, that the more educated couples had on the average a superiority, however slight, in the genetic capacity for developing intelligence, then, as between these groups and for this period of time, natural selection had gone into reverse.

This situation was in part temporary. The great post-war baby boom was largely a phenomenon of the white-collar classes. Their birth rate almost doubled between 1940 and 1960 and accounted for most of the rise in births in that period.[2] The

---

[1]Sherwood L. Washburn. Tools and Human Evolution, *Scientific American*, Sept. 1960.

[2]Calculated from Graybill, W. H., Kim, C. V., Whelpton, P. K., "Fertility of American Women," Wiley & Son, 1958. Tables 48-54. Cum. Series P20, No. 84, Table 5.

proportion of families of five or more children continued its long decline. Group differentials lost much of their significance. Within the major groups in our society we do not know what kind of people are having the most children or in what direction we are moving. There is evidence for genetic gains but also for genetic losses.

Meantime deaths have continued their rapid decline. In the United States today, over 95% of all children born alive live to reach their thirtieth year, which is the middle of their reproductive period. At such a low death rate, the chief selective effect of death is to keep down those genetic anomalies which result in early deaths. Selection for traits of personality and intelligence can no longer operate through selective deaths.

Thus there is no evidence now available which indicates that on balance, a man's genetic structure is changing in a direction which will make him more able to cope with his increasingly complex environment. It seems that natural selection for man's higher traits has for all practical purposes ceased to operate. The most important long-term job facing man today is to get natural selection back to selection for success or achievement in the human environment.

A Return to Natural Selection

In the past, natural selection operated mainly through deaths. Obviously we are not going to return to that method if we can help it. Our alternative is to enlarge the part played by birth selection. Voluntary birth control gives us a special opportunity to switch from death selection to birth selection.

It seems pretty certain that present methods of birth control, which are difficult to use and not always effective, will soon be superseded by physiological means of control which will prevent conception over considerable periods of time by means of a pill, a vaccine, or an immunization. The prospect of such a development is sufficiently good to justify our considering birth selection in terms of conditions of entirely voluntary parenthood, when parents will have children only after passing a deliberate judgment that they want them. We can get some idea of the effect of such a development by analysis of studies of couples using present methods of birth control.

Recent studies have been made on large-scale samples of the U.S. population, in which the couples using contraception most effectively have been studied separately from the others. For such couples there are indications of a trend toward size of family being related directly to income, that is, couples with incomes of six thousand dollars may, under certain conditions, be having more children than couples with incomes of five thousand dollars, and so on.[3] Indications of a similar trend

[3]Freedman, R., Campbell, A. A., Whelpton, P. K., "Family Planning, Sterility and Population Growth," McGraw-Hill, 1959, p. 293.

have appeared in a number of studies in this country and abroad. Thus, under conditions in which birth control is used and is effective, pressures of an economic sort have a direct effect on the distribution of births. It seems probable that if, with new methods of fertility control, parenthood became even more a matter of personal decision, similar tendencies would provide opportunities for the development of a whole set of pressures not only economic but also social and psychological, which would tend to a selection of births favorable to genetic improvement. "The response would be voluntary, a selection drawn out by the environment itself, so that, in so far as children tend to resemble their parents for either hereditary or environmental reasons, there would be in each generation a proportionate increase in the number of people best fitted to the environment."[4]

It is not likely that voluntary parenthood alone would be sufficient to produce a population best fitted to succeed in its social environment. But so long as birth control offers some tendency along selective lines, it should be easier to develop measures which would strengthen and confirm the tendency. Such measures should be developed from the knowledge obtained in studies of factors affecting family size among people of different kinds and different environments. This is a proper field for sociological study. To be carried out effectively, we require criteria for those qualities which are to be selected for survival.

Criteria for Natural Selection

The criteria must be of a sort which will define fitness in terms of success in the various social environments men live in today. For this we know nothing better than to accept the forms of recognition given by society itself. Every society offers its members rewards for the work they do. In our society the rewards vary in kind, including wages paid and prestige elements, such as Federal judgeships and inclusion in "Who's Who," a higher education, and many other evidences of achievement. But for the great majority of people the rewards are mostly in money, and this criterion will be necessary in measuring the larger groups. Money rewards can be scaled, which is an advantage to the investigator. The recognitions given by society provide a generally accepted value judgment of the individual's achievement of "success" in his environment. If the judgments are "wrong" the fault lies with society, and we should try to change the society.

It will be immediately argued that even in the best society many rewards go to the wrong people for the wrong services. But so was the reward of death in many instances in the past. All we can hope for is to find measures which most nearly, and for the greatest number of instances, define qualities valuable to society. The measures now proposed are probably more accurate in more instances than were

[4]Osborn, Frederick, "Preface to Eugenics." 2nd ed. Harper's, 1951, p. 240.

the measures used in the old processes of natural selection where pure chance played so large a part. Considerable elasticity is permissible in using the rewards given by society as a measure of value. We do not include rewards to those whose work is clearly anti-social, or those who have taken rewards in defiance of the law, as gamblers, racketeers, etc. Nor need we be too concerned that we are eliminating from procreation some valuable elements now not recognized by society. No man-made alterations in natural selection are going to work so well that man will thenceforth breed to a particular type or types and none other. All we can hope for, and indeed all that is desired, is that by and large, in general, man's genetic structure will be changing in the direction of making him more able to cope with his environment. Since man must operate in a great number and variety of environments, he would by natural selection be breeding not to one type but to as many different types as there are different environments and different types of human activities.

The proposed criteria would provide tools for measurement which do not involve value judgments on the part of those who are studying and making proposals. The factors to be studied would range from existing social institutions as they affect birth rates, to psychological pressures which may affect parental decisions. Most of the institutions needing study are already under pressure for change by people who want to improve social conditions.

A few of such situations may be listed as examples.

Social Sanctions Affecting the Distribution of Births

Tax Structures. There have been a number of studies recently on the attitudes of parents in regard to desired size of family. These all tend to confirm the impression that in the great middle-income group of families which includes the professional classes, the cost of rearing children and particularly the cost of higher education is a limiting factor in decisions as to size of family. Present income-tax load is particularly high for the parents of children in this group. Deductions for children have been made larger in recent years but are not nearly large enough to offset the cost of a college education. Most professional people spend an unusually long time in preparation before they begin to have earning power, and they marry late. But they cannot regain the lost ground. After paying their income tax, there is no sufficient margin for the later education of their children. These people are among the most useful and most needed in our society, and it has long been felt that changes should be made in the income-tax laws for their benefit as a matter of social justice and to encourage more people to enter these fields. An equally important reason for such changes is the effect of the present law on the distribution of births. Tax laws which put a disproportionate burden on couples in the middle-income groups when they have children tend toward an inverse relation between selection and achievement.

Illegitimacy. There are about 200,000 illegitimate children born in the United States each year.[5] Few people would affirm that bearing illegitimate children is in itself a mark of "fitness" in our society. Yet, apart from a diminishing social disapproval, society does little to discourage such parents in their ill-considered propagation. Society makes varying degrees of effort to care for the children, and this is right and proper. But these efforts should not relieve the parents of their responsibility, as is now too often the case. Our laws should be greatly strengthened, to seek out the father and require that he bear part of the expense and to hold the mother to her responsibilities to the child. Such changes are widely urged on the grounds that society needs them for its protection. Changes are also desirable for their effect on the distribution of births. A high frequency of illegitimacy is also a force tending to an inverse relationship between selection and an effective response to the environment.

Marriage Laws and Divorce. It is quite generally agreed that our marriage laws should be drawn to prevent, so far as possible, the marriage of people who are not prepared for the responsibilities of having children. Most states do not permit marriages at ages considered too young to undertake the responsibility of rearing children. But the state laws are not uniform in this matter. Few states proscribe marriage by individuals who are too feeble minded to give their children adequate care, or by alcoholics, and others whose children will in turn become social problems. In most states divorce laws do not sufficiently hold parents seeking divorce to their responsibility to their children. Because state divorce laws are not uniform and requirements for residence easy or easily evaded, most people think that divorce will be easy. But easy divorce tends to encourage careless marriage, and marriage, careless or not, promotes childbearing. Marriage and divorce laws should be drawn with more consideration for their effect on the distribution of births.

There are many other economic and legal conditions which could be changed to the immediate benefit of society, and such changes would at the same time bring us closer to the principles of natural selection for achievement.

Further studies may be expected to develop valuable information in the field of psychological pressures which affect couples in their decisions as to size of family. It is already known that couples are influenced by the opinions and attitudes of those in their immediate surroundings. They want to follow the fashion in regard to size of family, and the fashion is usually quite undiscriminating. During the depression, in the white-collar groups, strong social pressures were exerted to maintain the "fashion" of small families. Just now, in the same groups, the "fashion" is for large families. It should be possible to develop a fashion which is discriminating and encourages children among people who are making a recognized contribution to

[5]U. S. Dept. of Health, Education & Welfare, National Office of Vital Statistics, Vital Statistics of U. S. 1958, "General Characteristics of Live Births," Sect. 12, Table 38, p. 264, 1960.

society and who could be counted on to give their children responsible care, while it discourages children among couples less suited to parenthood. Psychological pressures of this sort are becoming increasingly important with the increasing spread and effectiveness of means of fertility control.

How Fast Might Change Be Effected?

Such proposals as those which we have indicated, and those we believe can be developed through further study, would change present trends. We would be coming back toward the principles of natural selection for an environment by which man became continually better fitted to handle an increasingly complex environment. It is time that this system of continuing genetic change be re-established. We seem to have developed a society so complex that it is beyond the ability of any but the most highly endowed of our people to handle it. We are only fooling ourselves if we think that we can sufficiently increase, by educational means alone, from the ranks of the less well endowed, the proportion of really able people we presently need.

Our hope lies in the fact that we do not have to wait for nature to develop a new and different type of man. We have only to increase the proportion we now have of people whose genetic endowment fits them to achieve the highest development. Scientific opinions differ as to how rapidly this could be done. Geneticists do not know how many genes are involved in the development of man's mental and emotional machinery. They do not know the method of their interaction, nor in which chromosomes they are located. Nor is it known whether the genes are separately and widely distributed throughout the population or are considerably concentrated in particular family lines. Lacking such knowledge, while geneticists do not reject proposals of the sort being considered, they cannot at present endorse them. There are, of course, exceptions. H. J. Muller, Nobel prize winner in genetics, supports such proposals and goes considerably further. He urges insemination with sperm from people generally recognized as having made important contributions to society. He suggests that the sperm might be frozen and used after the death of the donor so as to avoid the psychological difficulties arising when the donor is alive at the time of insemination. He would be using the soundest criteria of a person's value, namely, the judgment of society after death. It thus embodies to the highest degree the criteria of selection for achievement in an environment. But it is opposed to our present mores and may take some time for acceptance.

Paleontologists, with their long view of man's evolution, may find ground for approving the type of selection here proposed, but they are not in a position to tell us how rapidly it could begin to increase the proportions of able people. We might hope to find the best answers in analogies from the breeding of domestic animals, or from changes in gene frequencies in wild populations under the influence of changes in the environment. But animal breeders use methods incompatible with man's other needs, and this would invalidate the analogy. In the case of wild populations, the analogy might be fairly close if it were not for an important difference in the breeding habits of man compared to wild populations such as the fruit fly *Drosophila*. There is no evidence of selective breeding among fruit flies. They mate by chance, and geneticists call this system pan-mixia. But humans exercise considerable selection, and this selective or assortative mating tends to concentrate groups of genes in specific family lines. For a number of reasons which need not be considered here, there is probably a particularly high rate of assortative mating in the United States. In New England the able stocks have been inter-marrying since their arrival in this country; among the newer immigrants, the abler individuals have moved rapidly up into the industrial-management and professional classes, while the less able individuals have found mates in other groups. Even more significant is the probability that within, as well as between, each of our social and economic classes there has been a continuous process of assortative mating with a considerable sorting out of able strains into specific family lines. By this process the proportionate distribution of favorable genes would be somewhat heavier among the smaller well-educated groups, compared with a somewhat thinner distrubiton among the larger less well-educated groups, but the greater aggregate of genes for ability would still be found in the larger though less well-educated group. These assumptions appear to get some confirmation from the studies made by psychologists on the distribution of capacity for intelligence among the various socio-economic classes in this country.[6] Such findings are important for they indicate that any process of selection, to be effective, must be all-inclusive, operating within every social-economic and educational group, and including the whole population.

To the extent that specific types of genes tend to concentrate in specific family lines within every social class, a system of selection which was only moderately effective might still make a considerable increase in the proportion of able people within one or two generations. Able people, like all other people, develop out of conditions in which heredity and environment are so intertwined that their "relative" effects cannot be sorted out. The people who have responded most successfully to their environment have by and large the best genetic capacity for responding and live in home environments best fitted to bring out such a response. Their children, therefore, have a double reason, both genetic and environmental, for achieving the ability to respond successfully to the environment. This reasoning is supported by the extensive studies of Terman[7] and many others in the field. There are thus strong grounds for the belief that any system of birth selection which

[6]Anastasi, Anne, Differentiating Effect of Intelligence and Social Status, *Eugenics Quarterly*, Vol. 6, No. 2, June 1959, pp. 84-91.

[7]Terman, L. M., Genetic Studies of Genius. Vol. V, "The Gifted Child Grows Up," Stanford University Press, 1947.

improved genetic capacity would at the same time favor a better environment for the rearing of children. For this reason a return to the principle of selection for an environment should be favored by "environmentalists" as well as by "hereditarians."

Such principles of selection do not carry connotations of "superiority" or "inferiority" except with respect to relative success or failure in specific environments. A man might fail in business when he might have made a great success as a musician, and so on through all the categories of different kinds of service to society. The success of a system of selection using the criteria of success or achievement in an environment, would depend a great deal on people finding the right niche for their particular capacities. In our society this is increasingly possible, and desirable for many reasons. The kind of proposals here suggested for establishing principles of selection conductive to the improvement of the genetic endowment of man are not of the sort to arouse opposition. Hereditarians, equalitarians and environmentalists can join hands in their support, along with all those who are interested in the future condition of society. No stage could be better set for their adoption.

Yet today the only important group of men engaged in the problems of man's genetic future are the medical geneticists with practical applications in the field of heredity counseling. This is an important field for it deals with genetic factors which natural selection should certainly tend to reduce. As present the field is necessarily limited to the simpler genetic manifestations such as genetic abnormalities; as knowledge increases it will move on to more complicated formulations and finally reach consideration of the extreme complexities involved in the factors which underlie the higher traits.

It would be a long time to wait for geneticists to unravel the genetic complexities of man's higher traits. And it is not necessary to do so. The assumptions on which we can base the development of eugenic proposals have far more scientific justification than those assumptions which, in other fields, we are using to determine actions of great importance. It is time that social scientists should be undertaking the serious study of the social, economic and psychological factors which affect parents of various kinds and in various environments in their decisions as to size of family. On the basis of such studies we can embark with confidence on a return to those principles of natural selection which accompanied the evolution from ape to man, from the society of animal bands to the complex societies of civilized man.

25. What Genetic Course Will Man Steer?*

H. J. MULLER

Of course we—that is, humanity—will take our biological evolution into our own hands and try to steer its direction, provided that we, humanity, survive our present crises. Have we not eventually utilized, for better or worse, all materials, processes, and powers over which we could gain some mastery? And are there not means already by which we can influence our heredity, and other means that we are likely to gain? Some detractors may call the use of them tampering, tinkering, or even blasphemy. But this attitude is like that of the recalcitrants who still hold out against whatever foods, remedies, or measures they consider "unnatural" (e.g., the Amish who even today consider higher education to be a perversion of the mind). Any of these things can be disastrously misused, or they can be used to great advantage.

It is life's essence to utilize, wherever possible, more and more effective means of servicing itself. That implies doing things in ways that were previously unnatural. Moreover, the rise of man to ascendancy over all other forms of life has resulted from his having been so unusually successful in this very respect. This has been true in both his genetic and his cultural evolution. Indeed, man is the first organism for whom any culture except a trace has become natural. That is, increasingly, the ways and products of culture, the so-called artificial, have been naturalized by man. All this, however, is far from saying that every new or artificial way or product is a sound one. The so-called "primrose path" is readily found, and wider than the so-called "straight and narrow one." That is why there must be a most far-sighted, judicious, and beneficent steering by man in general of both his genetic course and his cultural one. These must in fact be intermeshed and complementary to one another.

SOURCE. Hemann J. Muller, "What Genetic Course Will Man Steer?" *Proceedings of the Third International Congress of Human Genetics* (Baltimore, Maryland: The Johns Hopkins Press, 1967), pp. 521-543.

*The writer acknowledges with thanks support in the preparation of this paper afforded by grant number GB 4764 from the National Science Foundation to the Indiana University Foundation in behalf of his work.

Life the Lucky

For our present purposes, we may define genetic advances as the gaining of abilities for making use of the environment more effectively, and for withstanding or even making use of circumstances that earlier would have been useless or hostile. By this measure, the totality of living things—that is, all species taken together—has certainly advanced enormously through the ages. For it has increasingly extended life's domain, increased its resources, and made it more secure. Moreover, certain lines of descent, most notably the one leading to ourselves, have ultimately advanced the most by these criteria, or, as we say, they have become the highest. They, and especially we, are the ones that can overcome the greatest difficulties, and the most adverse ones. And so, even though we certainly have our worries, these are largely of our own making. And we, the self-styled heirs of all the ages, constitute the very luckiest, the most improbably lucky, combination of trials of the whole lot.

The luck that allowed any line to advance genetically was of course based on the Darwinian natural selection of mutant types, and of combinations of them. Since the kind of mutation occurring cannot be influenced by the effect it will have, and since there can be ever so many more ways of harming than of improving any mechanism, vastly more mutations, and combinations of them, proved to be failures rather than successes in their influence on "survival." Here "survival" must be understood to include multiplication, and so it would be better to say "genetic survival," "net multiplication," or simply "fitness." It is the *multiplication* of the successful mutants that plays the key role in evolution, for it alone allows for additional successful steps. To permit room and resources for this multiplication there must of course be correlative reduction in numbers, or extinction of some less successful types, except when the new ones that succeed all go into virgin territory, or somehow make enough extra resources available for the others too.

Thus it is clear that "natural selection," or "the survival of the fittest," really means the *differentially* high net-multiplication, in some environment, of certain rare mutant types which we call "the fitter" or "fittest." This process has been made possible, even inevitable, by the gene-material's unique properties of replicating, as such, even its mutant forms, and accumulating them to an unlimited degree within one chromosome-set. Moreover, despite the rarity of serviceable mutations and the minuteness of the effects that individual serviceable mutations usually produce, the speed of evolution has been greater than these circumstances would seem to allow, for the getting together of the mutant genes has been greatly aided by sexual recombination. The eons of time over which they have accumulated has allowed for formation of such marvelous organizations, having such consummate integration, as protoplasm, and eventually man himself.

Our Forebearers' Run of Luck

Let us review briefly some clues to man's concatenation of luck having been so much greater than that of other organisms by focussing upon his ancestors of the last hundred million years, the primates. A long succession of events had already made the mammals the most advanced animals. Of the primates, remains of the most primitive known group, the prosimians, which today includes the lemurs, have been found in strata that also contain remains of dinosaurs. Prosimians must early have gained (if not inherited) such physical advantages for active life in trees as opposable first digits, improved vision, equipment for a somewhat omnivorous diet, and uniparity. But they were soon pushed into the background and thus hampered in advancing further by their more successful offshoots, the simians, that is, monkeys and apes. Thus, they failed to gain the simians' greater maneuverability, curiosity, and general intelligence.

However, Joyce's recent studies in the field show that lemurs do have a highly developed maternal solicitude—as demanded of animals that must take constant heed and care of their single young. There is a similarly high level of intra-group cooperation at all ages, which is a kind of extension of maternal love and empathy to companions in a permanent group whose members include individuals of both sexes. There is also play between adults. And so, what might be called social intelligence was enabled to flourish. It is important to note that this faculty probably preceded the kinds of intelligence concerned with inanimate objects and with other types of organisms. Moreover, there goes with this intelligence much learning, from elders and from playmates, of what behavior to adopt in given situations, what foods to seek, what to reject, and so on. The receptiveness to the attitudes and acts of others affords an important part of the basis needed for our own cultural evolution.

The bodily and psychological advances made by monkeys and apes gave a further basis for the advances afterward made by the apes' protohuman offshoot, which split off from the other apes some twenty million years ago. However, they have been too much discussed to be closely reviewed here. So have the factors that probably led to these traits having been favored by natural selection. Suffice it here to call attention, on the so-called physical side, to the constant view forward, with its opening up of wider opportunities, permitted by the apes' arm-mobility and consequent arm-swung mode of progression and, derived from the latter, their semi-erect posture even on the ground.

These traits put even more of a premium on broad awareness, versatility, and love of variety, hence too on curiosity concerning objects—both inanimate and animate—and general intelligence. The latter includes a higher ability to transfer lessons learned in a given field to another one, and to solve problems. This in turn allowed, at least in the chimpanzee, the making of very simple tools and some hunting of game.

Meanwhile, social intelligence, affection between companions, and cooperation had also developed. For the little groups had continued to be fairly permanent, and to include individuals of all ages, and could therefore profit by emphasis on these social traits. Moreover, the division into many small social groups must have promoted natural selection for the genetic bases of social intelligence, and of social traits in general. This is because genes that tend to extend maternal and brotherly feelings to other members of the closely related little group result also in mutual aid. By thus helping the group's survival, these genes actually foster their own survival even when they lead to self-sacrifice, since others of the tiny band tend to have the same genes. By the greater growth, followed by the resplitting of the more social little groups, the genetic groundwork of cooperation was increasingly strengthened in the species.

Man Improves on Luck

In these ways, the genetic structure must have been laid down for a line of descent which, branching off from that of other apes some twenty million years ago, could by virtue of both its bodily and mental traits get along increasingly well on the ground by defending itself better from predators and in various other ways. By some two million years ago, its members had already become fully erect and much like ourselves in form, except for their little more than ape-size brain and rather large jaws. Since their lairs contain abundant broken bones of fair-size game, as well as rough-hewn tools, they must not only have evolved much more initiative, including aggressiveness, than apes but also, and most important, they must already have accumulated a substantial amount of extra-genically transmitted experience. In other words, cultural evolution, a process so nearly unique in the human line, had begun in earnest.

Like the evolution of the genetic constitution, that of culture requires the arising, the transmitting, and the selecting of innovations. But since the cultural innovations are in thought and behavior, their transmission is by some form of imitation, not heredity, even though genes must afford the abilities for these processes. Of course this form of transmission allows a much more rapid spreading than that through differential multiplication. But in the earlier stages the acceptable innovations, like acceptable mutations, tended to be rare and apparently insignificant, for they arose in a rather haphazard way, in which foresight was extremely limited. Therefore, like mutations, they had to be selected after their origination, according to their helpfulness to the individual and his group.

But as culture very slowly accumulated by these means, the rate of development gradually increased. This was not only because culture itself affords means of producing more culture. It was also because in those times the use of culture afforded more opportunity for the natural selection of the genetic traits which allowed that culture to be utilized in ways serviceable to the particular user and the user's little group. Hence culture must have reacted, via natural selection, to enhance the genetic foundations of cooperativeness, initiative, general intelligence, and such more special faculties of mind and body as facilitated the use or the accumulation of culture itself, or caused adaptations to the conditions arising from culture. Foremost among these special genetic faculties was that of communication, especially speech, which in turn depends on a complex of unusual propensities. All these genetic advances allowed, in their turn, faster cultural evolution.

Thus there was for a very long time a "positive feedback," of reciprocal nature, between evolution of the genetic and the cultural types. So on the genetic side, it is not surprising that during the past two million years brain size (which is but one of many factors in mental ability) underwent a most drastic increase—about a tripling. And, as we all know, cultural evolution had, by some ten or twelve thousand years ago in some regions, reached such a rate that it resulted in the successive breakthroughs (so close together in terms of evolutionary time) of food-growing, then town-dwelling, and now our modern sciences and technologies.

During the course of these developments human foresight as well as hindsight became enhanced. Hence the initiation of cultural innovations gradually, and with the scientific technological breakthrough very rapidly, became less haphazard, unlike that of mutations. They could increasingly be preselected to advantage, more reliably and rapidly post-tested, and their transmission became faster and more diverse. Larger steps then became more feasible, and even necessary, and they were and are being taken. Imagination and foresight are supplanting luck through trial and error, although they can never do so completely.

Meanwhile, with the improved techniques of production and transportation of even the prescientific stages of urban life, the groups of people, originally so tiny, grew in density and area and merged increasingly, or were forced to do so. Thus great states emerged in diverse regions, and religions sprang up that emphasized the brotherhood of all men. Today, of course, it has become both possible and necessary, if civilization is not to founder, for all groups to federate together, within a surprisingly few scores of years, into one great community, all of whose people share comparably in the fruits of modern technology, in the modern scientific world-view, and in human dignity and opportunity. The convergence of attitudes and cultures has recently been admirably set forth in a statement, *Education and the Spirit of Science*, issued by the Educational Policies Commission of the National Education Association.

In this community, no place can be left for biases against races or social classes— *or else!* Hence, racial amalgamation will gradually and voluntarily but inevitably ensue. Hawaii demonstrates how peaceable, how successful, and how attractive in results this process can be. True, there are today conspicuous differences between the major races, but these are adaptations to conditions long different in their respective areas, and modern techniques give ready protection against these regional environmental difficulties. As for generalized genetic advances, although the details call for much more factual study, it is evident that the gene leakage between areas would have allowed those migrant genes which presented universal advantages (even if they had had different times of entry and possibly somewhat different ratios of selection after that) to have become selected virtually everywhere.

Similarly, social origins must not be regarded as valid clues to genetic level for classes with different status are all highly heterogeneous genetically. In the case of both ethnic and social or economic classes, however, the cultural differences are often so great as to give an utterly false appearance of genetic differences. At the same time, it must not be denied that the extent and importance of genetic differences between *individuals within any group* are often enormous though, here too, any given person will have been shaped by environmental differences to a degree which would elude reliable assessment today.

It is generally conceded that the advances of science and technology already carry the physical potential of bringing dignity, affluence, health, enlightenment, and brotherhood within the reach of all. It is also conceded that, because of the dearth of really integrative and cooperative thinking and the inertia of old ways, these very advances are misused to cause the desperate crises of fast mounting population, massive depletion of resources, mass pollution, maldistribution, mass want that knows it need not exist, inflexible privilege, mass miseducation along outgrown lines, mass deception, frenzied fanaticism, mass coercion, the threat or actuality of mass slaughter and the destruction of civilization.

Thus, the changes in social conditions depicted in the earlier paragraphs constitute, all taken together, no more than a now-forseeable larger cultural step forward which, as informed realists everywhere are aware, has become mandatory for the survival of civilization. It can bring no utopia—there will never be such a stasis, it is to be hoped!—but it will herald, in a sense, only a beginning of progress on a somewhat less insecure basis.

Man Undermines Himself

In considering these matters, we have not really strayed from our original subject. For, as we have seen, our genetics and our culture are inextricably interrelated. Only after such a general review could one informedly consider man's genetic future.

It might here be objected: "Why be concerned about our genetic future at all since, if civilization does continue, our scientific culture is likely to advance so fast as to much more than make up for genetic shortcomings—especially those affecting mental traits, in view of the enormous plasticity of man's mind? The means will be found," it is declared, "of suitably affecting gene action, immediate and remote, through methods of DNA and RNA repression and stimulation and other influences on the phenotype, or so-called euphenics, which should of course be considered as including ontogenetic influences, medicine, prosthetics, education, and applied sociology." In answer, I for one would thoroughly agree that efforts of these kinds ought to be actively pursued, and that some are bound to be highly rewarding both theoretically and practically. And certainly our minds do have immense hardly-tapped reserves that could be made far more available just by the more suitable organization of both our early and our later experiences. The most immediately promising of these fields is of course educational reform.

Nevertheless, it would be utterly unrealistic to ignore the genetic side, in relation to either body or mind. If in this argument the antigeneticists who pose as supergeneticists were right, they would have to admit that their future methods could, by considerable expenditure at least, also convert apes, or, for that matter, any forms much less advanced than apes, into the equivalent of the most advanced people. Moreover, if genetic defects and shortcomings were to be allowed to accumulate to an unlimited extent among us, as seems to be happening now, the condition would eventually be reached in which each person likewise would present an immense, yet in his case distinctive, complex of problems of diagnosis and treatment.

But why then start with organisms? As I have pointed out before, the designing, once and for all, and the manufacturing of robots of choice from inorganic materials should prove much simpler. It is a possibility not to be scoffed at in this context, especially if people who have achieved such advanced techniques still insist on the *mystique* that the younger generation in each human family must be physical continuations of the older ones and must carry, if not their defects, at any rate a random sample of their oddities.

Just as natural mutations had to be stringently sifted by natural selection if a population were to advance or even not to deteriorate, so, in species divided into many small groups the mutational combinations in each had to be sifted, by a longer-range natural selection, in the interests of the species as a whole. And again, genera with only one species had, other things being equal, less chance of surviving than did multispecific ones, since any single species is so likely to prove, in the still longer run, to have been a natural error. This is shown by the fact that such a minute per cent of species of the past have turned out to represent lines that persisted. In accord with this principle is the finding that the category with the highest per cent of survivals has been that of phyla, and that successively narrower categories have had a correspondingly decreasing survival rate.

All this might have been expected, since some evolutionary trends are in directions that are later blocked by changes in other species or in the physical environment. In addition, as both Haldane and I long ago pointed out independently, some individuals which are technically fitter, in that they multiply faster than the rest of the population, do so at the expense of the rest and thus sap that population as a whole. Here, a division of the species into many small groups would tend to save it by weeding such groups out before the trait spread.

In the case of man, it has been intrinsically dangerous for him to have so long existed as just one species. He has been saved not only by his unparalleled advantages but also by having until recently been divided into thousands of tiny bands of at most a few score members each. In fact, as we have seen, this condition was especially favorable for the genetic enhancement of cooperative traits—including, I might add, those promoting group initiative or even—to use a harsher word—aggression. Until some two hundred generations ago the population pattern remained like this over by far the largest portion of the area inhabited by man. However, the agricultural revolution resulted in larger, denser, fewer groups, and the urban revolution greatly intensified this trend, thus practically preventing further genetic advances based on intergroup competition and even, in all probability, threatening the maintenance of those previously gained. This must be all the more true in the world of today and tomorrow.

At the same time, intragroup natural selection, working via families and via individuals, is also counteracted, as much as our improving techniques can, by saving everyone whom they can for survival and for reproduction. They have already become highly effective in this job. This means that mutations having a net detrimental effect on body or mind may now be accumulating almost as fast as they arise. We can escape the inference that such mutations far outweigh any advantageous ones only by believing that mutations are designed by Providence for a species' direct benefit, but in the case we run contrary to the clear experimental results.

It has, however, been suggested that even without the "survival of the fittest" that operated on man in past times there are features of modern conditions which cause enough differential reproduction, not consciously directed, to maintain or possibly advance the genetic constitution. But the data offered as evidence, which deal with mental ability (as gauged by IQ) or psychopathic deviation, are unsatisfavorily meager in just those parts of their range—the extremes—where they seem to indicate such an effect. Moreover, out-of-wedlock offspring are not mentioned, and they might have far more than offset any result of the kind supposed.

Although larger amounts of more detailed data on these matters should certainly be sought, as by the kinds of statistics concerning whole populations urged by Newcombe, whatever trends may be found are likely to become outdated rather soon. For techniques, social conditions, and attitudes throughout the world are changing so complexly and rapidly as to give such trends a seemingly erratic course. More trustworthy conclusions regarding long-term trends may therefore be reached on the basis of theoretical considerations rather than of extrapolated statistics.

As we have seen, these considerations show that modern culture by maximal saving of lives and fertility, unaccompanied by a conscious planning which takes the genetic effects of this policy into account, must protect mutations detrimental to bodily vigor, intelligence, or social predispositions. Hence it must allow more accumulation of detrimentals in populations than would otherwise be the case. It appears wishful thinking to suppose that there is in our type of culture a built-in selective mechanism, not designed by us intentionally, which acts over a long period so as adequately to replace the earlier positive feedback whereby the genetic constitution was advanced.

Yet degeneration by passive accumulation of mutant genes is extremely gradual in its manifestation. This is the case even though approximately one individual in five can be reckoned to have received a detrimental mutant gene which arose in one of his parents and is added to the much larger number which they passed on to him from earlier generations. The reason for this creeping pace is that most mutant genes exert such minute effects, at least when the given gene has been received from only one parent. Slow also would probably be the effect of an automatic selection by modern culture in the direction of a lower level of physical or mental fitness. Hence these problems of creeping genetic deterioration are not acute in comparison with the fast-growing menaces presented by our cultural imbalances.

The much more important genetic problem arising out of modern cultural conditions lies in the need for a further *advance* in the genetic level of those psychological endowments which have already attained a height so distinctive of man. These, we have seen, are cooperativeness and general intelligence, including the creativity which arises from initiative working through intelligence.

Why More Intrinsic Cooperativeness is Urgent

Let us consider cooperativeness first. A stronger, broader cooperation is becoming imperative for adjusting to the relatively new conditions of life in large communities, and especially in the hoped-for world community of equal opportunities. Even in the scant two or four hundred generations since the ancestors of most people gave up living in tiny bands there may have been some significant passive accumulation of retrograde genetic changes that adversely affected one's brotherly feelings toward one's more distant associates. Under the dog-eat-dog mores prevalent, unofficially, within some of the larger, later communities there may have been an actual selection downward in such respects.

At any rate, the inadequacy for large communities of the level finally reached is indicated by the repeated and forceful entreaties in behalf of a broader brotherliness put forth by both past and present leaders of nearly every influential religion or ideology. We know of them from widely scattered regions, from soon after the time when diverse peoples there had been brought under a common rule. Yet, though they have fitted-in to some degree with most people's deep feelings, they have never in these millenia sunk-in thoroughly enough to be acted on without great inner resistance and outer friction.

So-called enlightened self-interest is no substitute. It can lead people in communities already having socially oriented practices to conform, outwardly at least, though it alone cannot initiate such community practices. But these same conformists, including those of high intelligence, may on feeling safe from exposure engage in unfair, cut-throat competition, covert fraud, or more extreme criminality.

More modern means of bringing up the young and of influencing the mind will doubtless be much more effective in the development of social consciousness and behavior, and the repression of antagonism. Yet we are far from knowing to what extent this influence would be able to rival or exceed a deep and broad warm-heartedness which was genetically built-in. That is, we cannot now estimate success in evoking from a man genuinely cordial reactions toward outsiders, utilization of his potentials, and affording him as much of a sense of fruition, as from one of those persons, at present so rare, whose genetic constitution has gone far in these respects. Meanwhile, the exigencies of recent culture call on us not to leave a stone unturned that could cause more of the population to be of this predisposition.

That differences in genetic constitution, not only in upbringing, between *individuals* of the same group are highly important in evoking social feelings and behavior has long been objectively evident. This is now reported to be marked even among wolves. Such differences between species, of types which practically lack culture, are enormous, as are those between artifically selected breeds of dogs and of other domestic animals. Of course, any character which shows differences between species or breeds discloses similar differences—some large but with a range of sizes, most of them small—between individuals of a given population. Moreover, Shields' studies of genetically identical twins reared apart, as compared with those reared together and with nonidentical twins or other sibs, demonstrate that in man this principle holds for the other personality traits which he investigated.

But the avoidance of disaster should be far from man's only motivation in seeking a stronger, broader brotherly love. Many of us realize the truth behind the saying "Love is what makes the world go round." Since such feelings and behavior have already been built into our genetic constitutions and built-up in our cultures to a considerable, although not now sufficient, degree, we do appreciate and seek them, even for their own sakes. Built-in and built-up also has been the related gratification at knowing ourselves and our associates to be co-spectators and co-workers forming a part of some much greater phenomenon than that of serving our individual self-interests. And, of course, both our basic feelings and our major cultural tendencies fit in with all trends which, like these, have long been of survival value in our line of ancestry. It is clear that we have in this way been brought to a consciousness that brotherly love will at the same time promote our survival, help to remove the aimlessness and sense of alienation so prevalent today, and afford man deeper inner fulfillment in working for his own vast community.

At the same time, we must maintain if not increase our comparatively high initiative or, in a sense, aggressiveness, but employ it hereafter in the form of independence of judgment and moral courage which, contrary to the ways of the so-called "organization man," braves social disapproval in order to start or support social and moral reforms. This initiative is also essential in our joint war to overcome the difficulties of nature. Moreover, it can certainly be exerted usefully in competition with one's fellows, so long as the competition is honest and constructive. It is evident, however, that its combination with sincere cooperation is indispensable in directing these activities toward the service of society rather than toward mere self-interest. When in combination with intelligence, initiative might better be termed creativity, but the combination of all three is required for the creativity that man needs.

On the Need for Increased Innate Intelligence

As for intelligence, consider how lost most people are today if they try to grapple realistically with out bewildering ideological, social, technical, or scientific problems. In all these areas more background, insight, and integrative ability are fast becoming required. Meanwhile, it is also becoming increasingly important that the politicoeconomic system be such as to seek from everyone his sincere, informed voice in the determination of policies affecting himself and his narrower group. These would of course include general policies too. Otherwise, fanatical individuals and cabals, avid for power, can too easily come to control and misuse mass methods of influencing the mind and of coercion, and by these means precipitate worldwide catastrophe. Yet a person's voice is worse than useless if it is not informed and understanding.

Here again better education, using the term in its broadest sense, can be of enormous help, and future biomedical and biochemical methods might also go a long way. But similar considerations apply here to those discussed in the case of cooperative feelings: the genetic constraints, though stretchable, are very real ones. Within any given social group of our own countrymen, the inter-individual differences in presently measurable features of intelligence are found to be based at least twice as much in genetic as in cultural differences, and to give a range from idiocy to generalized genius. But suppose we might someday exert considerable nongenetic influence to raise the level. Should we now count on this to satisfy our need? Personally, if I had an opportunity to gain greater intelligence or understanding I would take advantage of any and all means of doing so, except, like Faust, had I to sell my soul. Moreover, our species as a whole for a very long time made the same choice, even though unconsciously. Hence our own dominance.

In fact, in consequence of the long-continued genetic selection in that direction, intelligence and probably cooperativeness are traits which would allow artificial selection of their positive extremes without, or with minimal upsets in other respects. This conclusion is verified by the relatively high level of vigor and other valuable attributes which accompany these extremes and by the positive correlations among nearly all these traits.

But in spite of the continuous advance which natural selection has caused in our intelligence during the past two million years, at least as attested by one of its many factors—brain size—the rate of that advance, so amazingly rapid for mammalian evolutionary change, would be pitifully inadequate for us now if we could restore it. It is easy to reckon that, during the entire interval over which this rise in brain size was spread, the increase averaged not quite a tenth of one per cent per thousand years. This was also the percentagewise increase during that part of the interval which lasted from the man who is first known to have used fire—the Peking man of some 400,000 years ago—until the man of today. This means an absolute increase that averaged not quite one gram per thousand years, from Peking man to our modern man whose brain weighs 1,400 or so grams.

There seems no reason why there need be any limit, except that set by our intelligence, to the advances made in science and technology and to the creative powers they would allow us to exert. Nor do we now see any necessary limit to intelligence, although great increases in human intelligence would doubtless require, at times, breakthroughs released by anatomical and/or biochemical innovations in the brain or its accessories. Such innovations (e.g., the corpus callosum) have taken place in past mammalian genetic evolution. In culture, there have been analogous ones (e.g., writing).

The present advances in our techniques and the consequent increases in the complexity of our social organization have, as we have seen, created a situation which demands much more understanding than people at large can muster. Hence they tend, nowadays, to become mere cogs in the mechanism, as well as to be too dependent upon it. It has sometimes been held, however, that this situation is a necessary evil, since there is today no place for many highly intelligent people: they find themselves doing work that is too routine for them. But the contrary is potentially the case, now that so much of the routine can be left to machines. For by these means an increasing number of people can be left free both to attain greater understanding and also to carry on higher work in general, including varied creative activities. That is, the intelligent can find more ways to utilize automation and other complex technical and social organizations so as to proceed to higher enterprises, which hold for them more fulfillment.

Perhaps the person who even today finds his work most gratifying, and who least feels the need of distractions such as hobbies, is the scientist, provided he is not hounded by too much competitive pressure to obtain and exhibit, with maximum speed, masses of novel-appearing but perhaps ill-considered data. Intrinsically, man is made in such a way that if he is intelligent science can be play for him. For, after all, it is really a superb expression, developed in cultural evolution, of primate curiosity and love of variety. In the future, if people in general can come to have a greater amount of creative intelligence they can partake of this fun in larger measure, both on their own, in groups, and, by learning about the work of others, vicariously.

Even when scientists seem to be on their own they are able to pluck their fruits only because they are privileged to be standing at the top of a vast pyramid of mental athletes of past and present. Would it not be better if just about everyone were so constituted that he could share in the joy of understanding the great collective conquests of his species, such as mathematical relations, relativity and its development, cosmogony as known at the time, biochemical evolution, exobiology, mind-body relations, intra- and intermind workings, social and industrial organizations, the latest artificial mechanisms, and so on, instead of having such understanding necessarily confined to a rare few, and compartmented among them at that?

Science and its technologies are unprecedentedly great enterprises of cooperation in intelligent creativity resulting from the cultural evolution of our genetically gifted species. They not only foster, when duly integrated, man's greater expansion and security but also give deeper fulfillment to man's own inner urges, both as built in him genetically and as further shaped culturally. They should therefore be, for us, at the same time means of advancing our species and also their own excuse for being. Their development is soundest when the validity of either motivation by itself is recognized, as well as that of both working together. We would be false to ourselves if we did not try to promote throughout the population of future generations, by all possible means and without limit, the intelligence and the motivation which encourage the further advance of these activities. Therewith would also come the flowering of beauty: mainly in forms very unlike those of past or present but, utilizing the more advanced means and fitting in with the outlook, appropriate to the world which we will then be finding and creating, thereby enriching us both in our rest and our work-play.

Ways of Working Toward the Major Aims

The most basic way of working toward the major aims is to educate everyone not later than in high school in the main principles of biology, including especially genetic and cultural evolution and their lessons for ourselves. On the heels of this should be a sketch of world history, depicting the growing unity of man.

Even with all this background, most genetically less-fit individuals would not accept the judgment of their being so themselves, and then voluntarily engage in less than the average amount of reproduction. Nor, *vice versa*, would the more fit choose to make the career sacrifices today made likely by their having larger families. True, this difficulty could be taken care of on paper, by taxes and subsidies, but in a democracy the enforcement of such seeming discrimination would hardly be accepted. Also rejected as discrimination would be known arrangements whereby persons of types prejudged less fit were automatically shunted (even though induced by the "carrot method") into ways of life otherwise attractive to them but affording less opportunity to reproduce, while those of types prejudged fitter had occupations that had been adjusted to provide greater opportunity to have children at little or no sacrifice, or with a subsidy.

However, with the educational background outlined, increasingly large numbers of couples who were suffering from sterility in the husband would be eager to avail themselves of means of having one or more children derived on the male side from someone they both hold in deepest regard as a person physically by no means inferior while morally and mentally outstanding. There are perhaps ten thousand children a year produced in this country by artificial insemination with semen from donors chosen by the physician; but he does not select them according to such standards and he keeps their identity secret from everyone, including the couple. Well-endowed children would be far more desired if the couples were allowed to exercise the deciding voice in the choice of the genetic father after seeing the records concerning a wide range of possibilities, considering counsel concerning them, and judging which of them have shown more of the traits preferred by the couples themselves. Are not fertile couples nowadays expected to make their own choices of their partners in marriage, and are they not in that way allowed to choose also—even though with far less directness or likelihood of getting what they prefer than by the method here proposed—the kind of children whom they themselves want?

Openness of choice regarding donors would make it desirable that the semen had been stored, preferably for decades, until after the donors' decease. Thus the disclosure of the fact that a given person had been the donor could no longer handicap him nor present the possibility of personal entanglements between him and the recipient couple. Moreover, perspective could better be gained on the possible donors' phenotypically expressed merits and their genetic reliability in passing these along—information which would be invaluable in the making of choices.

Gradually, increasing numbers of nonsterile couples also would want to take advantage of so attractive an opportunity, for at least one child in their family. The previously mentioned resistance to self-condemnation which would interfere with voluntary eugenic action of an old-style kind would not operate here, since one would be comparing oneself with a personality whom one felt to be really great. The first participants would be those wanting a child without some particular defect of the husband's and idealistic realists who were far from subnormal. For the latter, clearly, quite open choices made voluntarily but after counseling and considering of the documentary evidence, would be essential. Then later, others would be proud to follow suit, letting it be known that they had done so.

There are additional reasons against secrecy. One is that adopted children usually find out that they have been adopted, as would "half-adopted" ones (to use Julian Huxley's term). The adopted child's attempt to discover his genetic derivation when (as is now usual) it is a closely guarded secret, commonly acts like a cancer in his life. On the other hand, in cases of the sort here described, knowledge of the facts would exert the opposite influence. Moreover, due appraisal of the data actually *requires* genetic recording of an open type. So does the making of genetic judgments about the future possibilities of an individual's germ cells, as well as the avoidance of incest, when the time comes for any given child to reproduce.

Of course the couples would be warned beforehand that genetic segregation and environmental influences allow the results of no human reproduction to be predicted, and that such selection as here depicted only *weights* the results in their favor. It would however be pointed out that outstandingly good performance has almost always required a combination of both favorable environment *and* favorable heredity; also that one-half of the child's non-sex-linked genes are those of the donor father. That the environment of these children also would tend to be favorable is indicated by follow-up studies on the families of those sterile couples who even today have resorted to artificial insemination, for their marriage and family life have turned out to be actually improved, on the average.

Nevertheless, there should be noted here the qualifying phenomenon of "regression toward the mean." This (as Dr. Crow kindly pointed out to me later) I failed to mention in my address. Measurements of all sorts of traits in all sorts of organisms have long been known to show that the magnitude of any trait in an offspring tends to be nearer to that of the population's mean than just halfway between that of the more extreme parent and the less extreme one. Considering only cases in which the more extreme parent is on the "plus" or "positive" (useful or beneficial) side, and the less extreme one either "plus" or "minus," the genetic reasons are probably as follows: (1) The genes of the more plus parent (whom we'll say was the male) formed a uniquely favorable *combination*, in relation to the environment in which he was raised and/or maintained. So much was this the case that a random sample of half of them would, oftener than not, tend to operate somewhat less favorably in relation to their environment. Hence they tend to afford somewhat less than their "expected" half share of the effect in the offspring. (2) It is likely that extremely deviant genes (whether plus or minus) tend to have less dominance than genes closer to the normal in their effects.

As far as environmental effects alone are concerned, even when the genetic composition is identical in both the parents and all their offspring, as in a homozygous stock, similar considerations operate, since the more plus individual has usually had a combination of environmental factors operating in his behalf, such that their dilution by half or by any given proportion would be likely to result in a somewhat less than proportionate plus effect. Among human beings, both the above genetic and the environmental mechanisms mentioned would be operating at once. In addition, the highly favorable matching with one another of the more extremely plus parent's genetic and the environmental combinations would tend to be less than half attained in the offspring, regardless of the other parent's contributions to the genetic and the environmental circumstances.

Natural and artificial selection, however, have both operated successfully, though somewhat more slowly than otherwise, using the performance test (even when that of the individual alone is used). Selection based partly on the relatives' performance succeeds with a good deal less regression. Nevertheless, in terms of gene content, the "upgrading" does remain 50 per cent, and very real progress is evident. In general the more it is a matter of individual dominant genes of large effect, the less regression there can be, and regression is also reduced when the given gene tends to express itself with about as much relative strength when the accompanying group of genes and/or environment have a generally favorable or unfavorable phenotypic effect. Here the gene in question does tend to vary in its expression, but in the main only *pari passu* with the genes in general. (Thus such a gene tends, at the *very least*, to be expressed with similar sign on "backgrounds" in general, as has been true of so many genes as to have maintained sexual recombination—a matter explained in other papers.)

At any rate, regression toward the mean may be regarded as traceable entirely to phenomena of the gene's *expression* varying with its environment. This may be its "exo-environment" or its environment of genes at other loci, or that determined by the kind of allele it has. In the genes the matter is purely one of pheno- or developmental genetics, not of heredity "proper." The time is exceedingly remote when the intricacies of all the specific interrelations here involved can be ascertained, even by the best computers. Meanwhile, we can *and must* improve greatly on nature's and the breeder's most successful attempts by using their performance criterion with the modern far more advanced techniques and the best pooled foresight that are now available to us.

Then, as the results, so favorable on the whole, of the relatively few first trials gradually become known, ever more couples will want to follow these pioneers' example; that is how new customs usually start. Previous taboos against the practice will dwindle. In their place, a new atmosphere of hope will emerge: hope both for the rewarding results likely to accrue to the couples themselves, and hope among them and others for mankind in general. Thus a genetic leaven will tend to diffuse through the population, and also a cultural, spiritual leaven. At last *human* resources, even on the genetic side, will begin to be enhanced, at an accelerating pace.

Admittedly, all this selection will be empirical, that is, based on performance rather than genetic analysis. After all, performance has been the criterion by which nature effected all our past evolution. Human discrimination, refined with the help of intelligent counselors, can however result in a far faster upgrading than nature has ever achieved in higher organisms, especially in regard to the two major characters that are being stressed. For considerable allowance can be made for the interfering effects of environment and of modifying genes, and heritability as indicated by close relatives can be taken into account. Knowledge of the actual genes concerned here is far from essential, however, and may be rather distant; each trait may well have many similarly acting major enhancers.

Whether, or how soon, we could supplant the chiefly empircal method with what I have termed "genetic surgery," we must not wait for this contingency to arrive, in view of our having the empirical method already at hand to some extent. For the application of the ultrasophisticated techniques to the kind of characters here being considered is likely to come much later than their more optimistic present supporters imply. Similarly, we should not wait for the success of so-called euphenics in this area.

Although success in upgrading the two major characters requires the high value of them to be recognized by the volunteering couples who engage in the project, they can and should have a great diversity of preference with regard to other characters. At the same time, they must be willing to take into consideration the counsel given them; one of the functions would be to see to it that there actually was enough diversity in their choices of other traits. In this way, moreover, such faddism as a flocking to choose as donors some then-popular group—such as certain types of entertainers, exhibitionists, sports champions, or demagogues—would be held in check.

Despite the differences in choice among couples, they would wish, and should be guided, to include some of the more special gifts or predilections which tend to support or channel the two major ones of cooperative disposition and general intelligence aimed at by all. Among these are: joy of life, strong feelings combined with good emotional self-control and balance, the humility to be corrected and self-corrected without rancor, empathy, thrill at beholding and at serving in a greater cause than one's self-interest, fortitude, patience, resilience, perceptivity, sensitivites, and gifts of musical or other artistic types, expressivity, couriosity, love of problem solving, and diverse special intellectual activities and drives. This list is very incomplete, the traits are complex, and many overlap and are interdependent. Some can be too extreme and should be held in temperate measure; some have bad names also (e.g., recklessness for fortitude). Many are greatly influenced by past environment. Physical traits also (e.g., longevity, late senility, vigor, good autonomic regulation, agility) should be given considerable place. No one has nearly all these mental and physical endowments, but that choice should be made which, while largely consistent with the counsel, best fits the couple's ideals. Initiative was omitted above because outstanding persons usually have much of it anyway, and those with too much "push" tend to become too strongly represented.

As these more special gifts become commoner in the population they can and should be more and more combined. This process will not ultimately reduce diversity. For the resulting population of more generally well-endowed individuals will of course branch out again diversely from the higher general level so attained. Thus it will gain still greater aptitudes of varied kinds in its different members.

Since really outstanding persons are relatively few, semen from the same donor would have to be used for correspondingly many people. Apprehension has been expressed that the genetic diversity of the population might become too much diminished in this way, but a little consideration shows that this supposition is incorrect. For one thing, the genetic spread of even a relatively small group of donors taken from among the general population would be just as great as in the population in general in regard to other characters than the two major ones agreed upon, except for representing more of the really desirable "minor" characters above referred to, and even in regard to these and the two "major" characters, the phenotypic convergence would not necessarily imply a great deal of genetic similarity.

Moreover, our modern populations can be regarded as having arisen by a rapidly repeated doubling that occurred in a relatively small number of generations; during that time the newly added mutations were very few as compared with the ones that were carried along from the past, when the populations were a small fraction of ours today. In other words, our present genetic diversity is really that of a fairly small population, but the diversity is many times repeated in essentially the same way. Thus, the population could be subjected to generations of back-crossing, even to very few individuals in each generation, before it began to have its original *potential* diversity substantially reduced. But long before it was, the aims of the preliminary upgrading would have been achieved, and the improved population thus resulting would have devised better methods, of varied kinds.

In getting this project started, it is of the utmost importance that rigorous precautions be taken to insure that the persons in the group or groups taking part genuinely understand and favor the two major aims previously stressed. Persons who favor what *they* consider genetic improvement are of course all agreed on the major value of intelligence. However, they are far from agreed on the need for more cooperativeness, and even of those who believe they favor it a large number are gravely mistaken about its nature. That is one reason it has here been placed first, before intelligence. Many persons would consider as desirable cooperation today joint actions that would give preference to their own race, or nation, or class, or institution, or religious or provincial group, rather than to mankind as a whole. I do not mean by this to imply that mankind as a whole might never be served by taking sides in a dispute—far from it—but I do mean that a consistent policy of favoring your own side just because it is your own is contrary to the kind of cooperation needed in today's world. I have plenty of evidence that people would try to get into a project like the one here outlined and then channel it into this narrower concept in which it would be actually defeating its original aims.

Thus the group of prime-movers, must be small and carefully chosen, and guided by rules that maximally safeguard their future observance of this interpretation of social values. They will have to see to it that both the tasks of getting the material and of choosing the counselors are carried out in the same spirit. Although it would doubtless be helpful to have governmental blessing for the major aims and methods of such a project, actual governmental direction or detailed control at this stage of world affairs would present too great a risk of partisan influence, and also subjection to standards of excellence which were too bureaucratic.

The group of prime-movers should of course have as participants not only persons specialized in genetics, in the physiology of reproduction in its theoretical and/or medical aspects, in psychology, and in social sciences but also representatives from other truly humanistic fields. In this connection, it is important to note that I have found not a few religious or ideological leaders, of diverse commitments, adopting a not unfavorable attitude toward this project when it was explained. Included were representatives of the Catholic, Methodist, Baptist, and Unitarian-Universalist denominations, of Judaism, official Humanism, and Free-thinking. Moreover, persons brought up to Buddhism and to Shintoism have expressed approval. Representation from several of these groups, widely spread, would be important both in its own right and for its influence on the public and its leaders.

Choice of counselors should of course be made by the same standards as those used in the choice of the prime-movers, but less rigorously applied. So should the selection of the outstanding donors. These results would automatically follow from a good choice of alert prime-movers.

As regards the attitude of the above groups toward intelligence, they should keep in mind that eminence and creative intelligence are far from the same thing, though usually confused. Truly creative intelligence is likely to break barriers that were previously observed. Therefore these creative, highly intelligent persons all too often fail to be recognized as such by their contemporaries, although they have a relatively better chance of being so recognized by the following or a still later generation. That is another reason for storing most of the semen for decades. But it also means that precautions should be taken to have in the group of those selecting the semen to be stored some persons who can better recognize creativity and who may therefore be somewhat suspect themselves. Moreover, not only majority opinion should count in the original selection of the semen donors but each person participating in the choosing should be allowed the option of selecting some donors even without the approval of any of his colleagues. None of the semen of donors in this latter group should be used currently; all of it should be stored. On the other hand, a specified small amount of semen should be available for immediate use, that is, during the life of the donor, at least at the beginning of the project. For in this way some persons already eager to use the method can set an early example, and some preliminary results can be obtained as a background of information and statistics.

Of course it is scientists—evolution-minded scientists—who will form the core of the prime-moving group. It is important for some of them to get behind such a project as soon as possible, recognizing and stressing the major aims. For it must be started soon enough to show by its example the importance of its purposes as distinguished from the aims of some other groups that will surely form very soon and try to get priority. There are all too many of the so-called social Darwinist and racist types, and those espousing a master-and-slave society, who are interested in starting such groups, or may already have done so, even in our country. In addition, projects of this kind will certainly spring up soon in some other countries, and not all will have the desirable breadth of outlook. Thus there is sure to be a "germinal race," parallel with our weapons race, and in this race at least we can and must set the highest possible standards. Here we would have nothing to lose, but mankind would have everything to gain by our winning.

Although no one should be excluded from the prime-moving group, or the supporting one, just because of his official politics or religion, nevertheless there are political and also religious positions, seldom adopted by an entire denomination or political party however, which are incompatible with the truly social outlook needed in the project. Here we should not hesitate to decide against admitting persons having such attitudes, even though this policy may sometimes lead to our being accused of bias. Incidentally, one of the best ways of proving that our own aims are not narrow or biased is to have included in the material stored semen from donors of varied races and social groups, and to raise no objection to any couple of a different race or group using such material if they want to, but they should never be pressured in either direction on this point. Of course the data gathered by those selecting the donors will include information as to race, social class, etc., for these matters often have much bearing on the environments they have had to contend with, or benefitted from, and therefore on the contribution from genetic sources.

The taking of extra precautions to insure a sound, forward-looking social attitude on the part of the prime-movers and supporters of the project of germinal choice is made especially important by the present mores of our American society. Although it is far advanced in social outlook and practices as compared with its condition of only half-a-century ago—as my personal recollections can vividly attest—it has not yet advanced far enough along this road to make "performance," as measured by mundane success in our present society, a reliable clue to the possession of the two major traits here stressed. The reason for this is not only that nonconformity is still frowned upon but also because of the persistence of strong over-competitive, unsympathetic attitudes in our commerical Western culture.

Although the chief seeds of Western progress do lie in its science, technology, education, and struggling democracy, its most conspicuous spirit is after all that of raucous, hypocritical, and often misleading salesmanship, aided by vulgar display, along with mass distraction, petty politics, and a growing militarism.

It is unfortunate that military strength should still be paramount. Yet without it the West along with all the world, would be practically certain to be submerged in the spread of the much more outright deception and outwardly unquestioning conformism so commonly imposed on large areas lying on the other side of the ideological fence. At the same time, it should be recognized that many of the minor regions connected with the West are guilty of similar falsification and repression. Moreover, on both sides of the fence the leaders of the major regions, no matter how mistaken about the means they should use, have ultimate aims which are constructive, and feel similar menace. Barring war, there should be gradual, salutary convergence.

In view of this situation, still so confused and subject to strong and dangerous currents and countercurrents, unusual vigilance will be indispensable for keeping the aims of the genetic betterment group here proposed from becoming perverted. Yet delay in beginning the project as an active operation becomes all the more dangerous. For rival groups will in any case get busy very soon, whose procedures use the same techniques but have aims which are already oriented in clearly non-social ways, or are inadequately guarded against perversion. It will be impossible to repress such groups everywhere and only a better organization that would at the same time command great respect from the socially more knowledgeable elements will fill the need thus created.

Techniques, Research, and Practical Aims

As is now so well known, human spermatozoa can be kept deep-frozen at the temperature of liquid nitrogen (and lower), without deterioration during prolonged storage, even though the processes of freezing and thawing still incapacitate a minority of them for fertilization. The addition of glycerin, and probably still better, DMSO, considerably reduces this undesired effect. At several places in this country, and in at least one abroad, banks of frozen sperm are already being kept. However, there has been no attempt so far as is known to procure for any of these banks the semen of donors who (at least by the standards here discussed) are outstanding. Moreover, the material is being used in a dictatorial, secret way by the physician, just as with the artificial inseminations that employ nonfrozen sperm. The infants from the deep-frozen sperm have been comparable in their normality with those from unfrozen (or quite untreated) sperm.

Deep-frozen sperm are less sensitive to radiation damage than those at room temperature, although it would of course be preferable to have them stored where there is likely to be minimal exposure to radiation and other potentially damaging influences. By means of modern cryogenic methods, the cost of maintnance of large numbers of samples over many years is becoming surprisingly low per sample. For this reason it will be all the more necessary to have the arrangements for keeping them provide for their permanent identifiability, free from the chance of their becoming confused with one another.

Research is badly needed concerning ways of "stretching" the amount of use possible for a given sample, since suitable diluents, long known for domestic animals, have not yet been found for man. Nor have ways yet been found of reliably fertilizing a human egg by a sperm *in vitro* since some kind of sperm "capacitation" is needed beforehand, which normally occurs in the Fallopian tubes. Another need is to find out how immature germ cells, that could of course be kept deep-frozen like other tissues, can be caused to develop into normal spermatozoa *in vitro*. This would make possible the unlimited use of a given sample of immature germ cells. In addition, of course, it will be desirable to investigate these problems in the case of female germ cells, which Sherman, using mice, was able to deep-freeze without injury—but in rare cases only—and now Burks, using human eggs, has applied the technique. Ways of flushing them out of the tubes without injury or operation must also be sought more actively, and so must answers to such related problems as those of parthenogenesis and nuclear transfer. Again it must be borne in mind that the groups with aims less guarded from social misdirection will surely be conducting research and development of a comparable kind.

The fear is sometimes expressed that since outstanding merit often fails to be recognized until late (if ever) in life, the object of the project might be defeated by the mutations accumulated in the sperm of the elderly. However, a little calculation readily shows that, at most, such mutations would be very few in comparison with the numbers already supplied to the sperm by earlier generations. Hence the positive merit of the individual chosen would usually much more than compensate for this relatively small increment.

However, the best way of meeting this as well as other problems of germinal choice would be to have sperm storage much more widely practiced by the youthful population, and this procedure would eventually make for wider range of choice. Moreover, if increasing knowledge had in the meantime resulted in any substantial changes in our standards of excellence the repositories of germinal material would then be extensive enough to meet this situation. Ideally, almost every young man should have some of his germinal material stored, if only for his own sake and that of his spouse.

This consideration would apply especially to persons likely to have the germinal material in their bodies exposed to more influence from mutagens than is usual. Self-interest should dictate prior storage for persons in any work involving the danger of over-exposure to radiation. This would include astronauts, the crews of supersonic planes, workers in nuclear submarines, etc. Even men not so threatened would often be glad to pay for such a service, just as a means of allowing them to be vasectomized for the achievement of reliable birth-control without loss of the potentiality of later reproducing if and when they so desired, and many wives would also favor such a method for their husbands. Thus the stores could eventually be increased enormously. However, a large, well-financed organization (though perhaps eventually supported by the fees of those who used these services) would in that case be necessary.

With changing mores regarding germinal choice these services would eventually present much more possibility of choice to recipient couples, and with more choice the mores would change more rapidly, in a self-accelerating cycle. Thus the preliminary choosing of donors would become largely unnecessary. All the more necessary, however, would be the adoption of means of insuring that the counselors, and those engaged in choosing germinal material for their families, recognize and truly understand the major aims here stressed. These points are being raised now because some recent developments have already presented them.

Prospects

If to some people such discussions seem "far out," it should be remembered that they deal with measures closer to realization than those of applied gene knowledge of the traditional kind, and ever so much closer than those of "genetic surgery." In fact, the latter procedures would also present the same weighty problem of aims to decide if they were not to be confined to matters which, though important, were by comparison trivialities. As we have seen, the germinal choice by empirical methods that is so much closer to realization still has to clear one or more technical hurdles of importance before it can be in very wide use. But these should prove readily negotiable if subjected to some concerted action.

In the empirical germinal choice project it is the matter of values that looms as paramount at present. This is especially the case because so many people who would like to be associated with the project fail to realize its importance, or what the major values should be, and they are therefore striving to get the techniques going, willy-nilly.

It may be objected that we should not glean values simply from what evolution has found an aid to genetic survival in our line of ancestry, but must seek them out where our hearts and souls, aided by our reason, inspire us to find them. The answer to that is that our hearts and souls and reason have been, in every sense, the most important products of our own evolution. This implies that ultimately, for man, it is true that *right makes might*. But it is also crystal clear that for our modern world the values here emphasized need to be further enhanced by all means available, both cultural and genetic. After we have succeeded substantially in this great undertaking we will be in a better position to make further decisions about our more distant genetic and cultural problems. Enough for the present to be aware that, as far ahead as we can now see, there is no necessary limit to our cultural or genetic evolution, even though man confronts such a crisis in these fields now, and probably will for quite a while to come. In this crisis he can destroy his civilization if he does not take the speediest yet most carefully considered action possible.

We, as geneticists concerned with man, should see it as a part of our own responsibility not only to enlighten the public but also to promote, in the meantime, the collection, documentation, and storage of superior germinal material. This would be that of men who, according to considered judgment, best represent the major aims of enhanced cooperativeness, based on more heartfelt, broader brotherly love and more creative and generalized intelligence. Only in this way can we meet the obligation we all have to the multitudes who have made us what we are—to use the insights they have afforded us in behalf of our successors.

We must avoid getting sidetracked into acceptance of the delaying procedure so prevalent in both academic and political circles, which declares: "This needs more study!" Of course it does, but it is clear that there are certain things which can and must be done at this point; also that some of us are the ones to do them, in collaboration with suitable persons in other fields, whom we must find and encourage. Chief among these immediate tasks is the starting of the practice of accumulating germinal stores and records derived from persons who so far as we can see embody the major traits here stressed. The example thus set will be the main feature of this starting effort.

By this stage of our discussion it should have become clear that there are certain dangers to be specifically guarded against in the choice of participants in the project and also in the selection of germinal material. One of the types most to be avoided as collaborators or as donors is that of the egotistical, paranoid personality who would certainly try to push or squirm his way in. No prime-mover, supporter, or donor should be a person who seeks the storage of his own germinal material. I know definitely of a physician who in practicing artificial insemination often used his own semen in the treatments. He remains safe under the cover of the secrecy that dominates this field among today's physicians, and he has also saved money in this way. It happens that this physician was in fact intelligent, but his procedure illustrates that intelligence is not enough. A related type to be avoided as a prime-mover or supporter is the person who accepts as valid the over-individualistic, asocial viewpoint that is still so prevalent. The violation of these principles by any group engaged in an alleged human betterment program should, *if verified*, be exposed.

On the whole, physicians, especially those concerned with reproduction and the urinogenital organs, who would be willing to give up dictatorship and secrecy as principles to be adhered to in inseminations carried out under the germinal choice project, would be important to have as participants in the plan. But they are extremely hard to find. Meanwhile the work of getting the project going must be undertaken as soon as possible, and both medical and legal aid will eventually be forthcoming. In this connection, however, it should be borne in mind that general medical approval and adequate legalization seldom are accorded until after the practice itself has been initiated and has won approval in other highly regarded but more progressive circles. A good example here is that of contraception.

Thus we should not let ourselves be discouraged by the temporary difficulties. We should not only bear in mind the urgent need for success, we should also recall that, after all, man has gone from height to height, and that he is now in a position, if only he *will*, to transcend himself intentionally and thereby proceed to elevations yet unimagined. Unintentionally, he no longer can do so. It is up to us to do our bit in this purposive process, and to use what we know constructively, rather than remain in that ivory tower which has the writing on its wall. Our reward will be that of helping man to gain the highest freedom possible—the finding of endless worlds both outside and inside himself, and the privilege of engaging in endless creation.

Summary

Our line of ancestry, after splitting off as the primate order from other mammals during the era of reptiles, was built for active tree life of a kind which gave an advantage in genetic survival to the little bands whose individuals felt more empathy for one another and better understood one another. Thus natural selection here favored genetic constitutions which resulted in more "social intelligence" and cooperation. The primate structure and mode of life also caused natural selection, apparently somewhat later, for more curiosity concerning objects, love of variety, and versatility, and thus for increased general intelligence. These aptitudes the apes carried still further, with their arm-swinging progression giving them a semi-erect posture and a wider view, and also greater mobility.

In this way a trend toward a fully erect life on the ground was facilitated. It was pursued in a line of apes that split off from the others, some twenty million years ago, in a more definitely human direction. On the ground there was even more advantage in manual ability, improvisation, and coooperation—as in predation and in defense from predators. Thus these selective tendencies were accentuated. They gradually built up the intelligence and the collaboration between associates to such an extent as slowly to allow the evolution of culture, including speech. The increasing culture, in turn, provided conditions for further genetic selection among the small bands, thus stimulating the ability to better utilize and improve the culture. The major attributes thus favored were, again, cooperativeness and intelligence, aided by initiative, i.e., creative intelligence. This reciprocal positive feedback between cultural and genetic evolution must have lasted for some two million years, until about the start of the agricultural age.

At that time culture became so successful as to allow the groups of people to increase so in size that intergroup competition became virtually inoperative. Recently, success in saving nearly everyone for survival, and probably the great majority for full reproduction, must have gravely undermined nearly all genetic progress in the bases of our physical, mental, and moral natures.

By far the worst aspect of our present genetic crisis lies in the urgent need for further advances in the development of our mental and moral natures. First in importance should be the need for a greater capacity on the part of people in general to extend genuine warmheartedness, that is, brotherly love, to all sectors of humanity, in a *world-wide community*. Second would come the need to have our general creative intelligence deepened and broadened. By the latter means, the complex problems raised by our modern sciences and technologies, and our resulting social organizations, could be far better resolved by us. They could also be better understood by mankind in general, so that people could live in awareness of the modern world which, after all, their forebearers and contemporaries had discovered and created. Thereby their alienation and feeling of frustrated enslavement to a vast machine would be relieved. They could rise to savor the deep joys of sensing the great universes outside of and within them. They could eagerly join in the creation of ever better worlds, and rise over their machines to higher, less routine activities.

Genetic progress of these kinds will no longer occur automatically. We must use the advances of our modern culture to bring them about intentionally. Moreover, this aim should be pursued by every possible means, and we should also progress on a more purely phenotypic (nongenetic) level toward the same goal. These two methods of advance should not be regarded as antagonistic but as complementary. Moreover, in genetic progress we cannot wait until we have mastered some refined "genetic surgery," nor even until we have better knowledge of the individual genes concerned in the attributes we are aiming for—although of course that skill and knowledge should meanwhile be sought actively. We should at least make a start by using the empirical method—being guided by the test of performance which has worked in natural selection.

For this purpose a beginning should now be made in germinal choice. This means, for the present, the affording to couples of an opportunity to have one or more children derived on the father's side from germinal material of their choosing, out of well-documented stores which had usually been kept for decades after the decease of those who donated them. The transactions must be recorded, not secret, and not dictated. Physical techniques are already available for conducting such an enterprise on a small scale. Areas of research to extend and improve these techniques, and the project as a whole, are outlined in this article.

It is important that means be employed, including counseling, to maximize the chance that the material chosen represents a genuine advance in the major genetic attributes we have stressed, while salutary diversity is maintained in other respects. Special precautions and vigilance are required to minimize subversion of the major aims, because our modern cultures are not yet sufficiently advanced to insure this. For this reason there is imminent danger of other groups (such as racists and espousers of a master-and-slave society), lacking awareness of social fallacies, soon making headway. Thus it becomes all the more necessary for practical work to be started at once by a more responsible, but nongovernment body, as a salutary example. What is called for is action by a socially-oriented group of geneticists, backed by aides from the fields of physiology, psychology, and medicine and by diverse humanistically minded persons of competence, with the aim of beginning to establish and maintain germinal repositories intended mainly for use decades later.

Amidst the justified present concern about "natural resources" in general, our *human resources*—rooted in man's most basic nature—should by no means be neglected. It is hoped that in his present critical period of rapid cultural transition man will thereby be enabled better to envision and advance to ever greater realms of knowledge, creation, and spirit.

Addendum

The works drawn upon in this article proved to be too numerous for justice to be done to them in a list of references herewith attached without delaying the manuscripts submission unduly and drawing invidious lines between contributors to the background material. However, I hope to note them in full later, along with other references, in a much more voluminous, although in the main popular, account.

Selected Bibliography Including References Cited

Part V

Bajema, C. 1963. Estimation of the Direction and Intensity of Natural Selection in Relation to Human Intelligence by Means of the Intrinsic Rate of Natural Increase. *Eugenics Quarterly*, 10: 175-187.

Crowe, B. 1969. The Tragedy of the Commons Revisited. *Science*, 166: 1103-1107.

Cruz-Coke, R. 1968. Birth Control and Human Evolution. *Lancet*, 2: 1249.

Darwin, C. 1956. Letter Concerning Osborn's Eugenic Hypothesis. *Eugenics Review*, 48: 131.

Darwin, C. 1958. *The Problems of World Population.* Cambridge Univ. Press, Cambridge. 42pp.

Davis, B. 1970. Prospects for Genetic Intervention in Man. *Science*, 170: 1279-1283.

Davis, K. 1966. Sociological Aspects of Genetic Control. pp. 173-204 in Roslansky, J. (ed.) 1966. *Genetics and the Future of Man.* Appleton-Century-Crofts, N.Y.

Dobzhansky, T. *The Biology of Ultimate Concern.* New American Library, N.Y.

Dubos, R. 1965. *Man Adapting.* Yale Univ. Press, New Haven.

Dobzhansky, T. 1967. Changing Man. Modern Evolutionary Biology Justifies an Optimistic View of Man's Biological Future. *Science*, 155: 409-415.

Eiseley, L. 1963. Man: The Lethal Factor. *American Scientist*, 71-83.

Etzioni, A. 1968. Sex Control, Science and Society. *Science*, 161: 1107-1112.

Fratantoni, J. 1969. New Approaches in Genetic Counseling. *Archives of Environmental Health*, 19: 613-615.

Glass, B. 1967. *What Man Can Be.* Paper presented at the American Association of School Administrators Convention, Atlantic City, N.J. 23pp.

Goodman, L. 1963. Some Possible Effects of Birth Control on the Incidence of Disorders and on the Influence of Birth Order. *Annals of Human Genetics*, 27: 41-52.

Hardin, G. 1959. *Nature and Man's Fate.* Rinehart, N.Y.

Hardin, G. 1962. Comments on Genetic Evolution. Pp. 453-454 in Hoagland, H. and Burhoe, R. (eds.) 1962. *Evolution and Man's Progress.* Columbia Univ. Press, N.Y.

Hardin, G. 1964. Rhythm and Natural Selection. Letter to the Editor. *Science*, 143: 995. Replies. *Science*, 144: 365-366. Rejoinder *Science*, 144: 1531.

Hardin, G. 1967. Mankind by Design: A Book Review of "Genetics and the Future of Man" edited by J. Roslansky. *Science*, 156: 797-798.

Higgins, J., Reed, E., and Reed, S. 1962. Intelligence and Family Size: A Paradox Resolved. *Eugenics Quarterly*, 9: 84-90.

Hoagland, H., and Burhoe, R. (eds.) 1962. *Evolution and Man's Progress.* Columbia Univ. Press, N.Y.

Hotchkiss, R. 1965. Portents for a Genetic Engineering. *Journal of Heredity*, 56: 197-202.

Huxley, J. 1962. Eugenics in Evolutionary Perspective. *Eugenics Review*, 54: 123-141. (Reprinted in Perspectives in *Biology and Medicine*, 6: 155-187.)

Istock, C. 1969. A Corollary to the Dismal Theorem. *BioScience*, 19: 1079-1081.

Lederberg, J. 1966. Experimental Genetics and Evolution. *Bulletin of Atomic Scientists*, 22 (October): 4-11.

Lederberg, J. 1970. Genetic Engineering and the Amelioration of Genetic Defect. *BioScience*, 20: 1307-1309.

McConnell, R. 1961. The Absolute Weapon. *AIBS Bulletin*, 11 (June) : 14-16.

Matsunaga, E. 1968. Birth Control Policy in Japan: A Review from Eugenic Standpoint. *Japanese Journal of Human Genetics*, 13: 189-200.

Mayo, O. 1970. On the Effects of Genetic Counseling on Gene Frequencies. *Human Heredity*, 20: 361-370.

Mirsky, A. 1964. Genetics and Human Affiars: A Review of "Essays of a Humanist" by Sir Julian Huxley. *Scientific American*, 211 (October): 135-138.

Morison, 1967. Where is Biology Taking Us? *Science*, 155: 429-433.

Muller, H. 1959. The Guidance of Human Evolution. *Perspectives in Biology and Medicine*, 3: 1-43.

Muller, H. 1965. Means and Aims in Human Genetic Betterment. Pp. 100-122 in Sonneborn, T. (ed.) 1965. *The Control of Human Heredity and Evolution*. Macmillan, N.Y. 127pp.

Muller, H. 1964. Better Genes for Tomorrow. Pp. 314-338 in Mudd, S. (ed.) 1964. *The Population Crisis and Use of World Resources*. Indiana Univ. Press, Bloomington, 563pp.

Nadler, H. L. 1969. Prenatal Detection of Genetic Defects. *Journal of Pediatrics*, 74 (Jan.): 132-143.

Neel, J. 1970. Lessons from a "Primitive" People. *Science*, 170: 815-822.

Osborn, F. 1940. *Preface to Eugenics*. First Ed. Harper's, N.Y.

Osborn, F. 1962. Overpopulation and Genetic Selection. Pp. 51-68 in Osborn F. (ed.) 1962. *Our Crowded Planet*. Doubleday, N.Y.

Osborn, F. 1968. *The Future of Human Heredity. An Introduction to Eugenics in Modern Society*. Weybright and Talley, N.Y. 133pp.

Ramsey, P. 1970. *Fabricated Man: The Ethics of Genetic Control*. Yale Univ. Press, New Haven, Conn.

Roberts, C. 1964. Some Reflections on Positive Eugenics. *Perspectives in Biology and Medicine*, 7: 297-307.

Rosenfeld, A. 1969. *The Second Genesis: The Coming Control of Life*. Prentice-Hall, Englewood Cliffs, N.J. 327pp.

Roslansky, J. (ed.). 1966. *Genetics and the Future of Man*. Appleton-Century-Crofts, N.Y.

Sinsheimer, R. 1969. The Prospect for Designed Genetic Change. *American Scientist*, 57: 134-142.

Smith, J. 1965. Eugenics and Utopia. *Daedalus*, 94: 487-505.

Stebbins, G. L. 1970. The Natural History and Evolutionary Future of Mankind. *American Naturalist*, 104: 111-126.

Tatum, E. 1966. The Possibility of Manipulating Genetic Change. Pp. 49-61 in Roslansky (ed.) 1966. *Genetics and the Future of Man*. Appleton-Century-Crofts, N.Y.

Williamson, F. 1969. Population Pollution. *BioScience*, 19:979-983.

Wolstenholme, G. (ed.). 1963. *Man and His Future*. Churchill, London.